Acta Physica Austriaca
Supplementum XXVI

Proceedings of the
XXIII. Internationale Universitätswochen für Kernphysik 1984
der Karl-Franzens-Universität Graz
at Schladming (Steiermark, Austria)
February 20th–March 1st, 1984

Sponsored by
Bundesministerium für Wissenschaft und Forschung
Steiermärkische Landesregierung
International Centre for Theoretical Physics, Trieste
Sektion Industrie der Kammer der
Gewerblichen Wirtschaft für Steiermark

1984

Springer-Verlag
Wien New York

Stochastic Methods and Computer Techniques in Quantum Dynamics

Edited by
H. Mitter and L. Pittner, Graz

With 37 Figures

1984

Springer-Verlag
Wien New York

Organizing Committee

Chairman

Prof. Dr. H. Mitter
Institut für Theoretische Physik
der Universität Graz

Committee Members

L. Pittner
L. Mathelitsch
W. Plessas

Secretary

Mrs. E. Neuhold
Miss E. Tandl

ISSN 0065-1559
ISBN-13:978-3-7091-8782-1 e-ISBN-13:978-3-7091-8780-7
DOI: 10.1007/978-3-7091-8780-7

CONTENTS

PREFACE

This volume contains the written versions of lectures held at the "23. Internationale Universitätswochen für Kernphysik" in Schladming, Austria, in February 1984. Once again the generous support of our sponsors, the Austrian Ministry of Science and Research, the Styrian Government and others, had made it possible to organize this school. The aim of the topics chosen for the meeting was to present different aspects of stochastic methods and techniques. These methods have opened up new ways to attack problems in a broad field ranging from quantum mechanics to quantum field theory. Thanks to the efforts of the lecturers it was possible to take this development into account and show relations to areas where stochastic methods have been used for a long time. Due to limited space only short manuscript versions of the many seminars presented could be included. The lecture notes were reexamined by the authors after the school and are now published in their final form.
It is a pleasure to thank all the lecturers for their efforts which made it possible to speed up publication. Thanks are also due to Mrs.Neuhold for her careful typing of the notes.

H. Mitter
L. Pittner

Acta Physica Austriaca, Suppl. XXVI, 3–52 (1984)

STOCHASTIC PROCESSES - QUANTUM PHYSICS⁺

by

L. STREIT
Universität Bielefeld
BiBoS
D-4800 Bielefeld, FR Germany

I. SOME INTRODUCTORY REMARKS

It is now about 20 years ago that I first spoke about Wiener Integrals at one of the earliest Schladming Winter Schools. As I was assembling my notes for those lectures one of my senior colleagues remarked that Functional Integrals provided a nice reformulation of Quantum Theory, but really there was hardly anything you could do with them that one had not already done otherwise.

We have come a long way in those twenty years. - I am not only thinking of the celebrated successes of Constructive Quantum Field Theory - it is proper to mention here the inspiration and insistence of K. Symanzik - but also such "practical" results as that of E. Lieb on the dependence of electronic binding energy on atomic separation, an old conjecture from chemistry finally proven with the means of functional integration.

⁺ Lecture given at the XXIII. Internationale Universitätswochen für Kernphysik, Schladming, Austria, February 20-March 1, 1984.

Here in these lectures I shall not reach out to sophististicated applications. Instead I have been asked to give an elementary introduction to Stochastic Processes. Hence I shall not proceed from Definitions to Theorems and Proofs, but rather will use some concepts with which you are well acquainted to acquaint you with some others that are central to those parts of the mathematical theory which play a role in quantum physics applications.

We shall start from simple quantum mechanics which you all know, make a short tour through probability and stochastic processes and finally return to the start: to quantum dynamics in a form that may be less familiar to you, i.e. in terms of Stochastic Processes.

We shall go on this excursion without attempting to cover the whole field but on a rather narrow track like a cross country skier, but I hope that the trail has been so arranged that we shall pass by a couple of interesting vistas.

What now are <u>Stochastic Processes?</u>

One author writes

"A Stochastic Process may be loosely described as a system running along in time and controlled by probabilistic laws."

But then:

What is NOT a Stochastic Process?

Examples: 1) The Dachstein?? (see however the illustrations in [1])

2) The shape of Austria?? (see however Fig.3)

3) God?? (no comment)

Stochastic Processes are

- o a vast field of research
- o of great unifying power.

<u>Some Examples</u>,

to which we shall return in the course of these lectures:

1. Noise bursts in information transmission

Quiet periods are found to be interrupted by bursts of noise. These noise bursts form clusters, which form larger clusters, which form larger clusters,... .

Q: Make a mathematical model!

2. "Phase boundary"

For T = O the phase boundary is a straight line. (Fig.1)

Q: Make a mathematical model for T>O!(Fig.2)

3. Length of the Austrian border (Fig.3)

Map of scale s,

$$L = L(s) = L_o s^{1-D} \quad .$$

Q: Make a mathematical model!

4. Coin toss: win/lose 1 öS

Q_1: What happens to your capital?

A: You will surely lose it!

Q_2: In what time?

5. Brownian Motion (R.Brown, 1827;[2])

Q_1: Make a mathematical model!

.
.
.
.

Q_n: What is the probability for the particle to wander from x_o to x in time t?

A: $p_t(x,x_o) = e^{-H_o t}\delta(x-x_o)$

$$H_o = -\frac{1}{2} D \sum_{v=1}^{3} \frac{d^2}{dx_v^2}$$

Q_m: Given x_o, x - how long will the particle take from x_o to x?

6. Let $H \equiv H_o + V(x)$

Q: Is there a corresponding stochastic process?
A: Yes, this is Quantum Dynamics in terms of Distorted Brownian Motion! [3,4]

II. SOME CONCEPTS FROM PROBABILITY THEORY

We recall that many probabilistic concepts are familiar from Quantum Mechanics:

Probability Theory:	(Q.M.)
Random Variable X	observable X
Expectation E(X)	$\langle X \rangle$
State Space Ω	spectrum $\sigma(X)$
Event $\chi_A = \begin{cases} 1, x \in A \\ 0, x \notin A \end{cases}$	projection Yes-No Experiment
Probability p(A)	$\int_A \lvert\psi(x)\rvert^2 dx$
P.-Density $\rho(x)$	$\lvert\psi(x)\rvert^2$
Moments $m_n = E(X^n)$ (mean, variance,...)	$\langle X^n \rangle$

The "variance", or "mean square deviation" measures the

uncertainty of X:
(For simplicity $m_1 = E(X) = 0$.)

Q: $p(|X|>\varepsilon) = ?$

Consider $X_\varepsilon = \begin{matrix} \varepsilon, & |X| > \varepsilon \\ 0, & |X| \leq \varepsilon \end{matrix}$.

We have then

$$X^2 \geq X_\varepsilon^2$$

$$m_2 = E(X^2) \geq E(X_\varepsilon^2) = \varepsilon^2 p(|X|>\varepsilon) .$$

A: $\varepsilon^{-2} m_2 \geq p(|X|>\varepsilon)$, this is the "Tchebychev Inequality".

The variance controls the probability p that the random variable X deviates by more than a given amount ε from the mean.

"Characteristic Functions"

$$C_X(\lambda) = E(e^{i\lambda X}) = \int_\Omega e^{i\lambda x} p(dx)$$

Note: $\rho(x) \overset{F.T.}{\leftrightarrow} C_X(\lambda)$

Characteristic functions are Fourier transforms of probability densities.
They are generating functions for the moments:

$$C(\lambda) = \sum_n \frac{(i\lambda)^n}{n!} m_n \quad , \qquad m_n = (-i\partial_\lambda)^n C(\lambda)_{\lambda=0} .$$

Example (GAUSS):

$$C(\lambda) = e^{-1/2(\lambda,K\lambda)+i(m,\lambda)} ,$$

K a positive n×n matrix, $m \in R^n$.

For $m = 0$, $\rho(x) = (2\pi)^{-n/2} ||K||^{-1/2} e^{-1/2(X,K^{-1}X)}$.

Note: $C^r(\lambda)$ $(r>0)$ is again a (Gaussian) characteristic function with

$$K \to r \cdot K \quad , \quad m \to r \cdot m \quad .$$

Characteristic functions C such that C^r are also characteristic functions (for all $r > 0$) are called "Infinitely Divisible". Why is that? What is the special property of "infinitely divisible random variables"?
We call two random variables "independent" if $E(f(X_1)f(X_2)) = E(f(X_1))E(f(X_2))$.
Everything factorizes!

$$p_{\vec{X}}(A_1 \times A_2) = \prod_i p_{X_i}(A_i)$$

$$\rho(\vec{X}) = \prod_i \rho_i(X_i)$$

$$C(\vec{\lambda}) = \prod C_i(\lambda_i)$$

Consider now a sum of independent random variables:

$$Y = \sum_i X_i$$

$$C_Y(\lambda) = E(e^{i\lambda Y}) = E(e^{i\lambda \sum_i X_i})$$

$$= E(\prod_j e^{i\lambda X_j}) = \prod_j E(e^{i\lambda X_j})$$

$$= \prod_j C_{X_j}(\lambda)$$

Note: Products of characteristic functions are characteristic functions of sums of random variables.
In particular: Integer powers of characteristic functions are characteristic functions of identically distributed independent random variables!

$$C_Y(\lambda) = (C_X(\lambda))^n$$

Now we understand what it means that $C(\lambda)$ is infinitely divisible: We have: $(C(\lambda))^{1/n} \equiv C^{(n)}(\lambda)$ is again a characteristic function,

hence $C(\lambda) = (C^{(n)}(\lambda))^n$

$$Y = \sum_{i=1}^{n} X_i^{(n)} \quad ;$$

for any n, an infinitely divisible random variable Y may be decomposed into a sum of n independent, equally distributed random variables $X_i^{(n)}$.
These Y are rather special, hence we can give a representation for all of them (Levy, Khinchin, ect.[5]).
The Khinchin representation for characteristic functions of infinitely divisible random variables is given by

$$\ln C(\lambda) = ia\lambda + b\int\mu(dx)\,\left(e^{i\lambda x}-1 - \frac{i\lambda x}{1+x^2}\right)\frac{1+x^2}{x^2} \quad ,$$

a real, b>0, μ a probability measure.
(Gaussian for $\mu(dx) = \delta(x)dx$)
Note: Sums of Gaussian random variables are again Gaussian. In particular

$$C(\lambda) = e^{-\frac{1}{2}K\lambda^2} = \left(e^{-\frac{1}{2}K\left(\frac{\lambda}{\sqrt{n}}\right)^2}\right)^2$$

$$= (C(n^{-1/2}\lambda))^n$$

$$Y \overset{d}{=} n^{-1/2} \sum_{v=1}^{n} Y_v \text{ where } Y_v \overset{d}{=} Y \quad \text{for all } v \quad ;$$

$\overset{d}{=}$ stands for "equally distributed".
More generally this works for

$$C_\alpha(\lambda) = e^{-K|\lambda|^\alpha} \qquad 0 < \alpha \leq 2$$

$$\text{with } Y \stackrel{d}{=} n^{-1/\alpha} \sum_v Y_v \qquad (Y_v \stackrel{d}{=} Y)$$

These are (all the symmetric) "<u>Stable Random Variables</u>" . For any n, they can be decomposed into n "<u>similar</u>" (= equal up to a factor) ones.
(Obtain their characteristic functions by imposing

$$C(\lambda) = (C(\gamma_n \lambda))^n$$

on the Khinchin representation.)
Applications: Systems that are composed of similar subsystems.
<u>Example</u>: The Holtsmark Universe [6,1,7,8].
Consider a component F of the gravitational force at random points in the universe:

$F = F_\rho$ with ρ the matter density.

Subdivide the surrounding universe into n "subuniverses" with density

$$\rho \rightarrow \frac{\rho}{n}$$

$$F_\rho = \sum_{v=1}^{n} F_{\rho/n}^{(v)} \quad .$$

But $\rho \rightarrow \rho/n$ is equivalent to $l \rightarrow n^{1/3}\, l$.
From $F \sim l^{-2}$ (Newton) we have $F_{\rho/n} = n^{-2/3} F_\rho$.

Insert this: $F_\rho \stackrel{d}{=} n^{-2/3} \sum_1^n F_\rho^{(n)}$,

i.e. the "exponent" $\alpha = 3/2$.
The characteristic function of the Holtsmark distribution then has to be

$$C(\lambda) = e^{-K|\lambda|^{3/2}} \quad .$$

A Related Idea: Kadanoff's concept of "Criticality" in Statistical Mechanics: At a critical point a system will be similar to its subsystems.
Simple Model: A one dimensional lattice

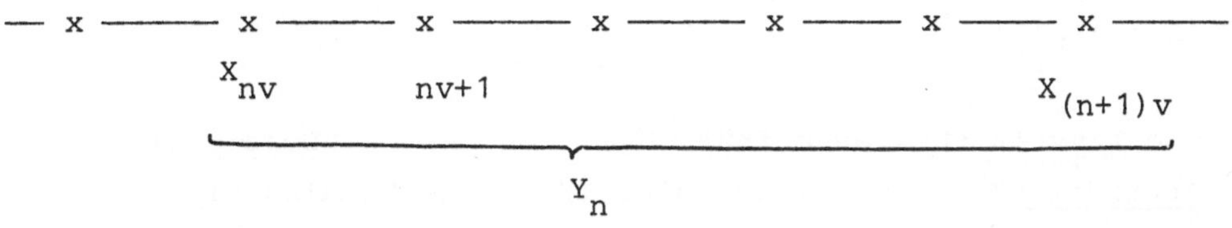

"Block Spin"

illustrates the
<u>"Renormalization Group"</u>:
1) Form "Block Variable"

$$Y_n^0 = \sum_{i=0}^{v-1} X_{nv+i} \quad ; \qquad E(X_v) = 0$$

for simplicity .

2) "Renormalize"

$$Y_n = n^{-1/\alpha} \sum_{i=0}^{v-1} X_{nv+i} \; .$$

I.e. the renormalization (semi-) group consists of these two steps and is given by

$$(R_v X)_n = v^{-1/\alpha} \sum_{i=1}^{v-1} X_{nv+i} \quad .$$

Exercise 1: Verify $R_v R_\mu = R_{v+\mu}$.

Critical Points are Fixed Points of the Renormalization Group. If one is lucky one can get them constructively from R: Assume that

$$\lim_{V\to\infty}(R_V X)_n = Y_n \equiv (R_\infty X)_n$$

exists. Then it will be a fixed point, since (formally) $R_V R_\infty = R_\infty$

$$R_V : (Y_n) \to (Y_n) \ .$$

One says in this case that (X_n) lies in the "<u>Domain of Attraction</u>" of the fixed point (Y_n). The Domains of Attraction for the Stable Distributions C_α are characterized by

$$\rho(x) \sim |x|^{-1-\alpha} \qquad \text{for } \alpha < 2$$

and

$$E(X^2) < \infty \qquad \text{for } \alpha = 2.$$

Hence we have the following alternatives for fixed points:

1. The random variables have infinite second moment ($\alpha < 2$).
2. Large sums tend to <u>Gaussian</u> distributions ($\alpha=2$).
3. The random variables are (<u>strongly</u>) <u>dependent</u>.

1. is not so nice for physical applications.
2. is the starting point for statistical thermodynamics, see e.g. [9].
3. implies "Long Range Correlations" at Critical Points [10].

A Simpl · Question: The "Law of Large Numbers".

Here we assume $E(X_V^2) = K < \infty$
and $E(X_V) = m$.

I.e. we will have a Gaussian Limit ($\alpha=2$).

But $R_n X = \frac{1}{\sqrt{n}} \sum_1^n X_v$

is not the "Average"

$$A_n \equiv \frac{1}{n} \sum_1^n X_v$$

("empirical average", do not confuse it with the mean).

Note $E(A_n) = \frac{1}{n} \sum_{v=1}^{n} m = m$,

whereas $E(R_n X) = \sqrt{n}.\, m$ blows up as $n \to \infty$ for $m \neq 0$. (This is why we chose m=0 in the previous discussion of renormalization group limits.) But with this precaution taken

$E((R_\infty X)^N)$ stays finite,

the contributions of the X_i partially cancel each other. Hence we expect the fluctuations of A_n around the mean value m to be strongly suppressed as $n \to \infty$. We verify this directly:

$$
\begin{aligned}
E((A_n - m)^2) &= \frac{1}{n^2} E((\sum_v (x_v - m)^2) \\
&= \frac{1}{n^2} \sum_v E((x_v - m)^2) \\
&= \frac{1}{n^2} \sum_{v=1}^{n} K = \frac{K}{n} \quad .
\end{aligned}
$$

So the Tchebychev inequality tells us

$$P(|A_n - m| > \varepsilon) \leq \frac{K}{n\varepsilon^2} \underset{n \to \infty}{\to} 0 \quad ,$$

i.e. the average of a large number of random variables will

approach a sharply defined value.
A final concept from the general theory: "<u>Conditional Expectation</u>".
Problem: Given two random variables X,Y - How much do I know about X if I have determined the value of Y?
<u>Example</u>: Let X be to-morrow's snowfall, and
Y to-morrow's noon temperature.
Assume I know to-morrow at 12^{00}, $T = -2^{\circ}C$, how does that affect my prediction of to-morrow's snowfall?
For simplicity we assume the existence of a positive probability density $\rho(x,y)$, i.e.

$$E(f(X)g(Y)) = \int_{\Omega} f(x)g(y)\rho(x,y)\,dxdy \ .$$

To get the conditional expectation we could fix Y through a δ-function:

$$E(f(X)\delta(Y-y)) = \int_{\Omega_X} dx f(x)\rho(x,y) \ .$$

But this is not good enough! For X independent of Y we want

$$E(f(X)|Y=y) = E(f(X)) \ .$$

"X does not care what value Y takes." Remember: E factorizes in this case!

$E(f.\delta) = E(f)E(\delta)$. Hence we set $E(f(X)|Y=y) =$

$$= E(f(X)\delta(Y-y)) / E(\delta(Y-y)) \ ,$$

and the Y-dependence will drop out if X and Y are independent. In terms of the density $\rho(x,y)$,

$$E(f(X)|Y=y) = \int dx\rho(x,y)f(x) \Big/ \int dx\rho(x,y) \ .$$

Note that $p_Y(dy) = \rho_Y(y)dy$

with $\rho_Y(y) = \int dx\rho(x,y)$

so that

$$E(f(X)) = \int_{\Omega_Y} E(f(X)|Y=y)p_Y(dy) \quad .$$

This is the "Law of Total Expectation". A slight generalization of this is

$$E(f(X)\chi_A(Y)) = \int_{A\subset\Omega_Y} E(f(X)|Y=y)p_Y(dy) \quad .$$

It is this equation which can serve as a rigorous (but not so transparent) mathematical definition of conditional expectations. Other notations to be found in the literature are

$$E(f(X)|Y)(y)$$

or just $E(f(X)|Y)$, we must then remember that this is a function on the state space Ω_Y.
A related notion is

$$E(f(X)|B) = \int_B E(f(X)|Y)p_Y(dy) \; .$$

Recall that one can obtain probabilities p from expectations E as follows:

$$f(x) = \chi_A(x) \rightarrow E(f(X)) = p_X(A) \quad .$$

We define likewise a "Conditional Probability"

$$p(A|Y) \equiv E(\chi_A(X)|Y)$$

with $p_X(A) = \int_{\Omega_Y} p(A|Y)p_Y(dy) \quad .$

This is the "Law of Total Probability".

III. STOCHASTIC PROCESSES

Recall: Random variables may be vector-valued:

$X = (X_1, \ldots, X_n)$.

Now we generalize to other index sets:

$X = (X_1, X_2, \ldots, X_n, \ldots)$

(application e.g. "time" T = 1,2,3,...)

or

$X = (\ldots, X_{-1}, X_o, X_1, \ldots, X_n, \ldots)$

(application eg.: "time" T extending to the infinite past; or: infinite 1-dimensional crystal)

or

$X = \{X_{\vec{n}} : n_1, \ldots n_d \text{ integers}\}$

(application e.g.: d-dimensional crystal) .

All of these are called "Random (or Stochastic) Processes with a Discrete Parameter" ("Time").

"Continuous Parameter Random Processes"

$\{X(t) : t \in R^d\}$

are also called "Random Functions" or, for d>1: "Random Fields".

How to characterize them?

We try again "Characteristic Functionals"

$$C(\lambda) = E(e^{i(\lambda,X)}) ,$$

but now with

$$(\lambda,X) = \sum_{v} \lambda_v X_v \quad \text{(discrete parameter)}$$

or $(\lambda,X) = \int\lambda(t)X(t)dt$ (continous parameter) .

<u>Example</u>: <u>Gaussian Random Process</u>

$$C(\lambda) = e^{-1/2(\lambda,K\lambda)+i(m,\lambda)}$$

where K is a positive linear operator on the space of the λ, (,) is a scalar product, and m a vector in the (dual) space. The Bochner-Minlos Theorem [5, 11] assures us that a probability measure p exists:

$$C(\lambda) = \int_{\Omega} e^{i(\lambda,x)}p(dx) .$$

Elements of the State Space Ω are now called "<u>Sample Functions</u>".

<u>Examples</u>: my capital as a function of time in a game of tossing coins, - or the various possible trajectories of a particle undergoing Brownian Motion.

What classes of functions occur? This asks for the "<u>support properties</u>" of stochastic processes.

Example 1: Gaussian White Noise

$$C(\lambda) = e^{-1/2\int\lambda^2(t)dt}$$

i.e. Gaussian with $m \equiv 0$ and

$$(\lambda,K\lambda) = \int d^2t\lambda(t_1)K(t_1,t_2)\lambda(t_2)$$

given by $K(t_1,t_2) = \delta(t_1-t_2)$.

We want $C(\lambda) = \int_{\Omega} e^{i(\lambda,x)}p(dx)$,

but what is the sample function space Ω?

We could try $\lambda(\cdot)\in H = L^2(R)$

$x(\cdot)\in H = L^2(R)$;

and could then introduce a fixed basis $\{e_n\}$ in H :

$$\lambda(t) = \sum_n \lambda_n e_n(t) \qquad X(t) = \sum_n X_n e_n(t)$$

where X_n are random variables: $X_n = (X, e_n)$.

This leads to $C(\lambda) = e^{-1/2\sum \lambda_n^2}$

$$= \prod_n e^{-1/2\lambda_n^2}$$

$$= \prod_n \int e^{i\lambda_n x_n} p_{Gauss}(dx_n) ,$$

i.e. the X_n are independent equally distributed (Gaussian) random variables.

We have $(X,X) = \sum_n^{\infty} X_n^2 = \lim_{n\to\infty} S_n$

$$S_n = \sum_{v=1}^{n} X_v^2 .$$

The Law of Large Numbers tells us:

$$\frac{1}{n} S_n \to E(X_v^2) = 1$$

i.e. $\sum_1^n X_v^2 \approx n$.

The sample functions of White Noise are "vectors on a sphere of radius $R = \sqrt{\infty}$",
i.e. <u>not</u> in the Hilbert space, in fact $\int_H dp(x) = 0$. Evidently we need to consider a <u>larger</u> space of sample functions X, hence a smaller one for λ.
A choice that works is Schwartz Space S ,

$$\lambda \in S \subset L^2 \subset S^* \ni X$$

λ	X
test functions	generalized functions .

Bochner-Minlos Theorem: (see [5,11]).
If C is positive definite and C(O) = 1 and C continous on Schwartz Space S, then $C(\lambda) = \int_{S^*} e^{i(\lambda,x)} p(dx)$.
We could have guessed this by a look at the covariance of White Noise:

$$E([X(t_1)-m(t_1)].[X(t_2)-m(t_2)]) .$$

For Gaussian processes this equals

$$K(t_1,t_2) .$$

Hence in particular for White Noise

$$E(X(t_1)X(t_2)) = \delta(t_1-t_2) .$$

White Noise is a <u>Generalized Random Function</u>,
i.e. $\int X(t)\lambda(t)dt$ is a random variable if $\lambda(\cdot)$ is a test function.

<u>Example 2</u>: <u>Brownian Motion</u>. <u>The Wiener Process</u>.
Brownian motion is the "macroscopic" result of many "microscopic" collisions. For these microscopic dynamics of Brownian motion see e.g. [12, 13].

<u>Postulates</u>:

I.	$B(t)$ is a Gaussian process	(sum of many effects)
II.	$B(o) = O$	(starts at the origin)
III.	$E(B(t+\Delta t)-B(t)) = O$	(is isotropic)
IV.	$E([B(t+\Delta t)-B(t)]^2) = \Delta t$	$(\Delta x)^2 \approx \Delta t$, as in the heat eg.: $(\frac{d}{dt} - \frac{d^2}{dx^2})\ u = O$

Calculate the Characteristic Functional of Brownian Motion:

$$C(f) = E(e^{i\int_0^\infty f(t)B(t)dt})$$

$$= i\int m(t)f(t)dt-\tfrac{1}{2}\int d^2tfKf \qquad (I)$$

where $m(t) = E(B(t))$

$$= E(B(t)-B(0)) \qquad \text{(II)}$$

$$= 0 . \qquad \text{(III)}$$

Hence the covariance K is simply

$$K(t_1,t_2) = E(B(t_1)B(t_2))$$

$$E(B(t_1)B(t_2)) = E(\tfrac{1}{2}B^2(t_1)+ \tfrac{1}{2}B^2(t_2)-\tfrac{1}{2}(B(t_1)-B(t_2)))$$

$$= \tfrac{1}{2}E(B^2(t_1))+\tfrac{1}{2}E(B^2(t_2))-\tfrac{1}{2}E([B(t_1)-B(t_2)]^2)$$

$$= \tfrac{1}{2}\, t_1 \qquad +\tfrac{1}{2}t_2 \qquad -\tfrac{1}{2}|t_1-t_2| \qquad \text{(IV)}$$

$$K(t_1,t_2) = E(B(t_1)B(t_2)) = \min(t_1,t_2) = t_1 \wedge t_2 .$$

The Characteristic Functional of Brownian Motion is

$$C(f) = E(e^{i\int_0^\infty fBdt}) = e^{-\frac{1}{2}\int d^2tf(t_1)t_1 \wedge t_2f(t_2)} .$$

Let us study some properties of Brownian Motion:

1) The Relation of Brownian Motion and White Noise:
From White Noise $X(t)$, we build a new process W:

$$W(t) = \int_0^t X(s)\, ds .$$

This is again Gaussian with mean zero .

Note: $E(e^{i\lambda.W(t)}) = e^{-\frac{1}{2}\lambda^2 t}$.

To calculate $C(f)$ for W we only need to compute the covariance

$$K(t,s) = E(W(t)W(s))$$

$$= \int_0^t d\tau \int_0^s d\sigma\, E(X(\tau)X(\sigma))$$

$= \int d\tau \int d\sigma \; (\tau-\sigma)$

$= t \wedge s.$

W(t) is a representation of the Wiener Process in terms of White Noise, and White Noise is the velocity of Brwonian Motion,

$X(t) = \dot{W}(t)$.

2) Next we consider the increment $\Delta B \equiv B(t+\Delta t)-B(t)$.
   Its variance (=2nd moment) gets small as $\Delta t \to 0$, recall $E((\Delta B)^2) = \Delta t$. The Tchebychev Inequality then tells us that

   $p(|\Delta B|>\varepsilon) \leq \varepsilon^{-2}\Delta t$.

   Large changes in short times are improbable. This suggests the continuity of Brownian sample paths, i.e. that we can realize Wiener measure on a space of continuous functions. Equivalently: the set of continuous path has measure $p = 1$. We will give general criteria for the continuity of sample paths later in these lectures.
3) All sample paths of $B(\cdot)$ start at zero:

$$B(0) = 0.$$

   Hence the Wiener process is not "stationary". We call a process $Y(\cdot)$ "stationary" if

   $C(\lambda_\tau) = E(e^{i\int\lambda(t+\tau)Y(t)dt}) = C(\lambda)$

   i.e. if Y is invariant under time translations.
   This implies in particular
   $m(t) = E(Y(t)) = \text{const}$
   $K(t,s) = E([Y(t)-m].[Y(s)-m]) = K(t-s)$.
   Processes with these two properties are called "weakly stationary" (for Gaussians: "weak" = "strict").

<u>Example</u>: White Noise is stationary.

4) How to get Brownian Motion to start from some point x?
This is easy: $B_x(t) = x+B(t)$

with $C(f) = e^{ix\int f(t)dt-\frac{1}{2}\int d^2t f(t_1)t_1 \wedge t_2 f(t_2)}$.

5) The "Brownian Bridge"
Consider now the process

$B^T(t) = B(t) - \frac{t}{T}B(T)$.

Its sample paths start at zero:

$B^T(0) = B(0) = 0.$

They all return to the origin at t=T!

$B^T = B(T) - \frac{T}{T}B(T) = 0.$

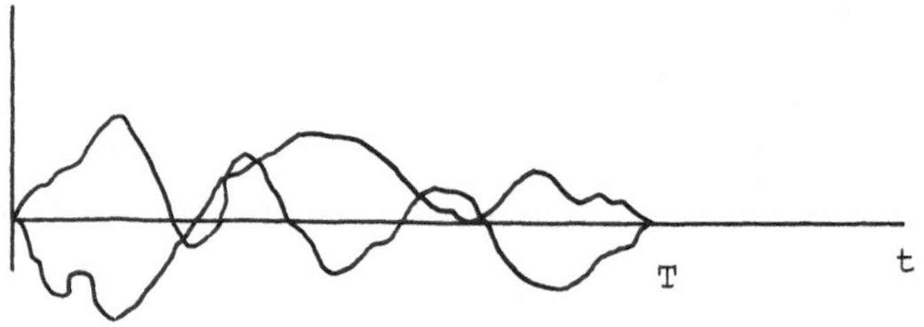

The Brownian Bridge occurs e.g. as Phase Boundary in the 2-dimensional Ising Model. [14] (Fig.2)

6) Another way to fix B(T) is by conditional expactations!

$E(e^{i(f,B)}) \to E(e^{i(f,B)} | B(T) = 0)$.

To calculate this, we use

$E(.|B(T)) = E(.\delta(B(T))) / E(\delta(B(t)))$.

We get the r.h.s. from the characteristic functional by using

$$\delta(B(T)) = \frac{1}{2\pi} \int dy \; e^{iyB(T)}$$

$$= \frac{1}{2\pi}\int dy \; e^{i(y\delta_T, B)} \quad ,$$

i.e. $E(e^{i(f,B)}\delta(B(T)) = \frac{1}{2\pi}\int dy\ C(f+y\delta_T)$

$$= \frac{1}{2\pi}\int dy\ e^{-\frac{1}{2}\int_0^\infty dt_1 dt_2 (f(t_1)+y\delta(t_1-T))t_1\wedge t_2(f+y\delta(t_2-T))}$$

$$= \frac{1}{2\pi}E(e^{i(f,B)})\int dy e^{-\frac{1}{2}y^2 T} e^{-y\int dt f(t) t\wedge T}$$

$$= \frac{1}{2\pi}E(e^{i(f,B)})\left(\frac{2\pi}{T}\right)^{1/2} e^{1/2\{\int dt f(t) t\wedge T\}^2 . T^{-1}} \quad .$$

In particular $E(\delta(B(T)) = \left(\frac{1}{2\pi T}\right)^{1/2}$ is the denominator (set $f \equiv 0$).

Finally we obtain in this way

$$E(e^{i(f,B)}\,|\,B(T)=0)=e^{-1/2(f,K_T f)}$$

with $K_T(s,t) = t \wedge s - \frac{1}{T}(t \wedge T)(s \wedge T)$.

Exercise 3: Compare this to the characteristic functional of the Brownian Bridge.

A little more generally:

For $E(e^{i(f,B)}\,|\,B(T)=x)$ we must use similarly $\delta(B(T)-x)$ and obtain

$$E(e^{i(f,B)}\,|\,B(T)=x)=e^{-\frac{1}{2}(f,k,f)+i(m,f)}$$

with $m(t) = \frac{x}{T}(t \wedge T)$.

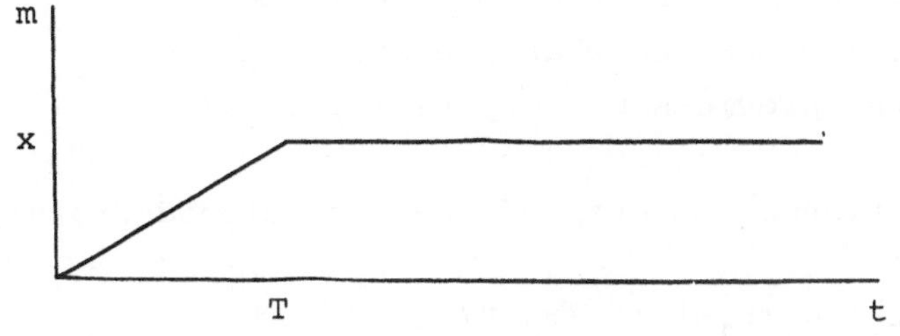

A closer investigation of the conditional expectation will tell us a few things about the properties of Brownian Motion.

1) For $t \leq T$ the conditioned process equals the Brownian Bridge:
   Let $f = f_- = 0$ for $t > T$,

   then $E(e^{i(f,B)} | B(T)=x) = E(e^{i(f_-,B_x^T)})$

   where $B_x^T(x) = x\frac{t}{T} + B^T(t)$.

2) Let $f = f_- + f_+$ $\quad (f_+ = 0$ for $t < T)$,
   then
   $$E(e^{i(f,B)} | B(T)=x) = E(e^{i(f_-,B_x^T)})E(e^{i\int_0^\infty dt f_+(t+T)(B(t)+x)}) .$$
   More generally we have the following independence (factorization) of past ($t<T$) and future ($t>T$):

   $$E(F_+(B)\cdot F_-(B) | B(T))$$
   $$= E(F_+(B)|B(T))\cdot E(F_-(B)|B(T))$$

   if $F_{\substack{+\\(-)}}$ depends only on $B(t)$ with $t \underset{(\leq)}{\geq} T$.

   Any such process is called a "Markov Process".
   Equivalently

   $$E(f_+(B)|B(T),\underset{s<T}{B(s)}) = E(F_+(B)|B(T)).$$

   In other words: Given the present $B(T)$, the future development $F_+(B)$ will not depend on the past $B(s)$, $s<T$. In any gambling game (without a "memory") your capital $K(t)$ for $t>T$ will depend on what you had at $t=T$, but knowing $K(T)$, the previous history $K(s)$ for $s<T$ is irrelevant.

3) Recall the covariance and mean of the conditioned Brownian Motion $(B(T)=x)$.
   We have $K_T(t_1,t_2) = (t_1-T) \wedge (t_2-T)$

   $$m(t) = x$$

   for times t, t_i later than T, i.e. the conditioned Brownian

Motion behaves as if starting afresh at time T from x, i.e. like B(t-T)+x.

4) Next we use this fact to calculate the probability distribution of the "First Hitting Time" of a point a:

$T_a = \inf\{t : B(t) = a\}$

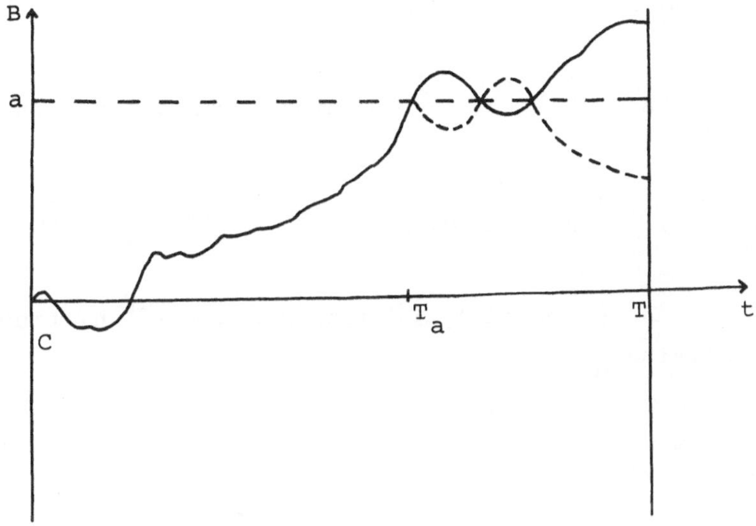

After a time $t = T_a$, the continuation of the trajectory has the same probability as its mirror image, hence

$$p(B(T)>a) = \frac{1}{2}p\{\max_{t \le T} B(t) \ge a\} .$$

This is the "Reflection Principle". It follows that

$$p\{\max_{t \le T} B(t) \ge a\} = 2\int_a^\infty \rho(x)\, dx$$

where $\rho(x)$ is the probability density for B(T). Recall that ρ is the Fourier transform of $C(\lambda)$:

$$C(\lambda) = E(e^{i\lambda B(T)}) = e^{-\frac{T}{2}\lambda^2}$$

$$\rho(x) = (2\pi T)^{-1/2}\, e^{-\frac{x^2}{2T}} .$$

Insert this to find

$$p\{\max_{[0,T]} B(t) \geq a\} = \frac{2}{\sqrt{2\pi T}} \int_a^\infty e^{-\frac{x^2}{2T}} dx \quad .$$

This is identical to the probability for the 1st hitting time T_a to be less than or equal to T,

$$p(T_a \leq T) = \frac{2}{\sqrt{2\pi T}} \int_a^\infty e^{-\frac{x^2}{2T}} dx \quad .$$

Recall that $p(T_a \leq T)$ is the distribution function of T_a:

$$p(T_a \leq T) = \int_0^T \rho_{T_a}(t)dt \quad ,$$

and get the probability density for the 1st hitting time by applying $\frac{d}{dT}$:

$$\rho_{T_a}(t) = \frac{a}{\sqrt{2\pi}} t^{-3/2} e^{-\frac{a^2}{2t}} \qquad (t \geq 0) \quad .$$

5) Recall that for $t \geq T$ the mean of the conditioned Brownian Motion is given by

$$m(t)=E(B(t)|B(T)=x) = x \qquad (t \geq T) \quad .$$

The location x where we fixed the process at time t=T determines its mean value for the future always to be x. Such processes are called "<u>Martingales</u>". One can visualize such processes as models for "fair games": If a gambler has a capital x at t=T, then that will be the average amount he will have in the future. Related concepts are

$m(t) \geq x$: "<u>super martingale</u>"
$m(t) \leq x$: "<u>submartingale</u>".

<u>Exercise 4</u>: Show that $M(t) = e^{B(t)-t/2}$ is a martingale.

6) "Transition Probabilities"
Recall one of the questions raised in the Introduction: What is the probability for a trajectory to reach a set

A at time T if it was at the point x at time T?
This is a conditional probability!

$$p(t,A;T,x) = E(\chi_A(B(t))|B(T) = x)$$

$$= \int_A dy\ E(\delta(B(t)-y))|B(T) = x)$$

$$= \int_{-\infty}^{\infty} dy\ \chi_A(y)p(t,y;T,x)$$

with $p(t,y;T,x) = E(\delta(B(t)-y)|B(T)=x)$ the "Transition Probability Density", i.e. the density of the measure P, $P(.,dy;.,.) = p(.,y;.,.)dy$.
We have seen how easy it is to calculate such conditional expectations:
Exercise 5: Verify that for Brownian Motion

$$p(t,y;T,x) = (2\pi(t-T))^{-1/2}e^{-1/2\frac{(x-y)^2}{t-T}} \quad .$$

Remarks and Results on Transition Probabilities.

1. For Brownian Motion p and P depend only on the time difference t-T. Such transition probabilities are called "stationary" and the process is called "homogeneous" (evidently a homogeneous process need not be stationary). Brownian Motion is called an "additive" homogeneous process, because p depends only on space differences x-y. Another such process is given by
2. Example 3: "Symmetric Random Walk" (on a one-dimensional lattice).

In each interval of time a unit step is taken either to the left or to the right, with equal probability.
Note that this models the coin toss game. The gambler's capital K(t) is controlled by
$p(t,x;O,K) = p(K(t) = x|K(O) = K)$.
K(t) is a "Markov chain".
Let us study the chances of winning!
The game ends if $K(n) = O$ (lost all)
or $K(m) = A$ (won all) .

Now define P_K = prob. to lose a capital K.
Note: The gambler has an equal chance in each game to increase or decrease K by one. Hence we must have

$$P_K = \frac{1}{2} P_{K+1} + \frac{1}{2} P_{K-1} \qquad \text{(by the law of total probability).}$$

Let us solve this difference equation with the Ansatz

$$P_K = a + bK$$

and the boundary conditions:

$P_o = 1$ (already lost all)

$P_A = 0$ (already won all).

We find $P_K = 1 - \frac{K}{A}$.

This says that one has no chance against a rich bank ($A \to \infty$).
"Gambler's Ruin":
You are sure to lose your money!
How long is a game expected to take?

$$T_K = E_K(T_{O,A})$$

where $T_{O,A}$ is the 1st hitting time of the set {O,A} for the random walk starting at K. Here we use the Law of Total Expectation:

$$T_K = E(T_{O,A} | \text{win } 1^{st} \text{ game}) \cdot 1/2$$

$$+ E(T_{O,A} | \text{lose } 1^{st} \text{ game}) \cdot 1/2$$

$$= (T_{K+1} + 1)1/2 + (T_{K-1} + 1)1/2 .$$

Ansatz: $T_K = a + bK + cK^2$
Boundary Conditions: $T_o = T_A = 0$ (game already decided)

$\to a = 0 \quad b = A \quad c = -1$

$$T_K = K(A-K)$$

As the bank is made infinitely rich

$(A - K \to \infty)$

it will - on the average - take infinitely long until you surely lose. Now we translate this back to the language of the symmetric random walk: Since we shifted the random walk to the origin, the "absorbing barriers" are $a = A - K$, $b = K$. The probability for absorption at b is then

$$P_b = 1 - \frac{b}{a+b} \, .$$

Note $P_a + P_b = 1$: Absorption is sure. The mean time for absorption to occur is

$$T = a \cdot b \, .$$

3. Symmetric Random Walk and Brownian Motion

For the random walk denote

$$P(t,x;o,0) \equiv p(t,x) \, .$$

It is the probability to find the walker at x at a time t after his departure from the origin. Obviously $p(o,x)=\delta_{x,o}$, and after the 1st step:

$$p(1,x) = \frac{1}{2}(\delta_{x,1} + \delta_{x,-1}) \, .$$

More generally

$$p(t+1,x) = \frac{1}{2}\, p(t,x+1) + \frac{1}{2}\, p(t,x-1) \, .$$

Rewrite this equation in the form

$$p(t+1,x)-p(t,x) = \frac{1}{2}\{(p(t,x+1)-p(t,x)-(p(t,x)-p(t,x-1))\},$$

generalize the time interval: $t+1 \to t + \Delta t$,
generalize the space interval: $x+1 \to x + \Delta x$, and choose

$$\Delta t = (\Delta x)^2 .$$

As a result

$$\frac{p(t+\Delta t,x)-p(t,x)}{\Delta t}$$

$$= \frac{\frac{p(t,x+\Delta x)-p(t,x)}{\Delta x} - \frac{p(t,x)-p(t,x-\Delta x)}{\Delta x}}{\Delta x} .$$

Now take the "continuum limit"

$$(\Delta x)^2 = \Delta t \to 0$$

to obtain

$$\frac{d}{dt}\, p(t,x) = \frac{1}{2}\frac{d^2}{dx^2} p(t,x) ,$$

the "Heat Equation". The Initial condition is $p(o,x) = \delta(x)$, and we integrate the differential equation by

$$p(t,x) = e^{-H_o t} p(o,x) , \qquad H_o = -\frac{1}{2}\frac{d^2}{dx^2}$$

$$= \frac{1}{2\pi}\int dp\; e^{-\frac{1}{2}p^2 t}\; e^{ipx}$$

$$= (2\pi t)^{-1/2}\; e^{-\frac{x^2}{2t}} .$$

This is nothing other than the transition density of Brownian Motion!

Brownian motion is the continuum ("scaling") limit of the Random walk! Figures 4-7 show clearly that $(\Delta x)^2 \sim \Delta t$: the amplitude scale (x) grows like the square root of the time scale (t).

4. A Remark on Fractals

It can be seen with a naked eye that the random walk on the large scale looks similar to that in a much smaller scale. Take, for instance, the points in time when the random walk returns to the origin. These zeros occur in clusters. Sets of this type have been used as a mathematical model for noise in communications channels where typically quiet periods will be interrupted by bursts (clusters) of noise [1] . If we look at such a cluster in the smaller scale, it would be composed of smaller clusters.

For the random walk, small clusters are grouped together to form big ones, those again form even bigger ones and so on ad infinitum. In the small however scaling will break down when one has dissolved the clusters all the way down the individual crossings of zero, which are spaced apart at least by the time step $2\Delta t$. Brownian motion arising if we take the limit $\Delta t \to 0$ should then be scaling invariant in both directions. Indeed $B_\lambda(t)$ which we define by

$$B_\lambda(t) = \lambda^{-1/2} B(\lambda t)$$

is a version of Brownian motion for all λ between zero and infinity!

So what can we say about zero sets of Brownian motion? Of course, for $t > 0$ they will be stochastic but otherwise much like a Cantor set; clusters contain similar clusters which contain similar clusters, ...
They are the prototypes of stochastic "fractal" sets. They are characterized by anomalous "fractional" dimensions (for definitions see [1] . While the set of "level crossings" will have dimension zero for a smooth curve, Brownian motion has

$$d = \dim\{t : B(t) = 0\} = \frac{1}{2}$$

for the level crossings. A symptom if not a proof of this fact is that

$$\int_{0}^{T} ds \int_{0}^{T} dt \frac{\delta(X(s)-X(t))}{|t-s|^{\beta}}$$

has finite expectation for all $\beta <$ dim $\{t:X(t) = 0\}$. This expectation is very easy to calculate for Gaussian processes with stationary increments (and zero mean). Their characteristic function is of the form

$$E(e^{i\lambda(X(t)-X(s)}) = e^{-\frac{1}{2}\lambda^2\sigma^2(t-s)} ,$$

and by the usual tricks of Gaussian integration we find

$$E(\delta(X(t)-X(s)) = 2\pi\sigma^2(t-s)^{-1/2} ,$$

so that we expect the dimension d to be the supremum of the β for which

$$\int \frac{d\tau}{\sigma(\tau)\tau^{\beta}}$$

is finite at the origin. Then for $\sigma^2(\tau) \approx \tau^{\alpha}$ this implies

$$d = 1 - \frac{\alpha}{2}$$

where $\alpha = 1$ is the Brownian Motion; for other α (0,2) we are dealing with "fractional Brownian motion". As one would expect $D = d + 1$ is the dimension of the graph of $X(t)$; in particular, one dimensional Brownian Motion traces out a non-rectifiable curve of dimension $\frac{3}{2}$: Such infinite length is typical for curves that look as jagged under a "looking glass" as they look in the large. Take for example the graph of a border line on the map (Fig.3). As more and more details are revealed on a finer scale the apparent length of the border becomes bigger and bigger:

$L = L_o\, S^{1-D}$.

We urge the reader to look into Mandelbrot's book [1] for stochastic models of landscapes and for much more on fractals and their applications.

5. Generalities on the Transition Probability for Markov processes

$$P(T+t,A;T,x) = P_t(A,x)$$

$$= \int p_t(y,x)\chi_A(y)\,dy .$$

We can associate a linear operator P_t with this:

$$P_t(A,x) = (P_t \cdot \chi_A)(x)$$

where

$$(P_t f)(x) = \int p_t(y,x) f(y)\,dy .$$

Recall for Brownian Motion:

$$p_t(y,x) = e^{-H_o t}\delta(x-y)$$

i.e. $P_t = e^{-H_o t}$ with $H_o = -\frac{1}{2}\frac{d^2}{dx^2}$.

The operators P_t of Brownian Motion form a semigroup!

$$P_{t+s} = P_t P_s \quad .$$

We write this in terms of the densities:

$$p_{t+s}(y,x) = \int_{-\infty}^{\infty} du\; p_s(y,u)\,p_t(u,x) .$$

This is the "Chapman-Kolmogorov Equation" (for the transition density of homogeneous processes). The factorization of Markov Processes! After the 1st time interval t is completed the further propagation depends only on the present value u of the process, not on past history.

6. To construct more general Markov processes with continuous sample functions (Diffusion Processes) from Brownian Motion, we have another look at the transition density of Brownian Motion:

$$p_t(y,x) = (2\pi t)^{-1/2} e^{-(y-x)^2/2t} \quad .$$

For $t \to 0$ this approaches a δ-function, more precisely:

$$\lim_{\Delta t \to 0} \frac{1}{\Delta t} \int_{|y-x| \geq \varepsilon} p_{\Delta t}(y,x)\,dy = 0 \;, \qquad \varepsilon > 0 \;.$$

We also have

$$\lim_{\Delta t \to 0} \frac{1}{\Delta t} \int_{|y-x| < \varepsilon} (y-x)\,p_{\Delta t}(y,x)\,dy = b = 0 \;,$$

i.e. the "Instantaneous Mean" $b = 0$, as well as

$$\lim_{\Delta t \to 0} \frac{1}{\Delta t} \int_{|y-x| < \varepsilon} (y-x)^2 p_{\Delta t}(y,x)\,dy = a = 1,$$

the "Instantaneous Variance" a equals one. The 1st of the three conditions, "Lindeberg condition", reflects Continuity of Sample Paths: it excludes large displacements in short times. All three equations make statements about the increment of the process (starting from the point x) in a short time t, i.e. about

$$\Delta Y(t) = Y(t+\Delta t) - Y(t) \;.$$

Generalization to $b = b(x) \neq 0$ and $a = a(x)$ can be viewed as a Brownian Motion in an inhomogeneous medium. Such "locally Brownian" random processes (with continuous sample paths!) are called "diffusions".
Let us try to build these processes from Brownian Motion. To produce a finite "instantaneous mean" (or "drift")

b(x) set

$$\Delta Y(t) = \Delta B(t) + b(Y)\Delta t \ .$$

This produces b(x) on the r.h.s. of the 2nd equation but will not change the instantaneous variance since it produces a term $\sim(\Delta t)^2$ there.
Similarly the replacement

$$\Delta B \to a^{1/2}(Y)\Delta B$$

will change the "instantaneous variance" from s to a(x), but not the mean.
Formally, the diffusions are thus characterized by the "stochastic differential equation"

$$dY = b(Y)dt + a^{1/2}(Y)dB$$

or

$$\dot{Y} = b(Y) + a^{1/2}(Y)\dot{B} \ .$$

A better defined expression is

$$Y(t) - Y(t_o) = \int_{t_o}^{t} b(Y(\tau))d\tau + \int_{t_o}^{t} a^{1/2}(Y)dB(\tau) \ .$$

This avoids the derivative of Y, but still:
dB not of bounded variation, the 2nd integral is a "Stochastic Integral" or "Ito-Integral", see e.g. Streater's, Zoller's lectures in this volume. Recall: In the case of Brownian Motion the transition density obeys the Heat Equation. Now we want to generalize this to other Diffusion Processes.
We start from the Chapman-Kolmogorov-Equation: (Markov property!)

$$p_{t+\delta}(y,x) = \int p_t(y,u)p_\delta(u,x)du$$

which gives us

$$p_{t+\delta}(y,x)-p_t(y,x) = \int\{p_t(y,u)-p_t(y,x)\}p_\delta(u,x)du \ .$$

Now we split the integral over u into regions of large and small $|u-x|$ and use a Taylor series for small $|u-x|$:

$$p_{t+\delta}-p_t = \int_{|x-u|\geq\varepsilon}\{\cdot/.\}p_\delta(u,x)du$$

$$+ \int_{|x-u|<\varepsilon}\{\partial_x p_t(u,x)(u-x) + \tfrac{1}{2}\partial_x^2 p_t(y,x)(u-x)^2\} \ .$$

$$\cdot \ p_\delta(u,x)du \ .$$

We divide by δ and let $\delta\to+0$ and find

$$\partial_t p_t(y,x) = (\partial_x p_t(y,x))b(x) + \tfrac{1}{2}(\partial_x^2 p_t(y,x))a(x) \ ,$$

the "Backward Equation" (since it operates on the initial position x). With respect to the "forward variable" y one finds similarly the "Forward" or "Fokker-Planck-Equation"

$$\partial_t p_t(y,x) = -\partial_y(b(y)p_t(y,x)) + \tfrac{1}{2}\partial_y^2(a(y)p_t(y,x)) \ .$$

Example 4: The "Ornstein-Uhlenbeck-Process"

$a = 1 \quad b(x) = -mx$

is a diffusion with instantaneous variance like Brownian Motion and a drift directed towards the origin. The corresponding stochastic differential equation is evidently

$$\dot{Y}(t) = -mY(t) + X(t)$$

where

$X(t) = \dot{B}(t)$ is White Noise, -

the "Langevin-Equation". The integration is easy:

$$Y(t) = \int_{-\infty}^{t} e^{-m(t-s)} X(s)ds.$$

This version of the Ornstein-Uhlenbeck-Process is called the "canonical" or "Hida representation". It depends causally on White Noise $X(s)$: $s \leq t$.
The O.-U.-Process Y is Gaussian:

$$Y = TX \quad \text{i.e.} \quad (f,Y) = (T^*f,X)$$

$$C_Y(f) = C_X(T^*f) = e^{-\frac{1}{2}(T^*f,T^*f)}$$

and has zero mean. We can also see this directly:

$$E(Y(t)) = \int_{-\infty}^{t} e^{-m(t-s)} E(X(s))ds = 0 .$$

Similarly we calculate the covariance K:

$$E(Y(t_1)Y(t_2)) = \int_{-\infty}^{t_1} ds_1 \int_{-\infty}^{t_2} ds_2 \; e^{-m(t_1-s_1)-m(t_2-s_2)} \cdot$$

$$\cdot \; E(X(s_1)X(s_2))$$

$$= e^{-m(t_1+t_2)} \int_{-\infty}^{t_1 \wedge t_2} ds e^{2ms}$$

$$= \frac{1}{2m} e^{-m|t_1-t_2|}, \quad E(X(s_1)X(s_2)) = \delta(s_1-s_2).$$

Note: Y is stationary!
Also: The O.-U.-Process is a Markov Process! To see this we consider the process

$$Z(t) = e^{-mt} B\left(\frac{e^{2mt}}{2m}\right) .$$

Z is Gaussian with $E(Z(t)) \equiv 0$.
Let us compute its covariance:

$$K(t_1,t_2) = E(Z(t_1)Z(t_2))$$

$$= e^{-m(t_1+t_2)}\left(\frac{e^{2mt_1}}{2m} \wedge \frac{e^{2mt_2}}{2m}\right)$$

$$= \frac{e^{-m|t_1-t_2|}}{2m} \quad .$$

→ Z is a version of the O.-U.-Process! The O.-U.-process Z inherits from B:

- the Markov Property
- continuity of Sample Paths.

Exercise 6: Is the O.-U.-Process Z(t) a Martingale? What is the canonical representation of the O.-U.-Process conditioned to Y(T) = x?

The version Z(t) is also useful to quickly calculate the transition density:

$$p(t,y;T,x) = E(\delta(Z(t)-y)\,|\,Z(T) = X)$$

$$= E\left(\delta\left(e^{-mt}B\left(\frac{e^{2mt}}{2m}\right)-y\right)\Big|\,B\left(\frac{e^{2mT}}{2m}\right) = x\,e^{mT}\right)$$

$$= e^{mt}\; p_{B.M}\left(\frac{e^{2mt}}{2m},\; y\,e^{mt};\; \frac{e^{2mT}}{2m},\; x\,e^{mT}\right)$$

$$= e^{mt}\left\{2\pi\left(\frac{e^{2mt}}{2m} - \frac{e^{2mT}}{2m}\right)\right\}^{-1/2} \cdot e^{-\frac{1}{2}\frac{(xe^{mT}-ye^{mt})^2}{\frac{e^{2mt}}{2m}-\frac{e^{2mT}}{2m}}}$$

$$= \left\{\frac{\pi}{m}\left(1-e^{2m(T-t)}\right)\right\}^{-1/2} e^{-m\frac{(x-ye^{m(t-T)})^2}{e^{2m(t-T)}-1}} \quad .$$

This is the "Chandrasekhar Formula". Note the difference between the canonical representation of the O.-U.process and Brownian Motion in terms of White Noise:

$$Y(t) = \int_{-\infty}^{t} e^{-m(t-s)} X(s)ds$$

$$B(t) = \int_{0}^{t} X(s)ds.$$

Stationarity of White Noise X, i.e. invariance of expectations under

$$X(\cdot) \to X_\tau(\cdot) = X(\cdot - \tau)$$

carries over to the O.U.Process:

$$Y_\tau(t) = Y(t-\tau) = \int_{-\infty}^{t-\tau} e^{-m(t-\tau-s)} X(s)ds$$

$$= \int_{-\infty}^{t} e^{-m(t-s)} X_\tau(s)ds = Y(t) .$$

This is not so for Brownian Motion: integration starts only at zero.

What if we consider the $m \to +0$ limit of the O.-U.-Process? Consider the characteristic functional:

$$C(f) = e^{-\frac{1}{2}(f,Kf)}$$

$$K(t_1,t_2) = (2m)^{-1} e^{-m|t_1-t_2|} .$$

In terms of Fourier transforms:

$$(f,Kf) = \frac{1}{2}\int dp\, \tilde{f}(p) \frac{1}{p^2+m^2} \tilde{f}(p) .$$

This will diverge for $m \to 0$ at $p = 0$ unless

$$0 = \tilde{f}(0) = \int dx f(x) .$$

Note this excludes considering the process at a fixed point t in time, for which $\tilde{f}(p) \sim e^{ipt}$. We are not even dealing with a generalized process on ϕ! A possible test function space [14] would be

$$S^{(1)} = \{f = F' \in S : F \in S\}$$

since $\int_{-\infty}^{\infty} f\, dx = F(x)\Big|_{-\infty}^{\infty} = 0$.

This is typical for scale invariant stationary processes;

$$\tilde{K} \sim p^{v}$$

gives rise to either infrared or ultraviolet divergences.

Note: $\tilde{K}(p) = \frac{1}{2} \frac{1}{p^2+m^2}$

of the O.-U.-Process looks like a "Two-Point Function" of Euclidean Quantum Field Theory (cf.eg. B. Simon); the O.-U.-Process is the Euclidean "free field" in one-dimensional space-time, i.e. the "Euclidean Theory" of the quantized harmonic oscillator.
To see this, let us now look at the semigroup of the process: Recall: To the stochastic differential equation

$$dY = bdt + a\, dB$$

there corresponds the (backward) equation for the probability density p_t:

$$-\partial_t p_t(y,x) = -a(x)\partial_x^2 p_t(y,x) - b(x)\partial_x p_t(y,x) .$$

Hence the semigroup $P_t = e^{-Ht}$ is generated by

$$H = -a(x)\partial_x^2 - b(x)\partial_x .$$

In our case $a = 1$, $b(x) = -mx$, so that

$$H_{O.U.} = -\partial_x^2 - mx\partial_x .$$

We diagonalize H:

$H\phi_n = \lambda_n \phi_n$ for $\phi_n = \text{const.} \cdot H_n(\sqrt{m}\, x)$. These are Hermite Polynomials, orthogonal in

$$\mathcal{H} = L^2(R, e^{-mx^2}dx) \ .$$

I.e. $P_t = e^{-tH}$, where H is the Harmonic Oscillator Hamiltonian in $\mathcal{H}$.

The occurrence of

$$\rho(x)dx = \psi_o^2(x)dx$$

where ψ_o is the ground state wave function is typical. Consider more generally

$$H = -\partial_x^2 - b(x)\partial_x \ .$$

We want this to be self-adjoint in

$$L^2(R, \rho(x)dx) \ .$$

How must we choose $\rho(x)$? Find the answer by comparison with

$$\hat{H} = -\partial_x^2 + V(x) \qquad \text{in } L^2(R, dx) \ .$$

Let $\hat{H}\psi_o = 0$, ψ_o the ground state. Write any other vector $\phi(x) = f(x)\psi_o(x)$. How does $\hat{H}$ transform f?

$$\begin{aligned} \hat{H}\phi &= -f''\psi_o - 2f'\psi_o' - f\psi_o'' + Vf\psi_o \\ &= (-f'' - 2\frac{\psi_o'}{\psi_o} f')\psi_o + f\,\hat{H}\psi_o \\ &= (Hf)\psi_o \text{ if we set } b(x) = 2\psi_o'/\psi_o \ . \end{aligned}$$

We have come to associate the following three structures with each other:

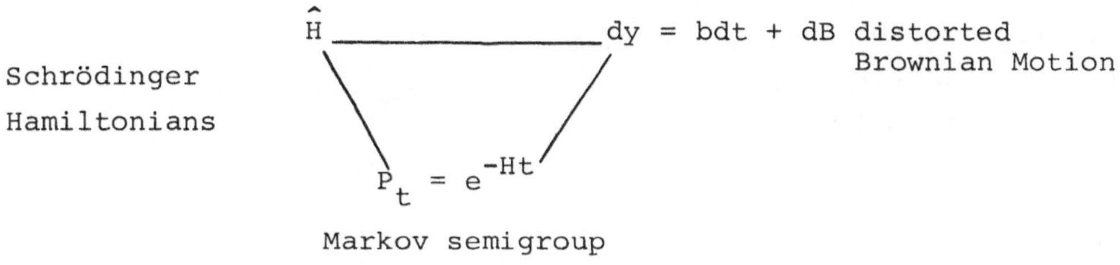

(1) Markov semigroup $P_t = e^{-Ht}$ with generators H self-adjoint in $L^2(R^n,\rho(x)d^nx)$.

(2) Schrödinger operators $\hat{H}$ in $L^2(R^n,dx)$
- unitarily equivalent to H
- such that $\hat{H}\psi_0 = 0$ ("ground state") and $|\psi_0^2(x)| = \rho(x)$
- formally $\hat{H} = -\Delta_x + V(x)$ with $V = \Delta\psi_{0/\psi_0}$ but existing for interactions much more singular than can be treated perturbatively!

(3) "distorted Brownian Motions" (diffusions Y(t))
- with a drift given by $b(x) = 2\nabla\psi_0(x)/\psi_0(x)$
- transition semigroup P_t
- invariant distribution $\rho(x)$ and
- under very general conditions for existence and uniqueness.

Many recent results in these three fields have sprung from their interrelation (see [3,4,15,16]); extensions for infinitely many degrees of freedom n have been studied by Albeverio et al.[17] and by Kusuoka.

FIGURE CAPTIONS

Fig.1.: Figure 1 shows a two phase system (Ising model, binary alloy, lattice gas) at temperature T = 0 and boundary conditions fixing the state at each lattice site to be + resp. - in the upper resp. lower half.

Fig.2.: Figure 2 shows the same system at nonzero temperature where fluctuations deform the phase boundary.

Fig.3.: The shape of Austria.

Fig.4.: One-dimensional random walk.

Fig.5.: One-dimensional random walk, time and space scaled by 10^{-1} resp. $10^{-1/2}$.

Fig.6.: One-dimensional random walk, time and space scaled by 10^{-2} resp. 10^{-1}.

Fig.7.: One-dimensional random walk, time and space scaled by 10^{-3} resp. $10^{-3/2}$.

REFERENCES

1. B. Mandelbrot, "Fractals. Form, Chance and Dimension" (Freeman, San Francisco, 1978).
2. T. Hida, "Brownian Motion",Applications of Mathematics, Vol.11 (Springer, Berlin, 1980).
3. S. Albeverio, R. Høegh-Krohn, L. Streit, "Energy Forms, Hamiltonians, and Distorted Brownian Paths", J. Math. Phys. 18 (1977) 907.
4. S. Albeverio, R. Høegh-Krohn, L. Streit, "Regularization of Hamiltonians and Processes", J. Math. Phys. 21 (1980) 1636.
5. J.M. Gelfand, N.Ya. Vilenkin ,"Generalized Functions", Vol.4 (Academic Press, New York, 1964).
6. J. Holtsmark, "Über die Verbreitung von Spektrallinien", Ann. d. Physik 58 (1919) 577.
7. W. Feller, "An Introduction to Probability Theory and its Applications", 2nd ed. (Wiley, New York, 1971).
8. S. Chandrasekhar, "Stochastic Problems in Physics and Astronomy", Rev. Mod. Phys. 15 (1943) 1.
9. A. Khinchin, "Mathematical Foundations of Statistical Mechanics" (1949).
10. J. Cassandro, G. Jona-Lasinio, "Many Degrees of Freedom-Field Theory" (Plenum, 1978), p.54 ff.
11. T. Hida, "Stationary Stochastic Processes" (Princeton University Press, Princeton, 1970).
12. E. Nelson, "Dynamical Theories of Brownian Motion" (Princeton University Press, Princeton, 1967).
13. C. DeWitt-Morette, K.D. Elworthy, "New Stochastic

Methods in Physics", Physics Reports 77 (1981) 121-382.
H. Ezawa, J.R. Klauder, L.A. Shepp, "A Path Space Picture for Feynman-Kac Averages, Ann. Phys. 88 (1974) 588.

14. Y. Higuchi, in "Field Theory - Algebras, Processes" (Springer, Vienna, 1980).
15. S. Albeverio, M. Fukushima, W. Karwowski, L. Streit, "Capacity and Quantum-Mechanical Tunneling", Comm. Math. Phys. 81 (1981) 501.
16. M. Fukushima, "Dirichlet Forms and Markov Processes" (North-Holland/Kodansha, Amsterdam/Tokyo, 1980).
17. S. Albeverio, R. Høegh-Krohn, "Topics in Infinite-Dimensional Analysis", in "Lectures Notes in Physics", Vol. 80 (Springer, Berlin, 1978).

Further references in this field are:

L. Breiman, "Probability" (Addison-Wesley, Reading 1968).

C. Carvalho, E. Ribeiro, L. Streit, "Stochastic Processes - Mathematics and Physics. An Introduction"(CFMC, Lissabon,1984).

J.L. Doob, "Stochastic Processes" (Wiley, New York, 1953).

E.B. Dynkin, "Markov processes" (Springer, 1965).

M.I. Freidlin, "Markov processes and differential equations", Progress in Math. III (Plenum Press, N.Y., 1969).

A. Friedman, "Stochastic differential equations and its applications", Vol.1 and II (N.Y., Academic Press, 1975).

N.G. van Kampen, "Stochastic Processes in Physics and Chemistry" (North Holland, Amsterdam, 1983).

P. Levy, "Processus Stochastiques et mouvement brownien", 2eme ed. (Gauthier-Villars, Paris, 1956).

E. Lukacs, "Characteristic functions", 2nd. ed. (Griffin, London, 1970).

M. Rosenblatt, "Random Processes" (Springer, Berlin, 1974).

B. Simon, "Functional Integration and Quantum Physics"(Academic Press, 1979).

J. Glimm, A. Jaffe, "Quantum Physics" (Springer,1981).

G. Jona-Lasinio, "The Renormalization Group - A probabilistic view", Nuovo Cim. 26B (1975) 99.

G. Jona-Lasinio, "Stochastic Dynamics and the Semi-Classical Limit of Quantum Mechanics", in "Quantum Fields - Algebras, Processes" (L. Streit, ed.)(Springer, Vienna, 1980).

E. Lieb, "Monotonicity of the molecular electronic energy in the nuclear coordinates", J. Phys. B15 (1982) L63.

M. Reed, "Functional Analysis and Probability Theory", in "Constructive Quantum Field Theory", G. Velo, A. Wightman, ed.,Lectures Notes in Physics, Vol. 25 (Springer, Berlin, 1973).

L. Streit, "Energy Forms: Schrödinger Theory Processes", Physics Reports. 77 (1980).

K. Symanzik, "Euclidean Quantum Field Theory", in "Local Quantum Field Theory", R. Jost, ed.,Academic Press, 1969).

R.L. Dobrushin, "Automodel generalized Random Fields and their Renormalization Group", Ann. Prob. 7 (1979) 1.

D. Dürr, S. Goldstein, J.L. Lebowitz , "Stochastic Processes Orginating in Deterministic Microscopic Dynamics", Journ. Statist. Phys. 30 (1983) 519.

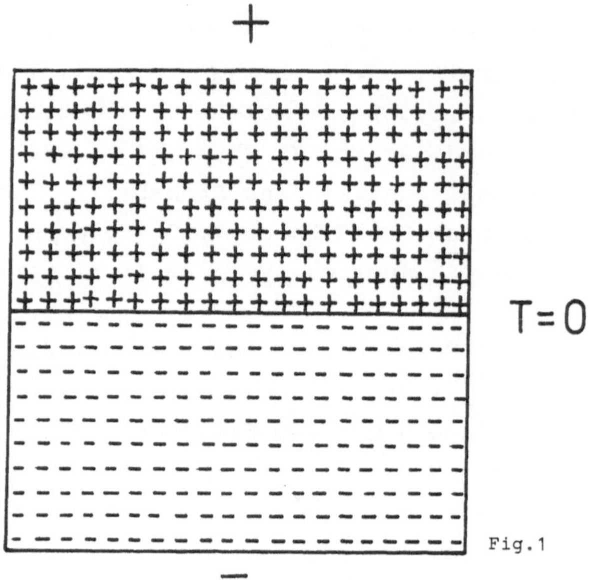

Fig.1

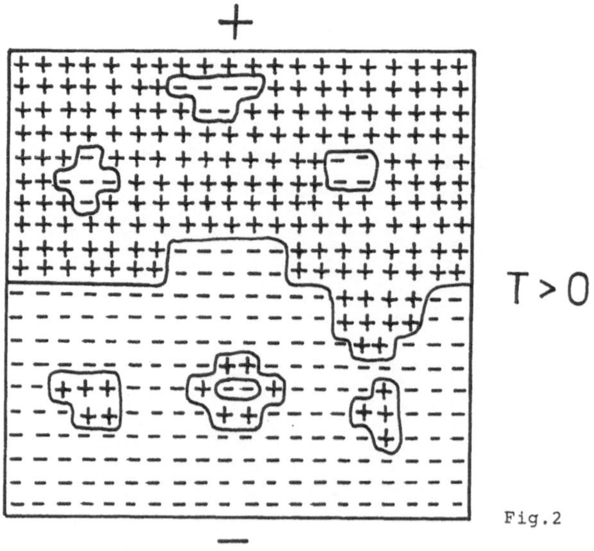

Fig.2

Fig.3

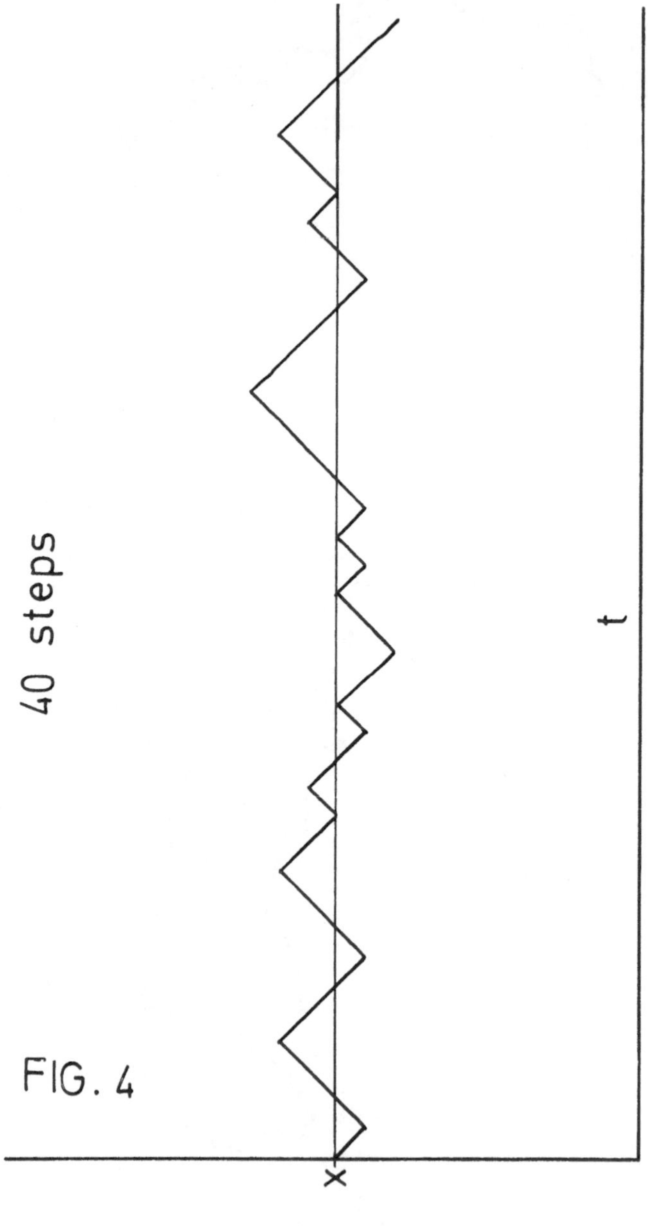

FIG. 4

40 steps

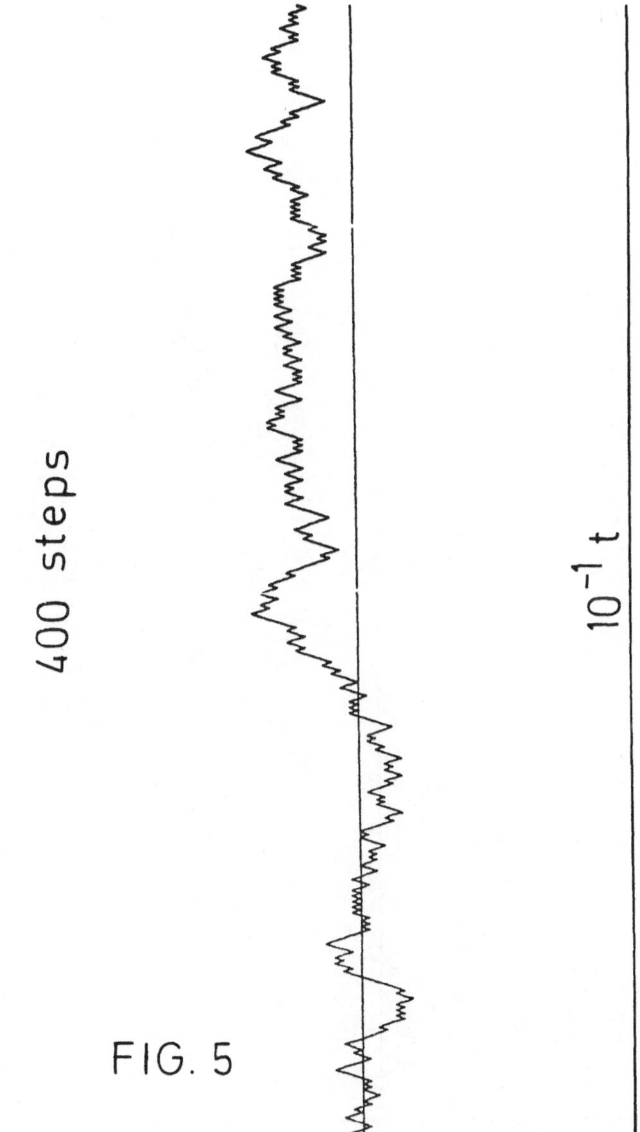

FIG. 5

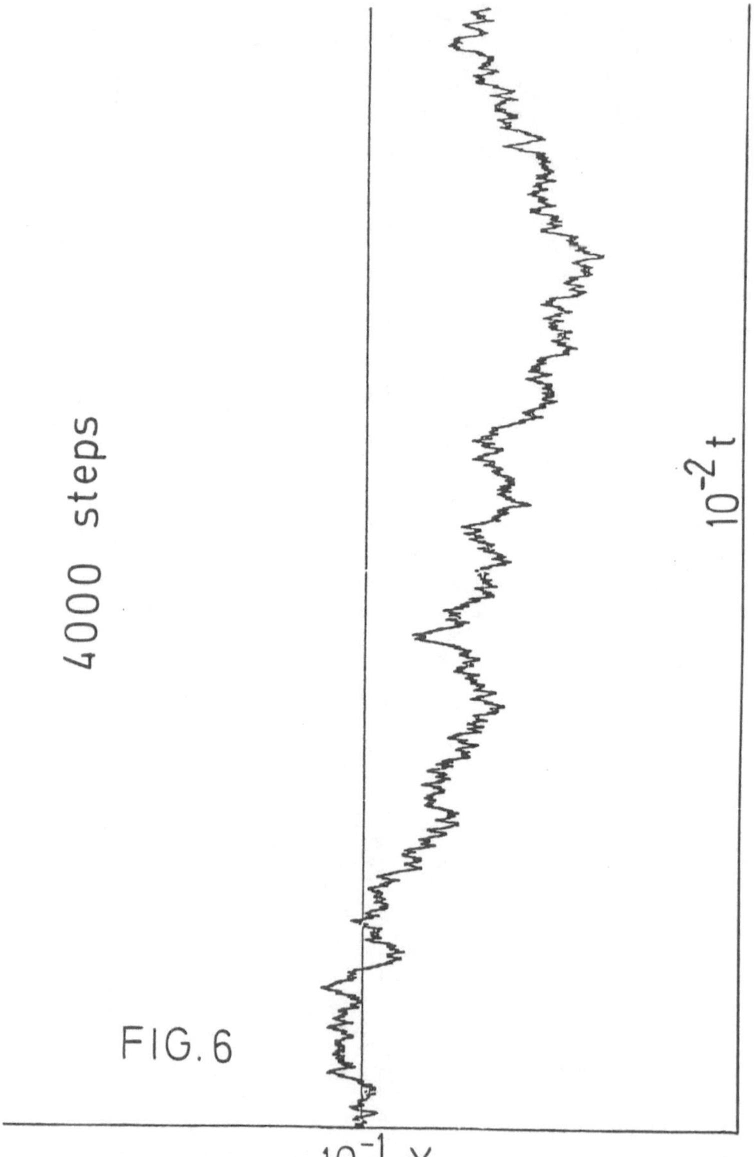

FIG. 6

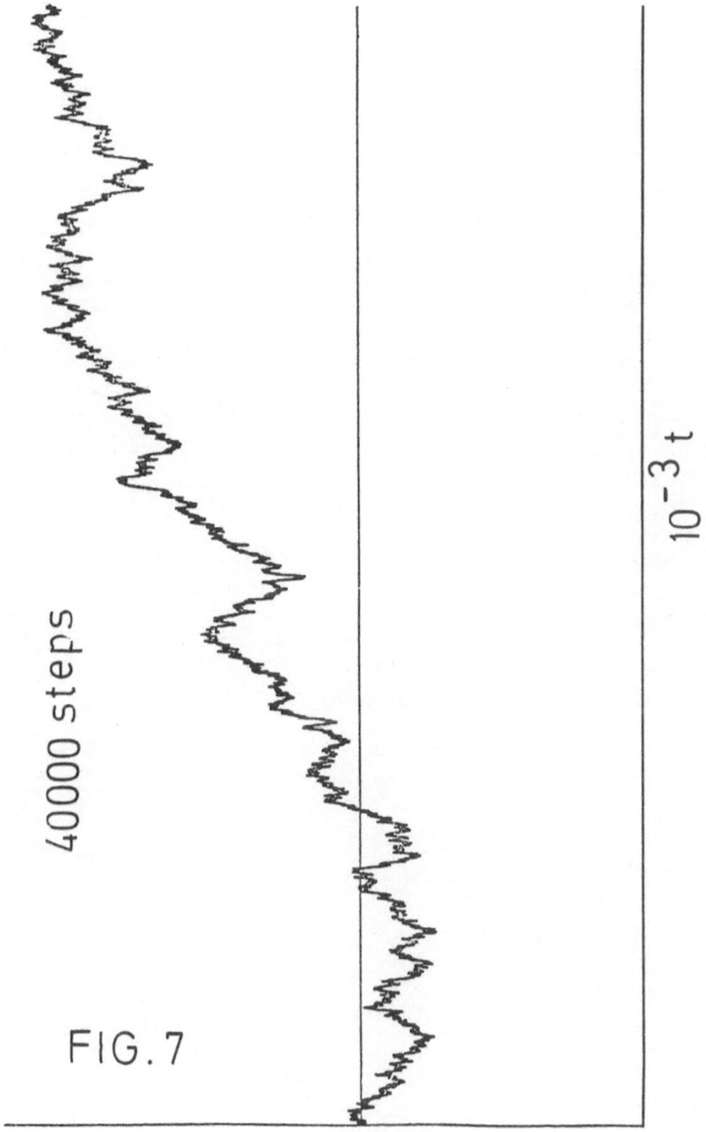

FIG. 7

Acta Physica Austriaca, Suppl. XXVI, 53–74 (1984)

QUANTUM STOCHASTIC INTEGRALS +

by

R.F. STREATER
Bedford College
Department of Mathematics
England

1. GENERAL REMARKS

There have been many attempts to set up quantum analogues of the theory of stochastic processes and stochastic differential equations. I should mention the many papers of M. Lax [1] on "quantum noise", and those of Senitzky [2]; these were inspired by the problem of describing a laser, and by quantum electronics. An account can be found in Haken [3]. Current-algebras led to my own research into continuous tensor products and infinitely divisible representations of algebraic structures [4], and to Araki's theory of factorizabel representations [5]. Survey articles (as at 1972) are [6,7]. Then there is the work of Evans and Lewis [8] and the book by E.B. Davies [9]. Independently, Hudson was developing a "quantum" version of Brownian motion and the Feynman-Kac formula [10]. The subject was stimulated by the work of Lewis and Thomas [11], and lately by

---

+ Lecture given at the XXIII. Internationale Universitätswochen für Kernphysik, Schladming, Austria, February 20-March 1, 1984.

the axioms of Accardi, Frigerio and Lewis [12]. These authors attempt to specify by means of axioms what a quantum stochastic process is. The axioms are determined by what they consider to be the essential physics of an open quantum system. Because of this, any model obeying these axioms must be quite realistic and probably not very easy to construct (cf. Maasen, thesis, Groningen 1982). Unrealistic but explicitly known theories like Brownian motion do not fall into their axiomatic scheme. We take a more lenient set of axioms which includes Brownian motion. In this our point of view is similar to that of Hudson and Parthasarathy [15] and Emch [16].

Classical diffusion is related to Brownian motion by the dilation problem. The quantum diffusions of Lindblad [17] and Gorini, Frigerio, Verri, Kossakowski and Sudarshan [18] can be similarly related to quantum stochastic processes in good cases [19,20].

The archetypical stochastic process is Brownian motion: for each time $t \geq 0$ we are given a random variable B_t, being the (random) x-coordinate of a particle at time t, given that is started at $x = 0$ at $t = 0$.

The process (B_t) can be characterized by

(1) $B_0 = 0$ with probability 1
(2) $(B_t)_{t \geq 0}$ is a Gaussian family

of mean 0 and covariance

$$E(B_s B_t) = \min(t,s) \quad . \tag{1.1}$$

As a consequence, (B_t) has independent increments: $X_t - X_s$ and $X_v - X_u$ are independent if $s<t<u<v$.

Brownian motion itself is not a very good model for the position of an actual particle in, say, a hot liquid. It is more realistic for the particle to suffer random jumps in its velocity, due to random impacts, than random jumps in position: the latter is tantamount to infinite velocity. This reflects the fact that B_t is, with probability 1, nowhere differentiable. These considerations led Ornstein and Uhlenbeck to propose that the position X_t of a particle in a hot liquid might be described by a "stochastic differential equation" or Langevin equation

$$m \frac{dX_t}{dt} = P_t$$

$$\frac{dP_t}{dt} = F_t - \gamma P_t + \lambda \frac{dB_t}{dt} \quad . \tag{1.2}$$

Here, F_t is the conservative force, $-\gamma P_t$ is the damping force, proportional to the velocity, and $\lambda \frac{dB}{dt}$ is the random impulsive force due to collisions: this term is called the <u>noise</u>. The fact that dB_t/dt is infinite with probability 1 leads to interesting problems of existence and meaning similar to the divergence problem of quantum field theory. In this case the problem is solved by stochastic integrals.

Before we come to that we need a word about "open" dynamical systems. In an isolated classical physical system, energy is conserved and the canonical variables p(t), q(t) at time t move according to Hamilton's equations of motion: p(t), q(t) are functions of p(0), q(0). If the system is not isolated we may nevertheless not wish to study more than the variables p(t), q(t) of the system, and to treat the effect of the environment, called the heat-bath, stochastically.

In an exact treatment, the variables p(t), q(t) are functions of p(0), q(0) and all the dynamical variables needed to describe the environment. In the approximate treatment, p(t), q(t) turn out to be functions of p(0), q(0) and the noise variables B_t from 0 to t. Thus p(t) and

$q(t)$ are random variables; one can study their means, variances and correlations, which are the things the stochastic theory predicts.

This picture gives us a strategy for constructing a quantum theory of an open system: at time 0 the system's variables $p(0)$, $q(0)$ are self-adjoint operators on the initial space of the system, $\mathcal{H}_o$ say. For each time t there will be a noise variable Φ_t, the quantum analogue of Brownian motion. Φ_t is a self-adjoint operator acting on a noise Hilbert space Γ. We then solve a quantum stochastic differential equation for operators $p(t)$, $q(t)$ on the space $\mathcal{H}_o \otimes \Gamma$ with initial values $p(0)\otimes 1_\Gamma$; $q(0)\otimes 1_\Gamma$. We expect the stochastic equation to be similar to the Langevin equation with $F(t)$ now given by Heisenberg's equation $F(t) = i[H,p]$, where H is a (conservative) Hamiltonian written as a function of $p(t)$, $q(t)$.

Quantum theory already is a probability theory: we do not need to add extra stochasticity. Thus Φ_t is to be an operator on Γ, not a random operator. In this, the spirit of the subject is different from that of the Schrödinger equation with random potentials.

We work in a Heisenberg-like picture: states do not move, and the dynamics is given by time-varying operators. The initial state (and so also the state at time t) is given by a density matrix on $\mathcal{H}_o\otimes\Gamma$. For example, if the environment is at temperature β^{-1}, the noise is described by a thermodynamic state ρ_β. If the system is initially described by the statistical state ρ and the system and noise are independent, then we would assign the state $\rho\otimes\rho_\beta$ to the combined system + noise. The interpretation of the theory is then as usual, e.g. the average position at time t is

$$\langle q(t)\rangle = \mathrm{Tr}_{\mathcal{H}_o\otimes\Gamma}((\rho\otimes\rho_\beta)q(t)) \quad .$$

At t=0, $q(t) = q(0)\otimes 1_{\Gamma}$, and this reduces to $Tr_{H_0}(\rho q(0))$, the system without noise at t = 0. As time goes by, the variables p(t), q(t) leap out into the unknown space Γ, picking up noise and forgetting the initial space H_0.

It will turn out that the Boson quantum noise Φ_t of Lax is actually identical to Brownian motion; there is a Fermion analogue Ψ_t, which we call the Clifford process. For these we can define the stochastic integral and using it, show the existence and uniqueness of solutions to stochastic differential equations.

2. REFORMULATION OF BROWNIAN MOTION

Any stochastic process has an underlying probability space Ω and a probability measure μ, which assigns to each measurable set $A \leq \Omega$ its measure $\mu(A)$, interpreted as the probability that the sample point lies in the set A: if so we say that the event A happened. We must then specify which sets are measurable: bitter experience tells us that not every subset of Ω can be assigned a probability. We call the collection F of measurable sets the Boolean σ-ring of the theory.

Although the concept of σ-ring of measurable sets was originally introduced for technical reasons, to eliminate pathology, the idea has proved to be very useful in specifying the observables of the theory: given the probability space (Ω,μ,F), an observable is an F-random variable i.e. a measurable function X from Ω to R: to each Borel set $E \leq R$, we require $X^{-1}(E)$ F, i.e. the inverse image of any measurable set in R is an element of F. We also say that X is F-measurable.

If we are given a collections of functions, $\Omega \rightarrow R$, then the σ-ring generated by S is to be the smallest σ-ring with respect to which each element of S is measurable.

In particular if we are given F and an F-random variable X_t for each $t \geq 0$, we can localize the observables in time. Thus, let F_t be the σ-ring generated by $S = \{X_s : 0 \leq s \leq t\}$. By construction $F_s \leq F_t \leq F$ if $s \leq t$. Any F_t-random variable can be regarded as an observable that can be measured in the time $[0,t]$. Usually, the increasing family $(F_t)_{t \geq 0}$ is fixed in advance, and we consider a <u>process</u> to be a random variable X_t for each $t \geq 0$, such that X_t is F_t-measurable for each t. We then say that (X_t) is <u>adapted</u> to F_t. An F-random variable X defines an operator on the Hilbert-space $L^2(\Omega, \mu F) = \Gamma$ by the formula

$$(X\phi)(\omega) = X(\omega)\phi(\omega) \quad . \tag{2.1}$$

If X is square integrable, it can also be regarded as an element of Γ. An important concept is the conditional expectation: let $G \leq F$ be a sub σ-ring. Then "the conditional expectation of X, given G", written $E(X/G)$ is the orthogonal projection of X onto $L^2(\Omega,\mu,G)$, at least, if X is square integrable. Note that $E(X/G)$ is a random variable.

For Brownian motion, $B_0 = 0$ and this generates the trivial σ-ring (containing ϕ,Ω only). If there are initial variables $p(0), q(0)$ say, we can describe them by random variables which generate the initial σ-ring F_0 say. The full σ-ring is then generated by F and F_0.

Now, classical probability is a special case of quantum probability (classical <u>mechanics</u> is not a special case of quantum <u>mechanics</u>), and the first step in our program is to rewrite the classical probability of Brownian motion as a commutative quantum theory.

Let $(\Omega,\mu,X_t,(F_t))$ be a stochastic process. We consider, in place of F_t, the (abelian) W^*-algebra $A_t = L^\infty(\Omega,\mu,F_t)$. By this I mean that A_t is the set of essentially bounded functions measurable relative to F_t, i.e. the bounded

F_t-random variables, regarded as multiplication operators on $L^2(\Omega,\mu,F_t)$. It is known that these form a weakly-closed $*$-symmetric algebra of operators, containing 1: i.e. a W^*-algebra. We can recover F_t (up to sets of measure 0) from A_t, because any self-adjoint projection in A_t is the indicator-function of a measurable set in F_t, and conversely. The function $1 \in L^2(\Omega,\mu,F)$ is a special vector: we recover all statistical properties of X_t from the relationship

$$E(y) = \langle 1, y1 \rangle = \int y(\omega)d\mu(\omega) \tag{2.2}$$

for every random variable y; the quantum expectation in the state 1 equals the classical expectation.

For Brownian motion, Ω is the set of continuous functions vanishing at t = 0. The σ-ring F is that generated by the "cylinder" sets". A cylinder set $A = \{(a_1,a_2), (a_2,b_2),\dots(a_n,b_n);t_1,\dots t_n\}$ is the set of $\omega \in \Omega$ such that $a_j \le \omega(t_j) \le b_j$, $1 \le j \le n$.
It is the set of slalom paths:

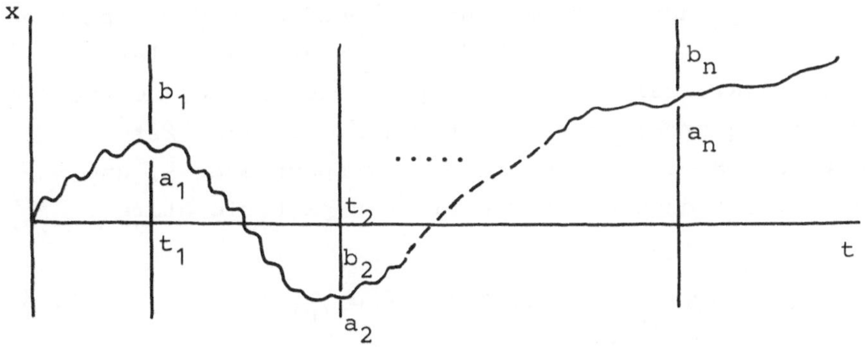

The Wiener measure μ assigns to A the probability

$$\mu(A) = \int_{a_1}^{b_1} dx_1 \int_{a_2}^{b_2} dx_2 \dots \int_{a_n}^{b_n} dx_n (4\pi t_1)^{-1/2}[4\pi(t_2-t_1)]^{-1/2} .$$

$$\ldots[4\pi(t_n-t_{n-1})]^{-1/2} .$$

$$\exp[-\frac{1}{4t_1}x_1^2-\frac{(x_1-x_2)^2}{4\pi(t_2-t_1)}-\ldots-\frac{(x_n-x_{n-1})^2}{4\pi(t_n-t_{n-1})}] . \tag{2.3}$$

This can be extended by taking limits to the σ-ring generated by the cylinder sets. The process known as Brownian motion is then $B_t(\omega) = \omega(t)$. For any interval (s,t) we can define the algebra $A(s,t)$ as the W^*-algebra generated by functions of ω that depend only on $\omega(t_1),\ldots\omega(t_n)$ for some n and $s = t_1 \leq \ldots \leq t_n = t$. One of the consequences of the independence of the increments is that the algebras $A(t_1,t_2)$, $A(t_2,t_3)$, ... are all independent in that

$$A(s,t) = A(t_1,t_2)\otimes \ldots \otimes A(t_{n-1},t_n) .$$

We say that the process is <u>factorizable</u>. Such processes are classified by the Levy-Khinchin formula. In [4] it is shown that any such process can be embedded in Fock space. For Brownian motion, this embedding is very easy; let $K = L^2([0,\infty),R)$, called the 'one-particle' space, be the set of real square-integrable functions of time. Let $K = C\otimes K\otimes(K\otimes K)_S\otimes \ldots$ be symmetric Fock space over K, and let $|\phi_1 \ldots \phi_n>$ denote the symmetric n-particle state

$$|\phi_1 \ldots \phi_n> = \frac{1}{\sqrt{n!}}\sum_{\pi\in S_n}\phi_{\pi(1)}\otimes\ldots\otimes\phi_{\pi(n)} . \tag{2.4}$$

Then define the creation operator $a^*(\phi)$, $\phi\in K$, by

$$a^*(\phi)|\phi_1\ldots\phi_n> = |\phi\phi_1\ldots\phi_n> . \tag{2.5}$$

Then $a^*(\phi)$, $a(\psi)$ obey the usual CCR

$$[a(\phi),\ a^*(\psi)] = \langle\psi,\phi\rangle\ 1\ . \tag{2.6}$$

The free quantized field Φ is defined by

$$\Phi(\phi) = 2^{-1/2}\ (a^*(\phi) + a(\phi)\ )\ .$$

Then we may define the following objects:

A_t = W^*-algebra generated by $\{\Phi_s,\ 0 \le s \le t\}$

where $\Phi_t = \Phi(\chi_{[0,t]})$. (2.7)

Since all Φ_t commute in the strong sense that all their spectral resolutions commute, A_t is abelian. Let A be the W^*-algebra generated by $\{A_t : 0 \le t\}$. Define an 'expectation' on A (a normalized positive linear form) by the quantum expectation in the Fock vacuum Ψ_F:

$$E(Y) = \langle\Psi_F, Y\ \Psi_F\rangle\ , \qquad Y \in A\ .$$

Then $\{\Phi_t\}$ is Gaussian and has all the properties of B_t, e.g.

$$E(\Phi_s \Phi_t) = \min(t,s)\ . \tag{2.8}$$

Indeed we have the following

Theorem. Let $(\Omega,(F_t),\mu,B_t)$ be Brownian motion. We regard B_t as an unbounded multiplication operator on $L^2(\Omega,\mu,F)$. Let $(H,(A_t),\Psi_F,\Phi_t)$ be the free quantum field over $K = L^2([0,\infty),R)$. Then there exists a unitary operator $V: L^2(\Omega;\mu,F) \to H$ such that $VB_tV^{-1} = \Phi_t$ for all $t \ge 0$, and $V1 = \Psi_F$.

To all intents and purposes, B_t and Φ_t are the same. It follows that $y \to \langle\Psi_F, |y|^2\Psi_F\rangle^{1/2}$ is a norm.

We see that Brownian motion is a commutative example of the following set-up, called a quantum stochastic process:

We are given a Hilbert space H, a vector $\Psi_0 \in H$ and a family of W^*-algebras $(A_t)_{t\geq 0}$ obeying $A_s \leq A_t$ if $s \leq t$, and a self-adjoint operator Φ_t, $t \geq 0$, whose spectral resolutions lie in A_t for each t. The quadrupole $\{H,(A_t)_{t\geq 0},\Psi_0,(\Phi_t)_{t \geq 0}\}$ is the quantum stochastic process.

This is a lenient view of what a process is; we do not require that the map $\Phi_s \to \Phi_t$ should be unitary or even an algebraic homomorphism; to do so would rule out Brownian motion.

The <u>Clifford process</u>, in which A_t are not abelian is constructed from the creation and annihilation operators $b^*(f)$, $b(f)$, $f \in L^2(0,\infty) = K$. These obey the anti-commutation relations

$$b(f)b^*(g) + b^*(g)b(f) = 2\langle f,g\rangle \ , \quad f,g \in K \ ,$$

and act on the anti-symmetric Fock space H_A with Fock vacuum Ψ_F. Let $\Psi_t = a^*(\chi_{[0,t]})+a(\chi_{[0,t]})$ and let A_t = W^*-algebra generated by $\{\Psi_s, s \leq t\}$. Then $\{H_A,(A_t)_{t\geq 0}, \Psi_F,(\Psi_t)_{t\geq 0}\}$ is the Clifford process. We note that Ψ_F defines a trace on A; this stems from the property

$$\langle\Psi_F, \Psi_{t_1} \cdots \Psi_{t_n} \Psi_F\rangle = \langle\Psi_F, \Psi_{t_2} \cdots \Psi_{t_n} \Psi_t, \Psi_F\rangle \ .$$

It follows that $F \to ||F\Psi_F||$, $F \in A$, is a norm.

It is natural to relax the condition that the process should be self-adjoint. We are then allowed to split up Φ_t or Ψ_t into creation and annihilation parts; the methods of quantum field theory then become available.

3. ITO'S ISOMETRY

The Clifford process, like Brownian motion, is not differentiable, in that $d\Psi_t/dt = \Psi'(t)$, the Fermion field at point t is not an operator (in the sense that $\delta(x)$ is not a function). This is because B_t and Ψ_t are not of bounded variation. Divergences therefore occur when we consider a stochastic differential equation

$$\frac{dX_t}{dt} = F(t,X_t)\frac{dB_t}{dt} + G(t,X_t) \tag{3.1}$$

or similar, with $d\Psi_t/dt$ replacing dB_t/dt. When we try to make sense of this equation by rewriting it as an integral equation

$$X_t = X_o + \int_o^t F(s,X_s)dB_s + \int_o^t G(s,X_s)ds \quad ; \tag{3.2}$$

the trouble now shows up as ambiguities in the definition of $\int_o^t FdB_s$: suppose we approximate F by a linear sum of step functions $F(s) = \sum_j \chi_{[t_j-t_{j-1}]}(s)F_j$ and the integral by the Riemann sum

$$\sum_j F_j(B_{t_j} - B_{t_{j-1}}) \quad ;$$

then the limit, as $t_j-t_{j-1}\to 0$, might not converge, or will depend on the choice of time $s_j(t_{j-1} \leq s_j \leq t_j)$ at which we evaluate F_j.

Ito suggested a resolution of this ambiguity: consider only <u>adapted</u> integrands, and always evaluate the integrand at the beginning of the time-interval, i.e. choose $s_j = t_{j-1}$. Then the increment dB points to the future: for a step-function the stochastic integral is thus

$$\int_0^t F(s)dB(s) = h_j[B(t_{j+1})-B(t_j)]$$

when $F(s) = h_j\chi_{[t_{j+1},t_j]}(s)$, $h_j\in A(t_j)$. (3.3)

We extend the definition, by linearity, to the linear span of adapted step-functions. We can also extend it to the completion of this space in the norm $||F||$,

$$||F||^2 = \int_0^t ds\ ||F(s)\Psi_F||^2 \quad . \tag{3.4}$$

This is the key result of the subject, and follows from
<u>Theorem: The Ito isometry</u>.

Let X_t be Brownian motion or the Clifford process. Let $F(s) = \sum F(t_j)\chi_{[t_j,t_{j+1}]}(s)$ be a sum of adapted step-functions. Then

$$||\sum_j F(t_j)(X_{t_{j+1}}-X_{t_j})\Psi_F||^2 = \int_0^t ds||F(s)\Psi_F||^2 \ . \tag{3.5}$$

<u>Proof</u>.If j>k, then $F^*(t_k)F(t_j)[X_{t_{j+1}}-X_{t_j}]\ \Psi_F$

is orthogonal to $(X_{t_{k+1}}-X_{t_k})\Psi_F$, since the latter contains a particle of wave-function $\chi_{[t_k,t_{k+1}]}$ not in the first state. Hence if j<k the contribution to the left-hand side of (3.5),

$$<\Psi_F,[X_{t_{j+1}}-X_{t_j}]\ F^*(t_j)F(t_j)[X_{t_{k+1}}-X_{t_k}]\Psi_F>$$

vanishes. Similarly it vanishes if j>k. Hence only the term j=k survives, and

$$||\sum_j F(t_j)(X_{t_{j+1}}-X_{t_j})\Psi_F||^2 =$$

$$= \sum_j <\Psi_F,[X_{t_{j+1}}-X_{t_j}]F^*(t_j)F(t_j)[X_{t_{j+1}}-X_{t_j}]\Psi_F> \quad .$$

In the Brownian case, everything commutes. In the Fermionic case, we can use the trace property of Ψ_F to bring the two δX's together, to get for this norm

$$\sum_j <\Psi_F,F^*(t_j)F(t_j)[X_{t_{j+1}}-X_{t_j}]^2\Psi_F> \quad .$$

Now $[X_{t_{j+1}}-X_{t_j}]^2 = \Phi(\chi_{[t_j,t_{j+1}]})^2$ or $\Psi(\chi_{[t_j,t_{j+1}]})^2$

$$=: \Phi(\chi_{[t_j,t_{j+1}]})^2 :+<\Psi_F,\Phi(\chi_{[t_j,t_{j+1}]})^2\Psi_F>$$

where: :denotes Wick ordering.

The Wick ordered term creates two particles of wave-functions later than t_j and so this term is orthogonal to the earlier times up to t_j. The second term is just $(t_{j+1}-t_j)1$. In the Fermi case, the first term is zero by Pauli's exclusion principle. In either case, then, we get

$$\sum_j <\psi_F,F^*(t_j)F(t_j)\Psi_F>(t_{j+1}-t_j)$$

which for step-functions is equal to

$$\int_0^t ds||F(s)\psi_F||^2, \quad \text{as required.}$$

The sum $\sum F(t_j)[X_{t_{j+1}}-X_{t_j}]$ is called the Ito integral $\int_0^t FdX_s$.

In [21] we show that any adapted function F for which $\int_0^t ||F(s)\Psi_F||^2 ds<\infty$ can be approximated in this norm by adapted step functions. Corollary: The Ito integral can be defined for all such functions as the (unique) limit of its value for approximating adapted step functions - the Ito isometry ensures that the map $F\to\int F dX_s$ is continuous.

Having defined $\int F(s)dX_s$ for a known adapted function F, we can formulate the integral equation (3.2), and iterate it by Picard's method:

$$X_o(t) = X_o$$

$$X_1(t) = X_o + \int_o^t F(s,X_o)dB_s + \int_o^t G(s,X_o)ds$$

$$X_2(t) = X_o + \int_o^t F(s,X_1(s))dB_s + \int_o^t G(s,X_1(s))ds$$

.... ;

if F and G obey Lipschitz conditions, this iteration converges to a solution of the integral equation; the solution is an adapted operator - valued function, i.e. a process. The same is true with Ψ_t replacing B_t [21].

The reason why the off diagonal terms in Ito's formula are zero is that B_t and Ψ_t are martingales, a fair game: if M_t is the size of our random gambling debts at time t, and $E(|F_s)$ is the conditional expectation, given all M_u, $0 \leq u \leq s$, then $E(M_t|F_s) = M_s$, $s \leq t$, holds for a martingale M: our future losses will average out leaving us with our latest actual known loss. By construction, $(B_{t_{j+1}}-B_{t_j})\Psi_F$ is orthogonal to all states created before t_j, so $E(B_{t_{j+1}}-B_{t_j}|F_{t_j}) = 0$ which is the martingale condition since

$$E(B_{t_j}|F_{t_j}) = B_{t_j} \quad .$$

A famous theorem of Doob and Meyer states that the square of a martingale M_t can be written as a martingale + increasing process; the latter is written $<M_t>$. The formulas

$$\Phi_t^2 = : \Phi_t{}^2 : + t\,1\,, \qquad \Psi_t^2 = 0 + t\,1$$

are the most famous examples. We have virtually used the fact that $:\Phi^2{}_t:$ is a martingale to evaluate the Ito isometry. The same method works in general, even in the quantum case [22]: for any martingale M_t, $\int_0^t F(s)\,dM_s$ can be defined for non-anticipating F and obeys the Ito isometry

$$\left|\left|\int_0^t F(s)\,dM_s\right|\right|_2^2 = \int_0^t ||F(s)||_2^2 \; d< M_s> \quad .$$

In the quantum case, the norm $||\ ||_2$ is taken to mean $<\Psi_0, A^* A\Psi_0>^{1/2} = ||A||_2$, and the conditional expectation $(E|F_t)$ is the projection onto the subspace spanned by $\{A_s\Psi_0,\ 0 \le s \le t\}$.
The first known non-abelina martingales were constructed by R.L. Hudson; his idea led us to the construction of copious examples as Wick ordered products [23].

Ito's method, using adapted integrands and pointing the increment of the noise to the future, resolves the ambiguity and has the neat consequence that any Ito stochastic integral is a martingale. Kunita and Watanabe show the converse - any martingale is a stochastic integral. We have a Fermion version of this [21]: given an L_2-martingale $(M_t)_{t\ge 0}$, it can be written

$$M_t = \int_0^t M_s' \, d\,\Psi_s$$

for some adapted function M', written symbolically as the stochastic derivative $\frac{\partial M}{\partial \Psi}$. This suggests the chain rule

$$dM_t = M_t' \, d\Psi_t \; ,$$

and indeed we prove that under wide conditions, we have

$$\int G(s)dM_s = \int G(s)M_s' d\Psi_s .$$

It is also possible to show that any bounded martingale is the conditional expectation of its "value at ∞".

The classical theory of martingales has many beautiful results, most of which are mirrored in the Clifford process. But the classical theory has moved away from martingales to the even more general theory of semimartingales. To see why martingales are not general enough to provide realistic noise, we turn to the simplest explicit example.

4. APPLICATIONS: DAMPED OSCILLATOR [24]

A Bose or Fermi oscillator is described by a creation operator $A^*(t)$ obeying

$$\frac{dA^*}{dt}(t) = i\omega\, A^*(t) \tag{4.1}$$

and acting on $H_0 = L^2(-\infty,\infty) \cong l_2$, which will serve as the initial space. Suppose that at $t = 0$ the oscillator moves into a damping medium. We could try adding a linear damping term $-\gamma A^*(t), \gamma > 0$, to (4.1), leading to the solution

$$A^*(t) = A^*(0)e^{(i\omega-\gamma)t} \quad , \qquad t \geq 0 \quad . \tag{4.2}$$

This leads to the commutation relations

$$[A(t),A^*(t)]_F = e^{-2\gamma t}1$$

which decays in time. Discussion of this decay has led to horrible contortions [25]. Senitzky noted that one should add noise to (4.1) as well as damping: the environment contributes a random force to the system, as well as damping. So a noise Hilbert space Γ is needed, with operators

$a^*(t)$, $a(t)$:

$$a^*(t) = \int dk\rho(k)e^{ikt}a^*(k) \tag{4.3}$$

where $[a(k),a^*(k')] = \delta(k-k')$. (4.4)

The weight-function ρ describes the physics of the noise. The quantum noise is the smeared field $X_t = a(\chi_{[0,t]})+a^*(\chi_{[0,t]})$. The damped equation with quantum noise is

$$\frac{dA^*(t)}{dt} = i\omega A^*(t) - \gamma A^*(t)+\lambda a^*(t) \quad . \tag{4.5}$$

We seek solutions $A^*(t)$, being operators on $H_0\otimes\Gamma$; $a^*(t)$ is short for $1\otimes a^*(t)$ acting on $H_0\otimes\Gamma$. The initial value is then $A^*(0)\otimes 1_\Gamma$. Equation (4.5) is similar to the Langevin equation (1.2) when written in terms of $P(t)$,$Q(t)$. The difference is that the damping and noise occur more symmetrically between P and Q in (4.5); this allows the decay caused by the damping to be cancelled by the increase in uncertainty due to the noise. Indeed we can solve (4.5) exactly,

$$A^*(t) = A^*(0)e^{(i\omega-\gamma)t} + \int_0^t ds\ e^{(i\omega-\gamma)(t-s)}a^*(s) \quad . \tag{4.6}$$

The requirement of no decay

$$[A(t),A^*(t)] = 1 \qquad \text{for all } t \geq 0 \tag{4.7}$$

then leads to an integral equation for ρ. One solution is [24]

$$|\lambda\rho(k)|^2 = \gamma/\pi \quad , \qquad -\infty < k < \infty \quad . \tag{4.8}$$

This is the "white" quantum noise of Lax [1], for which $X_t = \Phi_t$ or Ψ_t.

Senitzky recommended that only the $k \geq 0$ part of the oscillator's spectrum should be used - the requirement of positive energy. Then X_t is "coloured noise". See also Kubo [24]. I find all possible ρ such that (4.7) holds; one such is

$$|\lambda\rho(k)|^2 = \frac{2\gamma}{\pi}\,\theta(k-\omega) \quad .$$

There is a strong physical reason for prefering a model using only positive-energy noise - if the system is at zero temperature it cannot lose an indefinite amount of energy to the noise - as it could if the noise created negative - energy states. This is why Brownian motion (and the Clifford process) are not satisfactory as models of noise.

If the energy, k, of the noise is ≥ 0, we can discuss the thermodynamic state ρ_β of the noise

$$\rho_\beta = e^{-\beta H}/\mathrm{Tr}(e^{-\beta H}) \quad , \tag{4.9}$$

where $H = \int_0^\infty k\, dk\, a^*(k)a(k)$ is the second - quantized energy of the oscillators $a(k)$. Now $e^{-\beta H}$ is of trace class only if we impose periodic boundary conditions, so care is needed here.

Suppose the initial state of the system is ρ_0, and that this is uncoupled from the noise. Then the density matrix is $\rho_0 \otimes \rho_\beta$. A direct calculation from (4.6) gives for the expectation in this state

$$<A^*(t)A(t+\tau)>_\beta = \rho_0(A_0^*A_0)e^{-i\omega\tau-\gamma\tau-2\gamma t}$$

$$+ \int_0^\infty dk\, n(k)\rho^2(k)\int_0^t ds \int_0^{t+\tau} ds' e^{(i\omega-\gamma)(t-s)+iks-iks'-(i\omega+\gamma)(t+\tau-s')}$$

where $n(k) = (e^{\beta k} \pm 1)^{-1}$.

This is not a stationary process, in that it depends on t; but as $t \to \infty$ the t-dependence decays exponentially, and this converges to the "Wightman functions" of a generalized free field $A_\infty(\tau)$, with

$$<A_\infty^*(0)A_\infty(\tau)> = \lim_{\tau\to\infty} <A^*(t)A(t+\tau)>_\beta = \int_0^\infty \frac{dkn(k)|\rho(k)|^2 e^{-ik\tau}}{\gamma^2+(\omega-k)^2} \quad .$$

The Hilbert space of A_∞ is that built up by the Wightman reconstruction theorem, i.e. it is the unique cyclic generalized free field with this two-point function. The field $A_\infty(\tau)$ obeys equal-time commutation relations, and the limit state is stationary and is the thermodynamic state at temperature $1/\beta$, i.e. it obeys the Kubo-Martin-Schwinger conditions e.g.

$$<A_\infty^*(0)A_\infty(\tau+i\beta)> = <A_\infty(\tau)A_\infty^*(0)> \quad .$$

The solution $A(t)$ forgets its initial state ρ_0 exponentially fast, and moves to equilibrium at the same temperature as the heat-bath. The limit process exhibits a resonance at $k = \omega$ of width γ. The shape is not exactly Lorentzian even at zero temperature (except for the Lax model, where ρ is constant). The model describes an unstable system and gives a representation of the CCR or CAR at each time, and has well-defined many-time correlations (the Wightman functions).

This then is a satisfactory linear theory. The noise however is not a martingale. Indeed, coloured noise is not even adapted to the Wiener filtration $\mathcal{F}_t$ (or the Clifford filtration for Fermions). It is not even obvious that the future commutes with the past, and this is certainly not true of the A^* and A. Thus the axiom-builders should act carefully, in view of the importance of this model. We have been trying to move away from the martingale condition, and more to regularity conditions expressed in terms of Green's functions [26]. Of course, for the linear oscillator, no deep existence theory is needed - we can solve it exactly.

A non-linear model: $\lambda\phi^4$

We can introduce damping and noise into each momentum-mode of the free relativistic field of mass m, to get the equations

$$\frac{dA^*(\vec{k},t)}{dt} = i\sqrt{k^2+m^2}\;A^*(\vec{k},t)-\gamma A^*(\vec{k},t)+\lambda a^*(\vec{k},t)$$

where $a^*(\vec{k},t)$ is a generalized free field with two-point function (at zero-temperature):

$$\langle a(\vec{k},s)a^*(\vec{k},t)\rangle = \delta^3(\vec{k}-\vec{k}')\int_0^\infty \rho(\omega)e^{i(t-s)\omega}d\omega \quad .$$

This breaks Lorentz invariance, of course.
As time goes by, the stochastic differential equation churns the solution round in noise-space and imprints its influence on the rather bland properties of the noise; and in the stationary limit of large times we get a generalized free field $A_\infty(\vec{k},t)$ which has a resonance at $\omega = \sqrt{k^2+m^2}$ of width γ:

$$\langle A_\infty(k_0)A^*_\infty(k',\tau)\rangle = \delta^3(k-k')\int \frac{|\rho(\omega)|^2 e^{-i\omega\tau}}{\gamma^2+(\omega-\omega_k)^2} \quad .$$

We recover the relativistic free field as $\gamma\to 0$ if ρ^2 is proportional to γ.

D. Patel has studied the perturbation of this solution

by the non-linear term $:4g(A^*(\vec{x},t)+A(\vec{x},t)):^3$, in $\vec{x}$-space; in other words, he studies the damped $g\phi^4_4$ theory. To aid the convergence, he takes A_∞ rather than the free field as the "free" theory. We find that, in the Feynman-Dyson expansion (the Picard expansion), every graph is convergent. This is because each propagator has an extra two powers of $\vec{k}$ coming from the propagator $|\rho(\omega)|^2/(\gamma^2+(\omega-\omega_k)^2)$ which, as $\gamma \to 0$, loses its power to assist convergence in the $d\vec{k}$ integrations. It is likely, then, that the solution to $g\phi^4_4$ exists as operators in the noise-space generated by $A_\infty(\vec{k},\tau)$. This solution is non-relativistic but formally gives Lorentz invariance as $\gamma \to 0$. We might argue that this is the "correct" limit for any field theory: the damping and noise represent the coupling to other fields that are omitted, and are needed for the stability.

REFERENCES

1. M. Lax, Phys. Rev. 145 (1965) 111-129.
2. I.R. Senitzky, Phys. Rev. 119 (1960) 670, ibid. A3 (1970) 421.
3. H. Haken, Optics, Handbuch der Physik 25/20 (1970)p.43-44, Springer.
4. R.F. Streater, Current Commutation Relations, Continuous Tensor Products and Infinitely Divisible Group Representations (1969), in Local Quantum Theory (R. Jost, Ed.), Academic Press.
5. H. Araki, Reports. Res. Inst. Math. Sci. Kyoto, 5 (1970) 361-422.
6. A. Guichardet, Springer Lecture Notes in Mathematics 261 (1972), Springer.
7. K. Parthasarathy and K. Schmidt, Springer Lecture Notes in Mathematics (1972) 272, Springer.
8. D.E. Evans and J.T. Lewis, Commun. Dublin Institute for Advanced Studies A 24 (1977).
9. E.B. Davies, The Quantum Theory of Open Systems, Academic

Press.

10. R.L. Hudson, K.R. Parthasarathy and P.D.F. Ion, Commun. Math. Phys. 83 (1982) 261-280.
11. J.T. Lewis and L.C. Thomas, Ann. Inst. H. Poincaré A22 (1975) 241-248.
12. L. Accardi, A. Frigerio and J.T. Lewis, Publ. RIMS . Kyoto Univ. 18 (1982) 97-133.
13. Maasen, thesis, Groningen (1982).
14. R.F. Streater, Quantum Stochastic Processes. Rome II Conference on Quantum Probability (L. Accardi, A. Frigerio and V. Gorini, Eds.).
15. R.L. Hudson and K.R. Parthasarathy, Quantum Diffusions, in: Theory and Applications of Random Fields, ed. Kallianpur. Lecture Notes in Control & Information Sciences 49, Springer (1983).
16. G.G. Emch, The minimal K-flow associated to a Quantum Diffusion, in Physical Reality and Mathematical Description, Dordrecht (1974) 477-493, Ed. C. Enz and J. Mehra.
17. G. Lindblad, Commun. Math. Phys. 48 (1976) 119.
18. V. Gorini, A. Frigerio, A Kossakowski, M. Verri, and E.C.G. Sudarshan, Rep. Math. Phys. 13 (1978) 149.
19. H. Hasegawa and R.F. Streater, J. Phys. A (L) 16 (1983) L697-L703.
20. R.L. Hudson and K.R. Parthasarathy, Noncommutative Semimartingales and quantum diffusion processes adapted to Brownian motion, Nottingham preprint.
21. C. Barnett, R.F. Streater, and I.F. Wilde, J. Functl. Anal. 48 (1982) 172.
22. C. Barnett, R.F. Streater, and I.F. Wilde, Math. Proc. Camb. Phil. Soc. 94 (1983) 541-551.
23. R.L. Hudson and R.F. Streater, Phys. Lett. 85A (1981) 64; ibid. 86A (1981) 277.
24. R.F. Streater, J. Phys. A15 (1982) 1477-1485; R. Kubo, J. Phys. Soc. Japan 26 (1969) Suppl.
25. H. Dekker, Phys. Repts. 80 (1981).
26. C. Barnett, R.F. Streater, and I.F. Wilde, Quantum Stochastic Integrals under Standing Hypotheses. Submitted to J. Lond. Math. Soc..

Acta Physica Austriaca, Suppl. XXVI, 75–100 (1984)

STOCHASTIC DIFFERENTIAL EQUATIONS+

by

P. ZOLLER
Institute for Theoretical Physics
University of Innsbruck
A-6020 Innsbruck, Austria

1. INTRODUCTION

This is a brief introduction to Langevin equations(stochastic differential equations (SDE) with white noise terms)[1-3], with particular emphasis on its use as a calculational tool. We also discuss recently developed (matrix) continued fraction methods for solving certain types of stochastic differential equations and their associated Fokker-Planck equation [4-6].
Finally, we point out recent developments in quantum optics where nonclassical states of the radiation field can be described in terms of stochastic differential equations [3,7,13,14].

+Lectures given at the XXIII.Internationale Universitäts-wochen für Kernphysik,Schladming,Austria,February 20 - March 1, 1984.

2. A SUMMARY OF THE ITO AND STRATONOVICH CALCULUS [1-3]

In its simplest form a Langevin equation is a differential equation for a quantity $x(t)$,

$$\frac{dx(t)}{dt} = a(x(t),t) + b(x(t),t)\xi(t) \quad , \tag{1}$$

with $\xi(t)$ a "rapidly varying, fluctuating function of time", whose origin can either be internal noise (thermal fluctuations) or external noise originating in the environment of the system. Examples for such Langevin equations are the equation for the velocity of a Brownian particle, or an equation for the laser intensity, where the fluctuating term describes contributions to the light intensity from spontaneous emission of laser atoms.

A simple possible model for the "rapidly varying irregular function" $\xi(t)$ (which we require to be continuous as a function of time) is an Ornstein-Uhlenbeck process. This is a Gaussian stochastic process with mean value $\langle\xi(t)\rangle = 0$ and correlation function $\langle\xi(t)\xi(t')\rangle = 1/(2\tau_c)\ \exp(-|t-t'|/\tau_c)$ where the angular brackets denote an average over the possible realizations of $\xi(t)$. The coherence time τ_c gives us the time scale of fluctuations of $\xi(t)$. Since we are interested in the limit where the time scale of change of $x(t)$ is much slower than τ_c, we are led to idealize the concept of a rapidly varying, highly irregular function by white noise, defined as a Gaussian process with a δ-function as its auto-correlation function:

$$\langle\xi(t)\rangle = 0 \ , \qquad \langle\xi(t)\xi(t')\rangle = \delta(t-t') \quad , \tag{2a}$$

$$\langle e^{i\int\phi(t)\xi(t)dt}\rangle = e^{-1/2\int\int dt dt'\phi(t)\phi(t')\langle\xi(t)\xi(t')\rangle} \quad , \tag{2b}$$

where the characteristic functional of a Gaussian process (2b) gives us higher order correlation functions for $\xi(t)$. The problem with white noise is that in view of its infinite variance it does not exist as an ordinary stochastic process. One possible way around this difficulty is to interpret the Langevin equation (1) as an ordinary differential equation leaving τ_c finite and taking the limit $\tau_c \to 0$ only after calculating the averages of interest. Although this procedure comes often close to the physicists understanding of a Langevin equation, it is in many cases impractical as a calculational tool. An alternative way which is the basis of the Ito and Stratonovich calculus is to introduce the Wiener process $W(t) =$ " $\int_0^t \xi(t')dt'$" (which is mathematically well defined) and to interpret the Langevin equation (1) as an integral equation

$$x(t)-x(t_0) = \int_{t_0}^{t} a(x(s),s)ds + \int_{t_0}^{t} b(x(s),s)dW(s) \tag{3}$$

with the second integral a "stochastic Stieltjes integral" and $dW(t)$ a Wiener increment.

There are different ways of defining the stochastic integral in Eq. (3). The mathematical literature almost exclusively deals with the Ito Integral, because of its special mathematical properties; but unfortunately for physicists the Stratonovich definition is often the more natural choice.

Below we summarize the basic ideas of the Ito calculus and develop its connection with the Stratonovich formalism.

2.1 The Wiener Process

We defined the Wiener process as $W(t) =$ "$\int_0^t \xi(s)ds$" $(t \geq 0)$ with $\xi(t)$ white noise. This leads immediately to the following properties:

(i) $\langle W(t)\rangle = 0 \qquad \langle W(t)W(t')\rangle = \min(t,t')$ (4a)

(ii) The probability density of finding a particular value W realized at a given time t is

$$P(W,t) = \langle \delta(W - \int_0^t \xi(s)ds)\rangle \equiv \int_{-\infty}^{+\infty} \frac{d\lambda}{2\pi} e^{i\lambda W} \langle e^{-i\lambda \int_0^t \xi(s)ds}\rangle$$

$$= (2\pi t)^{-1/2} e^{-W^2/(2t)} \quad . \tag{4b}$$

(iii) The conditional density to find W at time t given W_0 at t_0 is $(t \geq t_0)$

$$P(W,t|W_0,t_0) = \langle \delta(W - (W_0 + \int_{t_0}^t \xi(s)ds))\rangle$$

$$= (2\pi(t-t_0))^{-1/2} e^{-(W-W_0)^2/[2(t-t_0)]} \quad . \tag{4c}$$

(iv) The n-th order probability density of finding W_1, W_2, ..., W_n at times t_1, t_2, ..., t_n respectively is, assuming a time ordering $t_1 \geq t_2 \geq \ldots \geq t_n$,

$$P(W_1,t_1;W_2,t_2;\ldots;W_n,t_n) =$$

$$P(W_1t_1|W_2,t_2)P(W_2,t_2|W_3,t_3)\ldots P(W_{n-1},t_{n-1}|W_nt_n)P(W_n,t_n) \, . \tag{4d}$$

Note that Eq.(4d) implies that the Wiener process is a Markov process. Furthermore it is a process with independent increments, i.e., $W(t_2)-W(t_1)$ and $W(t_4)-W(t_3)$ with $[t_1,t_2]$ and $[t_3,t_4]$ non-overlapping time intervals are statistically independent.

(v) The sample paths of the Wiener process are continuous with probability one (which is closely related to the assumption that $\xi(t)$ is Gaussian and not for example, a Poisson process). They are non-differentiable with probability one, however, in view of

$$\text{Prob}\{|\frac{W(t+h)-W(t)}{h}| > k\} = 2\int_{hk}^{\infty} dW(2\pi h)^{-1/2} e^{-W^2/2h} \xrightarrow[h\to 0]{} 1,$$

i.e., $\xi(t)$ does not exist.

2.2 Stochastic Integrals

We define the stochastic integral (compare Eq. (3)) $\int_{t_o}^{t} G(s)dW(s)$ with $W(s)$ a Wiener process and $G(s)$ a class of reasonable smooth stochastic functions (see below) as a limit of Riemann-Stieltjes sums of the form

$$S_n = \sum_{i=1}^{n} G(\tau_i)[W(t_i)-W(t_{i-1})] \tag{5}$$

where we have divided the integration interval according to $t_o \leq t_1 \leq \ldots \leq t_{n-1} \leq t$ with $t_{i-1} \leq \tau_i \leq t_i$. Obviously, the limit $n \to \infty$ depends on the choice of intermediate points, as illustrated by the example

$$\langle S_n \rangle = \langle \sum_{i=1}^{n} W(\tau_i)[W(t_i)-W(t_{i-1})] \rangle$$

$$= \sum_{i=1}^{n} (\tau_i - t_{i-1}) = \alpha(t-t_o)$$

with $\tau_i = t_{i-1} + \alpha(t_i + t_{i-1})$ and $0 \leq \alpha \leq 1$.
The Ito integral is now defined by choosing $\tau_i = t_{i-1}$ in Eq. (5),

$$I_n = \sum_{n=1}^{n} G(t_{i-1})[W(t_i)-W(t_{i-1})] \tag{6}$$

and taking the limit

$$I = \text{ms} - \lim_{n\to\infty} I_n = \int_{t_o}^{t} G(s)dW(s) \tag{7}$$

where by the mean square limit (ms-lim) we mean

$$\lim_{\to\infty} \langle (I-I_n)^2 \rangle = 0. \tag{8}$$

We expect that a physically reasonable solution of the SDE (1) at a given time t, x(t), is statistically independent of the Wiener increments W(s)-W(t) for s > t, i.e., we expect x(t) to depend on the past and not to anticipate the future (causality). Accordingly, the stochastic integral is defined for so-called non-anticipating functions G(t) which for t < s are statistically independent of W(s)-W(t). The simplest example of a nonanticipating function is W(t). Note that a function b(x(t),t) with x(t) nonanticipating is again a nonanticipating function.

It can be shown that the stochastic integral $\int_{t_0}^{t} G(s)dW(s)$ exists for nonanticipating and continuous functions G(s) [1,2]. In addition we have the following properties with G(t) and H(t) nonanticipating functions:

(i) the mean value formula

$$\langle \int_{t_0}^{t} G(s)dW(s) \rangle = 0 \tag{9}$$

(which is plausible in view of $\langle \sum_i G_{i-1}\Delta W_i \rangle = \sum_i \langle G_{i-1} \rangle \langle \Delta W_i \rangle = 0$ with $G_{i-1} = G(t_{i-1})$ and $\Delta W_i = W(t_i)-W(t_{i-1})$;

(ii) the correlation formula

$$\langle \int_{t_0}^{t} G(t_1)dW(t_1) \int_{t_0}^{s} H(t_2)dW(t_2) \rangle = \int_{t_0}^{\min(t,s)} \langle G(t')H(t') \rangle dt', \tag{10}$$

which again can be understood from $\langle (\sum_i G_{i-1}\Delta W_i)(\sum_j H_{j-1}\Delta W_j) \rangle = \sum_i G_{i-1}H_{i-1}\Delta t_i$ where we used

$\langle \Delta W_i^2 \rangle = \Delta t_i$.

(iii) As a function of the upper integration limit,

$$x(t) = \int_{t_o}^{t} G(s)\,dW(s)$$

is a nonanticipating function and has continuous sample path (with probability one).

2.3 Stochastic Differentials

The stochastic differential

$$dx(t) = a(t)dt + b(t)dW(t) \tag{11}$$

is an abbreviation for

$$x(t) = x(t_o) + \int_{t_o}^{t} a(s)ds + \int_{t_o}^{t} b(s)dW(s) \tag{12}$$

with $a(t)$ and $b(t)$ nonanticipating functions.
The key to applications of the Ito calculus are the formal "multiplication rules" for differentials

$$dW(t)^2 = dt, \quad dW(t)^{2+N} = 0, \quad dW(t)dt = 0, \quad \text{etc.} \tag{13}$$

$(N = 1,2\ldots)$

which can be summarized by saying that "$dW(t)$ is a differential of order 1/2 and that in calculations infinitesimals of order higher than one are discarded". Thus, changing variables from the stochastic differential (11) for $x(t)$ to an arbitrary $f(x(t),t)$ we have, by expanding $f(x(t),t)$,

$$df(x(t),t) = f_t(x(t),t)dt+f_x(x(t),t)dx(t)+\tfrac{1}{2}f_{xx}(x(t),t)dx(t)^2+.$$

where f_t, f_x and f_{xx} denote partial derivatives with respect to t and x. Using (11) together with (13) we find Ito's formula

$$df(x(t),t) = f_t(x(t),t)dt + f_x(x(t),t)[a(t)dt + b(t)dW(t)]$$

$$+ 1/2\ f_{xx}(x(t),t)b^2(t)dt\ . \tag{14}$$

It is simple to generalize the previous discussion to an m-variable Wiener process $W(t) = (W_1(t),W_2(t),\ldots,W_m(t))$ where $W_1(t),W_2(t),\ldots$ are statistically independent. Instead of Eq.(3) we have now

$$dW_i(t)dW_j(t) = \delta_{ij}dt\ . \tag{15}$$

2.4 Stochastic Differential Equations

The solution x(t) of an Ito SDE

$$dx(t) = a(x(t),t)dt + b(x(t),t)dW(t) \tag{16}$$

is a nonanticipating function obeying the integral equation

$$x(t)-x(t_o) = \int_{t_o}^{t} a(x(s),s)ds + \int_{t_o}^{t} b(x(s),s)dW(s) \tag{17}$$

with a(x(t),t) and b(x(t),t) nonanticipating functions and $x(t_o)$ a nonanticipating initial condition. The meaning of (16) is best illustrated by the discretized version

$$x(t_{i+1}) = x(t_i) + a(x(t_i),t_i)(t_{i+1}-t_i)$$

$$+ b(x(t_i),t_i)[W(t_{i+1})-W(t_i)] \tag{18}$$

with t_1, t_2, ... mesh points. The change of x(t) with time is then the sum of a deterministic drift and a stochastic term involving the Wiener increment. Eq. (18) can be used to simulate sampel paths x(t) on a computer by repetitive calls of a Gaussian random number generator for the Wiener increments. Conditions for existence and uniqueness of the nonanticipating solution $x(t)$ with continous sample paths (with probability one) are given in Ref. 1.

2.5 The Markov Property and the Fokker-Planck Equation

We want to show that the solution $x(t)$ of the SDE (16) is a Markov process. Consider an arbitrary function $f(x(t))$ with $x(t)$ a solution of (16). We can write the average $\langle f(x(t))\rangle$ as an integral over a probability density,

$$\langle f(x(t)\rangle = \int dx\; f(x)P(x,t) \quad , \tag{19}$$

where - at least formally - $P(x,t) = \langle\delta(x-x(t)\rangle$. The change in time of $\langle f(x(t)\rangle$ is according to Ito's formula (14) and the mean value formula (9)

$$\frac{d}{dt}\langle f(x(t))\rangle = \int dx\; f(x)\,\frac{\partial}{\partial t}P(x,t)$$

$$= \int dx[a(x,t)f(x,t) + \frac{1}{2}\,b^2(x,t)f_{xx}(x,t)]P(x,t) \quad .$$

Integrating by parts and discarding surface terms, the probability density $P(x,t)$ obeys, since $f(x)$ is arbitrary, the diffusion equation

$$\frac{\partial P(x,t)}{\partial t} = [-\frac{\partial}{\partial x}\,a(x,t) + 1/2\,\frac{\partial^2}{\partial x^2}\,b^2(x,t)]P(x,t) \tag{20}$$

with $a(x,t)$ a drift and $b^2(x,t) \geq 0$ a diffusion term. In particular we note that the above derivation also holds for the conditional density (transition probability) $P(x,t|x_o,t_o) = \langle\delta(x-x(t))\rangle \xrightarrow{t\to t_o} \delta(x-x_o)$ with the sample paths $x(t)$ satisfying the fixed initial condition $x(t_o) = x_o$. Eq. (20) for $P(x,t\;|x_o,t_o)$ is called a Fokker Planck equation (FPE).

In a similar way multitime averages

$$\langle f(x(t_1),x(t_2),\ldots x(t_n))\rangle = \int dx_1 \ldots dx_n \; f(x_1,\ldots,x_n) \times P(x_1,t_1;\;\ldots;\;x_n,t_n) \;, \qquad (21)$$

with f an arbitrary function of the solution of the SDE (16) $x(t_1)$, $x(t_2),\ldots,x(t_n)$ at times $t_1 \geq t_2 \geq \ldots \geq t_n$ respectively, can be expressed as integrals over an n-th order probability density

$$P(x_1,t_1;\;\ldots,\;x_n,t_n) = \langle \prod_{i=1}^{n} \delta(x_i - x(t_i))\rangle \;.$$

From the time evolution of the average for $t_1 \geq t_2$, keeping $t_2 \geq \ldots \geq t_n$ fixed, we conclude that

$$P(x_1,t_1;x_2,t_2;\ldots;\;x_n,t_n) = P(x_1,t_1|x_2,t_2)\;P(x_2,t_2;\;\ldots;\;x_n,t_n) \;, \qquad (22)$$

so that the n-th order density can be written in the form

$$P(x_1,t_1;\;\ldots;\;x_n,t_n) = P(x_1,t_1|x_2,t_2)\;\ldots\;P(x_{n-1},t_{n-1}|x_n,t_n)P(x_n,t_n) \qquad (23)$$

for $t_1 \geq t_2 \geq \ldots \geq t_n$. A stochastic process with probability densities obeying Eq.(23) is called a Markov process. The solution x(t) of a SDE, therefore, defines a Markov process. The meaning of the factorization in time implied by Markov property (22) is that the statistics of the solution $x(t_1)$ of the SDE at time t_1 is completely specified by the initial condition $x(t_2)$ at time t_2, and does not depend on information for earlier times (i.e., the "system has no memory"). This obviously is related to the fact that the

Langevin equation is first order in time and that the Wiener process has independent increments.

2.6 The Stratonovich SDE

The Stratonovich SDE for x(t)

$$(S)\ dx(t) = \alpha(x(t),t)dt + \beta(x(t),t)dW(t) \tag{24}$$

is an abbreviation for the integral equation

$$x(t) = x(t_o) + \int_{t_o}^{t} ds\alpha(x(s),s) + (S)\int_{t_o}^{t} \beta(x(s),s)dW(s) \tag{25}$$

where the Stratonovich integral is defined by

$$(S)\int_{t_o}^{t} \beta(x(s),s)dW(s) =$$

$$\underset{n\to\infty}{\text{ms}-\lim}\ \sum_{i=1}^{n} \beta\left(\frac{x(t_i)+x(t_{i-1})}{2},t_{i-1}\right)[W(t_i)-W(t_{i-1})]\ . \tag{26}$$

This differs from Ito's integral (10) by a symmetric choice of approximating step functions.

To establish the connection between the Stratonovich and Ito version of a SDE, suppose that x(t) is a solution of the Ito equation (16). Then we can write for the partial sums in Eq. (26)

$$\sum_{i=1}^{n} \beta\left(\frac{x_i+x_{i-1}}{2},t_{i-1}\right)\Delta W_i \simeq \sum_{i=1}^{n} \beta(x_{i-1},t_{i-1})\Delta W_i$$

$$+ \sum_{i=1}^{n} \beta_x(x_{i-1},t_{i-1})(1/2)[a(x_{i-1},t_{i-1})\Delta t_i+b(x_{i-1},t_{i-1})\Delta W_i]\Delta W_i$$

where the first term on the right hand side is an Ito integral which together with the second line gives in view of Eq.(13)

$$(S)\int_{t_0}^{t} \beta(x(s),s)dW(s) = \int_{t_0}^{t} \beta(x(s),s)dW(s)$$

$$+ (1/2) \int_{t_0}^{t} \beta_x(x(s),s)b(x(s),s)ds . \tag{27}$$

Comparing Eqs.(17), (25) and (27) we see that the Ito SDE

$$dx(t) = a(x(t),t)dt + b(x(t),t)dW(t) \tag{28}$$

is the same as the Stratonovich SDE

$$(S)\ dx(t) = [a(x(t),t)-(1/2)b(x(t),t)b_x(x(t),t)]dt$$

$$+ b(x(t),t)dW(t) . \tag{29}$$

Note that for additive noise, when $\beta(x(t),t)$ or $b(x(t),t)$ are independent of $x(t)$, the Stratonovich and Ito interpretation of a SDE are equivalent. The rules for changing variables in a Stratonovich SDE are the same as in ordinary calculus.

Whenever a SDE for a quantity $x(t)$ is given, it must be clearly stated whether this equation has to be interpreted as a Stratonovich or Ito equation. In external noise problems, where white noise is usually only an approximation for fluctuations with finite correlation time, as a rule the Stratonovich interpretation is the correct one. There is, however, an Ito-Stratonovich dilemma for internal noise problems[11].

2.7 Multivariable Equations

The n-variable generalization of the SDE for $x(t) = (x_1(t),..,x_n(t))$ is

$$dx_i(t) = A_i(x(t),t)dt + \sum_{j=1}^{m} B_{ij}(x(t),t)dW_j(t)$$

$$(i = 1,...,n) \tag{30}$$

where $(W_1(t),...,W_m(t))$ is an m-variable Wiener process. The FPE for the conditional density is

$$\frac{\partial}{\partial t} P(x,t|x_o,t_o) =$$

$$\left(-\sum_{i=1}^{n} \frac{\partial}{\partial x_i} A_i(x,t) + \frac{1}{2} \sum_{i,j=1}^{n} \frac{\partial^2}{\partial x_i \partial x_j}[B(x,t)B^T(x,t)]_{ij}\right)P(x,t|x_o,t_o) \tag{31}$$

Note that the FPE (31) depends on the matrix B only through the combination BB^T, i.e., replacing B by BS with S orthogonal leaves the FPE unchanged.

The Ito equation (30) corresponds to the n-variable Stratonovich equation

$$dx_i(t) = (A_i - (1/2) \sum_{k=1}^{n} \sum_{j=1}^{n} B_{kj} \frac{\partial}{\partial x_k} B_{ij})dt$$

$$+ \sum_{j=1}^{m} B_{ij}dW_j \ . \tag{32}$$

3. METHODS OF SOLUTION:
A MATRIX CONTINUED FRACTION APPROACH

A variety of techniques have been developed in the literatur to solve SDEs or the associated FPE [3,4]. Only few problems can be solved exactly, apart from single variable equations, or FPEs satisfying "detailed balance" conditions. Approximate methods include linearization (small noise expansions) of SDEs or adiabatic elimination of "fast variables". These analytical methods are complemented by numerical methods such as computer simulation [10] of SDEs or numerical integration of the FPE [4].

Here we confine ourselves to a discussion of recently developed (matrix) continued fraction expansions of averages or correaltion functions for certain classes of SDEs and FPEs [4-6]. The basic ideas will be illustrated below by two examples: the solution of a multiplicative SDE with colored noise; and the solution of the non-linear Langevin equation for a laser near threshold.

(i) Linear multiplicative SDEs of the form

$$\frac{d}{dt} u_\mu(t) = (A_{\mu\nu} + x(t) B_{\mu\nu}) u_\nu(t) \tag{33}$$

for a vector u(t) with A and B noncommuting matrices are encountered when the matrix governing the time evolution of a linear system contains a rapidly fluctuating (external) noise term x(t).[5] When the colored external noise x(t) is approximated by white noise, we have to interpret Eq. (33) as a Stratonovich equation (see below) which - when converted to Ito form - reads

$$du_\mu(t) = (A + \frac{1}{2} B^2)_{\mu\nu} u_\nu(t) dt + B_{\mu\nu} u_\nu(t) dW(t) \tag{34}$$

Suppose we are interested in the mean values $<u_\mu(t)>$. Taking averages we find

$$\frac{d}{dt}<u_\mu(t)> = (A + \frac{1}{2} B^2)_{\mu\nu}<u_\nu(t)> . \tag{35}$$

Equations for $<u_\mu(t)u_\nu(t')>$(and in a similar way higher order moments) can be derived by changing to $u_\mu(t)u_\nu(t)$ as new variables in the Ito equation (34) and then taking averages as above. Equations for correlation functions follow by similar arguments.
More complicated is the problem when x(t) is modelled by a colored Gaussian stochastic process, i.e., x(t) has the correlation function $<x(t)x(t')> = <x^2>e^{-\gamma|t-t'|}$. This can be described by adding to Eq. (31) the Langevin equation for the Ornstein Uhlenbeck process:

$$du_\mu(t) = (A_{\mu\nu}+x(t)B_{\mu\nu})u_\nu(t)dt , \tag{36a}$$

$$dx(t) = -\gamma x(t)dt + \sqrt{2\gamma<x^2>}\, dW(t) . \tag{36b}$$

Note that only u(t) together with x(t) defines a Markov process. Since the diffusion coefficient multiplying the Wiener increment in Eq. (36b) is constant, there is no Ito-Stratonovich ambiguity in this case. Instead of attempting to solve the nonlinear SDE (36) we introduce a a new stochastic variable $u(t,\lambda) = e^{i\lambda x(t)}u(t)$ with λ a parameter. The average $<u(t,\lambda)> = <e^{i\lambda x(t)}u(t)>$ may be called a marginal characteristic function; in particular we have $<u(t)> = <u(t,\lambda = 0)>$. From Eq. (36) we have (in matrix notation)

$$du(\lambda,t) = (A+x(t)B-i\lambda\gamma x(t)-\lambda^2<x^2>\gamma)u(\lambda,t)dt +$$
$$+ i\lambda\sqrt{2\gamma<x^2>}u(t,\lambda)dW(t) ,$$

so that the average $<u(\lambda,t)>$ obeys

$$\frac{\partial}{\partial t} \langle u(\lambda,t)\rangle = (A - i \frac{\partial}{\partial\lambda} B + \gamma\lambda \frac{\partial}{\partial\lambda} - \langle x^2\rangle\gamma\lambda^2)\langle u(\lambda,t)\rangle \qquad (37)$$

which, at the expense of introducing partial derivatives with respect to λ, is a linear equation.

The form of Eq. (37) suggests an ansatz

$$\langle u(\lambda,t)\rangle = \langle e^{i\lambda x(t)}\rangle g(\lambda,t) = e^{-\frac{1}{2}\langle x^2\rangle\lambda^2} g(\lambda,t) \quad ,$$

where the coefficients of a power series expansion of $g(\lambda,t)$,

$$g(\lambda,t) = \langle e^{\frac{1}{2}\lambda^2\langle x^2\rangle} e^{i\lambda x(t)} u(t)\rangle =$$

$$= \sum_{n=0}^{\infty} (-i\sqrt{\frac{\langle x^2\rangle}{2}}) \frac{1}{n!} \langle H_n(\frac{x(t)}{\sqrt{2\langle x^2\rangle}}) \; u(t)\rangle \quad , \qquad (38)$$

describes cross correlations between Hermite polynomials in $x(t)$ and $u(t)$. Inserting Eq. (38) into (37) we find the infinite system of differential equations

$$(\frac{d}{dt} + n\gamma - A)\langle H_n u\rangle = (n+1)\sqrt{\frac{\langle x^2\rangle}{2}} B\langle H_{n+1} u\rangle$$

$$+ \sqrt{2\langle x^2\rangle} B\langle (H_{n-1} n\rangle (1-\delta_{no}) \qquad (39)$$

for $n = 0,1,\ldots,$ which we want to solve for $\langle u(t)\rangle = \langle H_o u\rangle$ with the initial condition $\langle H_n u(t=0)\rangle =$ $= \langle H_n\rangle \; \langle u(t=0)\rangle \equiv \delta_{no}\langle u(t=0)\rangle$.

Taking the Laplace transform of Eq. (39) an explicit solution for the average

$$\langle u(t)\rangle = \frac{1}{2\pi i}\int ds\; e^{st}\,\frac{1}{s-A-\hat{K}(s)}\langle u(t{=}0)\rangle \tag{40a}$$

can be given in terms of a matrix continued fraction

$$\hat{K}(s) = B\,\cfrac{\langle x^2\rangle}{s+\gamma-A-B\,\cfrac{2\langle x^2\rangle}{s+2\gamma-A_{\ddots}}B}\,B\;. \tag{40b}$$

The matrix $K(\tau) = \frac{1}{2\pi i}\int ds\; e^{s\tau}\hat{K}(s)$

may be identified with the kernel in an integrodifferential equation for $\langle u(t)\rangle$,

$$\frac{d}{dt}\langle u(t)\rangle = A\langle u(t)\rangle + \int_0^t d\tau\; K(\tau)\langle u(t-\tau)\rangle \quad , \tag{41}$$

where $K(\tau)$ describes memory effects associated with the finite correlation time $1/\gamma$. The matrix continued fraction expansion may be considered as a summation of a "fast fluctuation" ($\sqrt{\langle x^2\rangle}\,||B||\ll\gamma$) perturbation expansion. In the limit of large γ, $\hat{K}(s)$ can be truncated after the first step, so that Eq.(41) takes on the form

$$\frac{d}{dt}\langle u(t)\rangle = A\langle u(t)\rangle + \\ + \langle x^2\rangle\int_0^t d\tau\; Be^{(-\gamma+A)\tau}\,B\langle u(t-\tau)\rangle\;. \tag{42}$$

This result can also be obtained if Eq.(33) is solved by decorrelation assumptions $\langle x(t)x(t')u(t')\rangle \overset{?}{\simeq} \langle x(t)x(t')\rangle\langle u(t')\rangle$. Note that in the white noise limit $\gamma\to\infty$ ($\langle x^2\rangle = \gamma/2$) Eq.(42) reduces to the Stratonovich result (35). There are obvious convergence questions regarding matrix continued fraction expansions of the form (40). Only

few convergence theorems are availabe, some of them derived only quite recently [12]. Convergence of the solution of (40) has so far only been proved for special cases where the form of the matrices A and B in Eq. (39) allowed further elimination of some of the vector components $\langle H_n u_\mu \rangle$, resulting in matrix continued fractions of lower dimensionality [5].
Obviously the present method can be extended to more general polynomial couplings $x^n(t)$ in Eq. (33)[5,6].

(ii) As our second example illustrating matrix continued fraction expansions we study the nonlinear Langevin equation

$$dx(t) = x(a-x^2-y^2)dt + \sqrt{2}\, dW_1(t) \ ,$$

$$dy(t) = y(a-x^2-y^2)dt + \sqrt{2}\, dW_2(t) \ , \tag{43}$$

which describes a laser near threshold [4,8,9]. The variables x(t) and y(t) are scaled components of the (slowly varying) comple: electric field amplitude $\varepsilon(t) = x(t) + iy(t)$; a is the laser pump parameter which is positive (negative) above (below) the laser threshold. The deterministic part of the Langevin equation (42) is a sum of a linear gain ($a > 0$) and a saturation term. The additive white noise terms correspond to spontaneous emission of laser atoms. If we identify x(t) and y(t) with the coordinates of a fictitious particle, we can visualize the time development of x(t) and y(t) as the strongly damped motion of a particle in the cylindrically symmetric potential $V(x,y) = -\frac{1}{2}(x^2+y^2)(a-\frac{1}{2}x^2-\frac{1}{2}y^2)$, subjected to a random force. Note that for $a<0$, $x=y=0$ is a stable point of the potential, while for $a>0$, $\sqrt{x^2+y^2} = a$ are the stable equilibrium positions with $x = y = 0$ unstable, i.e., we have a situation reminicent of a second order phase transition [8]. Fluctuations are particularly important near the transition point $a = 0$, where the system behaves in an intrinsically nonlinear way.

For the following it will be convenient to use polar "coordinates" $x(t) + iy(t) = \sqrt{I(t)}\, e^{-i\phi(t)}$ with $I(t)$ the light intensity and $\phi(t)$ the laser phase. According to Ito's formula the transformed equations (43) are

$$dI(t) = [2I(a-I)+4]dt + \sqrt{8I}\, dW_I(t) \quad ,$$

$$d\phi(t) = \sqrt{\frac{2}{I}}\, dW_\phi(t) \quad , \tag{44}$$

where

$$\begin{bmatrix} dW_I(t) \\ dW_\phi(t) \end{bmatrix} = S(t) \begin{bmatrix} dW_1(t) \\ dW_2(t) \end{bmatrix} \quad \text{with } S = \begin{bmatrix} \cos\phi(t) & -\sin\phi(t) \\ \sin\phi(t) & \cos\phi(t) \end{bmatrix}$$

are again Wiener increments.

Suppose we are interested in the stationary laser spectrum, for example, defined as the Fouriertransform of the amplitude correlation function

$$S(\nu) = 2 \operatorname{Re} \int_0^\infty d\tau\, e^{i\nu\tau} \langle \varepsilon^*(t+\tau)\varepsilon(t) \rangle \quad (t \to \infty) \quad . \tag{45}$$

As a preliminary remark to our calculation below we observe that the eigenfunctions $f(x)$ of

$$a(x)\frac{\partial f}{\partial x} + \frac{1}{2}\, b^2(x)\, \frac{\partial^2 f}{\partial x^2} = -\,\lambda f \tag{46}$$

with $a(x)$ and $b(x)$ time independent and λ an eigenvalue obey the Ito SDE

$$df(x(t)) = -\lambda f(x(t)) + b(x(t)) f_x(x(t)) dW(t) \quad ,$$

i.e., we have

$$\frac{d}{dt}\langle f(x(t))\rangle = -\lambda\langle f(x(t))\rangle \ .$$

For the one-dimensional Ornstein-Uhlenbeck process (36b), for example, these eigenfunctions are Hermite polynomials $f_n = H_n(x/\sqrt{2\langle x^2\rangle})$ with $\lambda_n = n\gamma$ (n=0,1,...).
For a two-dimensional Gaussian Process we obtain as eigenfunctions Laguerre polynomials [5]

$$f_{n\nu}(I,\phi) = \frac{n!}{(n+|\nu|)!} I^{|\nu|/2} L_\nu^{(n)}(I)\ e^{-i\nu\phi} \ ,$$

$$\lambda_{n\nu} = (2n+|\nu|)\gamma \qquad (n = 0,1,\ldots;\nu = 0,\pm 1\ldots) \tag{47}$$

with $I = x^2 + y^2$, $\phi = \arctan(y/x)$.

Let us return now to the Langevin equation (43). We are no longer able to solve the eigenvalue problem (46) analytically. We note, however, that the nonlinearity in Eq.(43) contains only powers of x and y. In view of the recursion relations for Laguerre polynomials the averages $C_{n\nu}(t) = \langle f_{n\nu}(x(t))\rangle$, with $I(t) = \lambda x(t)$ and λ an arbitrary scaling parameter, obey the four-term recursion relation [4] ($C_{00} = 1, \nu \geq 0$)

$$\begin{aligned}\dot{C}_{n\nu} &= (2n+\nu)(n+\nu+1)\lambda C_{n+1,\nu} \\ &+ [(2n+\nu)a-2n(3n+3\nu+1)\lambda-\nu(\nu+1)]C_{n\nu} \\ &+ n[(6n+3\nu-2)\lambda-2a-4/\lambda]C_{n-1,\nu}+2n(1-n)\lambda C_{n-2,\nu}\end{aligned} \tag{48}$$

which in the limit $t \to \infty$ becomes a bandstructured system of linear equations. In a similar way an equation for the spectrum (45) may be derived. Consider the correlation functions

$$g_n(t+\tau,t) = \langle f_{n1}(x(t+\tau))x^{1/2}(t)e^{i\phi(t)}\rangle$$

$$\xrightarrow{\tau\to 0} C_{no}(t)-C_{n+1,o}(t) \quad . \qquad (49)$$

The Laplace transforms $\hat{g}_n(s) = \int_0^\infty d\tau\ e^{s\tau} g_n(t+\tau,t)$ obey in analogy to Eq.(48) an inhomogeneous four term recursion formula

$$(2n+1)(n+2)\lambda\hat{g}_{n+1}+[(2n+1)a-2n(3n+4)\lambda-2-s]\hat{g}_n$$

$$+n[(6n+1)\lambda-2a-4/\lambda]\hat{g}_{n-1}+2n(1-n)\lambda\hat{g}_{n-2} = C_{n+1} - C_n \qquad (50)$$

whose solution gives us the spectrum $S(\nu) = 2\lambda\ \mathrm{Re}\ \hat{g}_o(-i\nu)$. In terms of a vector $\hat{x}_n(s) = (\hat{g}_{2n}(s),\ \hat{g}_{2n+1}(s))^T$ Eq.(50) can be rewritten as a triadiagonal vector recursion

$$Q_n^-\ \hat{x}_{n-1}(s)+(Q_n - s)\hat{x}_n(s) + Q_n^+\ \hat{x}_{n+1} = -x_n(\tau = 0) \qquad (51)$$

with $Q_n^\pm$, Q_n two by two matrices and $x_n(\tau = 0)$ inhomogenities. Eq.(51) is now solved by the ansatz

$$\hat{x}_{n+1}(s) = S_n^+(s)\hat{x}_n(s) + a_{n+1}(s)(n \geq 0) \ ,$$

$$\hat{x}_o(s) = a_o(s) \ . \qquad (52)$$

This leads to matrix continued fraction expansions for $\hat{g}_o(s)$ in terms of

$$S_n^+(s) = [s - Q_{n+1} - Q_{n+1}^+\ S_{n+1}^+(s)]^{-1}\ Q_{n+1}^- \ ,$$

$$a_n(s) = [s - Q_n - Q_n^+\ S_n^+(s)]^{-1}\ [x_n(\tau = 0) + Q_n^+\ a_{n+1}(s)], \qquad (53)$$

which in the present case are known to converge quite rapidly. Details of the solutions are summarized in [4]. The value of the present approach is its applicability to multivariable SDEs [4]. For example, we can construct joint solutions of equations of the type (35a) coupled to nonlinear Langevin equations (43) [5].

4. QUANTUM FLUCTUATIONS IN QUANTUM OPTICS PROBLEMS [3,7]

In quantum optics we often have problems of the type that modes of the radiation field (in a cavity) are coupled in a nonlinear way by the atomic medium. The quantum statistical properties of these modes can be described by a master equation for the reduced density operator ρ of the field, as a function of (harmonic oscillator) annihilation and creation operators a, a^+ for each of these modes. All physical observables are obtained from normally ordered moments or correlation functions of a and a^+. In practice it is often simplest to represent ρ by quasi-probability function expanding ρ with the aid of coherent states $|\alpha>$ (eigenfunctions of the destruction operator: $a|\alpha> = \alpha|\alpha>$). An example is the Glauber-Sudarshan P-representation for a single mode defined as an expansion in diagonal coherent-state projection operators:

$$\rho = \int d^2\alpha |\alpha><\alpha| \; P(\alpha,\alpha^*) \; . \tag{54}$$

Normally ordered operator products

$$<(a^+)^n a^m> = \int \alpha^{*n}\alpha^m \; P(\alpha,\alpha^*) d^2\alpha \tag{55}$$

can then be written in formal similarity to averaging with a classical probability distribution.
The above P-representation may be used to obtain time-development equations in a c-number representation from the

master equation for the density operator ρ. Quite often we find by appropriate reordering that the quantum mechanical master equation can be written in the form ($\alpha^{(1)} = \alpha$, $\alpha^{(2)} = \alpha^*$; $\mu = 1,2$)

$$\frac{\partial\rho}{\partial t} = \int |\alpha><\alpha| \; \frac{\partial P(\alpha,\alpha^*)}{\partial t} \; d^2\alpha$$

$$= \int \{[A^\mu(\alpha,\alpha^*)\frac{\partial}{\partial\alpha^\mu} + \frac{1}{2} D^{\mu\nu}(\alpha,\alpha^*)\frac{\partial^2}{\partial\alpha^\mu\partial\alpha^\nu}]|\alpha><\alpha|\}P(\alpha,\alpha^*)d^2\alpha \quad . \tag{56}$$

Integrating by parts and neglecting boundary terms, solutions for ρ (Eq.(54)) can be constructed by solving the Fokker-Planck type equation

$$\frac{\partial P(\alpha,\alpha^*)}{\partial t} = (-\frac{\partial}{\partial\alpha^\mu} A^\mu + \frac{1}{2} D^{\mu\nu} \frac{\partial^2}{\partial\alpha^\mu\partial\alpha^\nu})P(\alpha,\alpha^*) \quad , \tag{57}$$

which immdiately suggest that we can use the formalism of SDEs to calculate physical observables (55).
$P(\alpha,\alpha^*)$, however, is not a true probability distribution but belongs to a class of quasi probability functions. While $P(\alpha,\alpha^*)$ exists for thermal light (a Gaussian distribution) and coherent laser fields (compare the Langevin Equations (43)), for fields with nonclassical photon statistics $P(\alpha,\alpha^*)$ is not necessarily a well-behaved positive function. Examples for such nonclassical effects are antibunching [13] of photons in time as measured by photon coincidence counting, and squeezed states of the radiation field. For antibunching to be observed we require

$$<:[(a^+-<a^+>)(a-<a>)]^2:> = \int|\alpha-<\alpha>|^2 P(\alpha,\alpha^*)d^2\alpha < 0$$

(with ::denoting normal ordering of a and a^+), while for

squeezed states [14] of the radiation field the in-phase or out-of-phase normally ordered variance

$$<:(a_{1,2}-<a_{1,2}>)^2:> = \int(\alpha_{1,2}-<\alpha_{1,2}>)^2 P(\alpha,\alpha^*)d^2\alpha < 0$$

with $a = a_1 + ia_2$ and $\alpha = \alpha_1 + i\alpha_2$ becomes negative. The investigation of such nonclassical states of the radiation field is presently one of the main streams of active research in quantum optics [13,14]. The description of light statistical properties with the help of a FPE for $P(\alpha,\alpha^*)$ meets the difficulty that the diffusion matrix is not necessarily positive semidefinite.

In order to treat problems with nonclassical photon statistics, a class of generalized P-representations were introduced by Gardiner and coworkers [7]. An example is the positive P-representation $P(\alpha,\beta)$ of the density operator

$$\rho = \int d^2\alpha \int d^2\beta \, \frac{|\alpha><\beta^*|}{<\beta^*|\alpha>} \, P(\alpha,\beta) \tag{58}$$

where contrary to (Eq.55) α and β vary independently over the whole complex plane, and which allows normally ordered averages to be written in the form

$$<(a^+)^n a^m> = \int d^2\alpha d^2\beta \beta^n \alpha^m P(\alpha,\beta) \quad . \tag{59}$$

The positive P-representation has the properties [7]:

(i) $P(\alpha,\beta)$ always exists for a physical density operator;

(ii) it can be chosen positive (as a genuine probability).

(iii) Provided a FPE exists for a Glauber-Sudarshan representation, a corresponding FPE for $P(\alpha,\beta)$ exists with a positive semidefinite diffusion coefficient. This enables SDEs to be derived, i.e., we have a correspondence between quantum Markov processes (as defined by a quantum mechanical master equation) and diffusion processes. Note, however, that this is accompanied by a doubling of the dimensions of the variables.

As an example consider a single mode interferometer with a nonlinear absorber [3]. The master equation for the mode (a,a^+) with frequency ω is in the interaction picture

$$\frac{\partial\rho}{\partial t} = [\varepsilon a^+ - \varepsilon^* a, \rho] + \frac{1}{2}K[2a^2\rho(a^+)^2 - (a^+)^2a^2\rho - \rho(a^+)^2a^2] \,, \qquad (60)$$

where the first term describes the coupling of the cavity mode to an external classical field ε while the second term corresponds to nonlinear absorption. Using the P-representation we would obtain

$$\frac{\partial}{\partial t}P(\alpha,\alpha^*) = [-\frac{\partial}{\partial\alpha}(\varepsilon - K\alpha^2\alpha^*) - \frac{1}{2}\frac{\partial^2}{\partial\alpha^2}K\alpha^2 + \text{c.c.}]P(\alpha,\alpha^*) \qquad (61)$$

which by transforming to real variables $\alpha = x+iy$ can be seen to have a nonpositive semidefinite diffusion matrix. For the positive P-representation we again find a FPE which can be obtained from Eq. (61) by the replacement $\alpha^* \to \beta$ and $P(\alpha,\alpha^*) \to P(\alpha,\beta)$ and corresponds to the Ito equation

$$\begin{bmatrix} d\alpha \\ d\beta \end{bmatrix} = \begin{bmatrix} \varepsilon - K\alpha^2\beta \\ \varepsilon - K\alpha\beta^2 \end{bmatrix} dt + i\sqrt{K} \begin{bmatrix} \alpha dW_1(t) \\ \beta dW_2(t) \end{bmatrix} \,. \qquad (62)$$

The advantage of a SDE formulation of these nonlinear quantum fluctuation problems is that not only powerful analytical techniques are available for solving SDEs but computer simulation of these equations provides a convenient numerical tool for calculating correlation functions and treating multivariable systems.

REFERENCES

1. L. Arnold, Stochastic Differential Equations (Wiley-Interscience, New York, 1974).
2. I.I. Grihman and A.V. Skorokod, Stochastic Differential Equations (Springer, Berlin, Heidelberg, New York, 1972).
3. C.W. Gardiner, Handbook of Stochastic Methods (Springer, Berlin, Heidelberg, New York, Tokyo, 1983); and unpublished lecture notes.
4. H. Risken, The Fokker-Planck-Equation (Springer, in press).
5. P. Zoller, G. Alber and R. Salvador, Phys. Rev. A 24 (1981) 398; P. Zoller and J. Cooper, Phys. Rev. A 28 (1983) 2310.
6. K. Wodkiewicz, J. Math. Phys. 23 (1982) 2179.
7. P.D. Drummond, C.W. Gardiner and D.F. Walls, Phys. Rev. A 24 (1981) 914.
8. H. Haken, Synergetics, An Introduction (Springer, Berlin, Heidelberg, New York, 1978).
9. H. Haken, Laser Theory, Encyclopedia of Physics XXV/2c, eds. S. Flügge and L. Genzel (Springer, New York, 1970).
10. J.M. Sancho, M. San Miguel, S.L. Katz and J.D. Gunton, Phys. Rev. A 26 (1982) 1589, and references cited.
11. H.G. Van Kampen, Stochastic Process in Physics and Chemistry (North-Holland, Amsterdam, 1981).
12. H. Denk and M. Riederle, J. Approx. Theory 35 (1982) 355, and references cited.
13. D.F. Walls, Nature 280 (1979) 451.
14. D.F. Walls, Nature, in press.

Acta Physica Austriaca, Suppl. XXVI, 101–170 (1984)

FEYNMAN PATH INTEGRALS*

From the Prodistribution Definition to the Calculation of Glory Scattering

by

Cécile DeWITT-MORETTE++
Department of Physics
University of Texas, Austin, Texas 78712,USA

and

Zentrum für Interdisziplinäre Forschung
Universität Bielefeld
D-4800 Bielefeld 1, FR Germany

I. INTRODUCTION

In these lectures I shall present a path integral calculation, starting from a global definition of Feynman path integrals and ending at a scattering cross section formula. Along the way I shall discuss some basic issues which had to be resolved to exploit the computational power of the proposed definition of Feynman integrals.

I propose to compute the glory scattering of gravitational waves by black holes. What is glory scattering? And why choose this problem to develop the theory of path integration? Recall that the classical cross section for the scattering of a beam of particles in a solid angle

---

+ Lectures given at the XXIII. Internationale Universitätswochen für Kernphysik,Schladming,Austria,February 20-March 1,1984.
++Supported in part by NSF grant no. PHY 81-07381.

$d\Omega = 2\pi \sin\theta \, d\theta$ by an axisymmetric potential is

$$d\sigma_{cl}(\Omega) = 2\pi B(\theta) dB(\theta) \quad (1a)$$

where the deflection function $\Theta(B)$ giving the scattering angle Θ as a function of the impact parameter B is assumed to have a unique inverse +)$B(\Theta)$. We can write

$$d\sigma_{cl}(\Omega) = B(\Theta) \left.\frac{dB(\Theta)}{d\Theta}\right|_{\Theta=\theta} \frac{d\Omega}{\sin\theta}$$ which will be abbreviated to

$$d\sigma_{cl}(\Omega) = B(\theta) \frac{dB(\theta)}{d\theta} \frac{d\Omega}{\sin\theta} \quad . \quad (1b)$$

It can happen that, for a certain value of B, say B_g,

$$\Theta(B_g) = 0 \quad \text{or} \quad \Theta(B_g) = \pi \quad \text{while } B(\Theta) d\Theta/dB\Big|_{B=B_g} \neq 0 \; . \quad (2)$$

Then the classical cross section for forward scattering $\Theta = 0$ or backward scattering $\Theta = \pi$ is infinite. Unusual properties of forward or backward scattering of electromagnetic waves have long been observed. It has been called "the specter of the Brocken" after the name of the highest peak in the Harz mountains where it is often observed. It is now more commonly called "glory" [1]. It was to observe glories that C.T.R. Wilson built the first cloud chamber.

The formula derived by Mie in 1908 for the intensity of light scattered by a sphere is often used to compute glory scattering. This formula is obtained from a partial wave decomposition of the electromagnetic wave and the summation of many terms is needed to obtain glory scattering cross sections with a modicum of accuracy. The situation has been improved by

---

+) If not $d\sigma = 2\pi \sum_i B_i(\theta) dB_i(\theta)$ for $\{B_i(\theta)\}$ the inverses of $\Theta(B)$ at $\Theta = \theta$.

using complex angular momentum techniques for Mie scattering [1d].

The glory scattering of particles by a potential has been analyzed by Pechukas [9] in a path integral formulation of semiclassical scattering theory.

In a well known paper, Ford and Wheeler [1e] computed the quantum mechanical cross section for forward and backward scattering from a partial wave decomposition of the wave function (i.e., from the Rayleigh-Faxen-Holtsmark formula for the scattering amplitude), and by making several approximations obtained an analytical cross section. One of these approximations consists of replacing the phase shift by its WKB approximation,and the Ford and Wheeler cross section is sometimes called the semiclassical glory cross section. Berry [1f] derived the quantum mechanical glory cross section by a uniform approximation of formulae valid near and far from the glory.

A few years ago Matzner and Handler [1g] computed the scattering of gravitational waves by black holes by numerical calculations following a partial wave decomposition; their cross sections displayed oscillations reminiscent of glories. Were they indeed glories? Could one gain qualitative understanding of the numerical calculations and relate easily the amplitude and periodicity of the cross section to characteristics of the black hole and/or the scattered waves?

I thought I could answerthese questions easily using path integration techniques [2],[11]. Indeed it seemed that the only hope for an intuitive analytical calculation was a direct semiclassical calculation circumventing the partial wave decomposition-sum-resummation. - A strict WKB calculation would not be sufficient because it gives infinite cross sections. Path integration seemed to be just the right tool. It gives powerful semiclassical approximations valid even when the critical points of the action are degenerate, i.e., when strict WKB breaks down. In addition, this was a problem which could not be solved by analytical continuation of results obtained via Wiener integrals [3].

Although I thought I had the answers close at hand, several problems of interest in their own right had to be solved before a glory cross section could be computed:

i) A Feynman-Kac formula for momentum-to-momentum transitions other than the appropriate Fourier transforms of position-to-position, or momentum-to-position Feynman-Kac formulae [4a]. Indeed Fourier transforms are not available on curved spacetime. Moreover the one or two extra integrations imposed by Fourier transforms would have made the fully analytical calculation extremely difficult, if not impossible.

ii) Momentum-to-momentum transitions when the time interval becomes infinite, so as to have a clean identification of scattering cross sections [4a]. The early formulae for such transitions were not suitable for actual calculations [2e],[5].

iii) Momentum-to-momentum transitions when the corresponding initial and final classical states are constrained by conservation laws [4b].

iv) A real understanding of glories - i.e., a classification of various types of possible degeneracies of the critical point of the actions [4d],[4c].

v) Polarized glories [4f]. The problem here was to generalize to wave scattering the results obtained for particle scattering.

This work has been done jointly with Bruce Nelson and Tian-Rong Zhang. The end result of our calculation is the WKB approximation for the backward polarized glory cross section for a massless wave of spin s and wave length λ:

$$d\sigma_{WKB}(\Omega) = 4\pi^2\lambda^{-1}B^2(\theta)\,\frac{dB(\theta)}{d\theta}\,J^2_{2s}(2\pi\lambda^{-1}B(\theta)\sin\theta)d\Omega \qquad (3)$$

where J_{2s} is the Bessel function of order 2s.

For $s = 0$, $d\sigma_{WKB}$ (at glory) $= 0$,

for $s \neq 0$, $d\sigma_{WKB}$ (at glory) $\neq 0$.

Using $B(\theta)$ for the scattering of gravitons by Schwarzschild black holes obtained by Darwin [10], Matzner has computed (3) and compared it with his original numerical results [4e]. Both cross sections are shown on figure 1.

The path integral calculation of the glory scattering is so simple that one can, with hindsight, give a heuristic [4e,9] derivation of the cross section (3). To contrast the path integral calculation with earlier calculations I shall quote a recent book (1980) on "Rainbows, Halos, and Glories " by Robert Greenler [1a]:

> In one sense the glory is now fairly well understood. A mathematical theory (Mie scattering theory) enables us to calculate the intensity variation in the glory pattern. Unfortunately, it gives us little physical insight into the process that produces the rings. Nussenzveig has developed another mathematical treatment, in which he attempts to identify the paths of rays that interfere to produce the diffraction rings. His model is not simple, however, and we cannot use it to predict the size of the rings without the complicated mathematical treatment. Simple models exist for all the other phenomena described in this book - models that give a physical understanding of the phenomena. I wonder if there is no simple model containing the physical essence of the explanation of the glory.

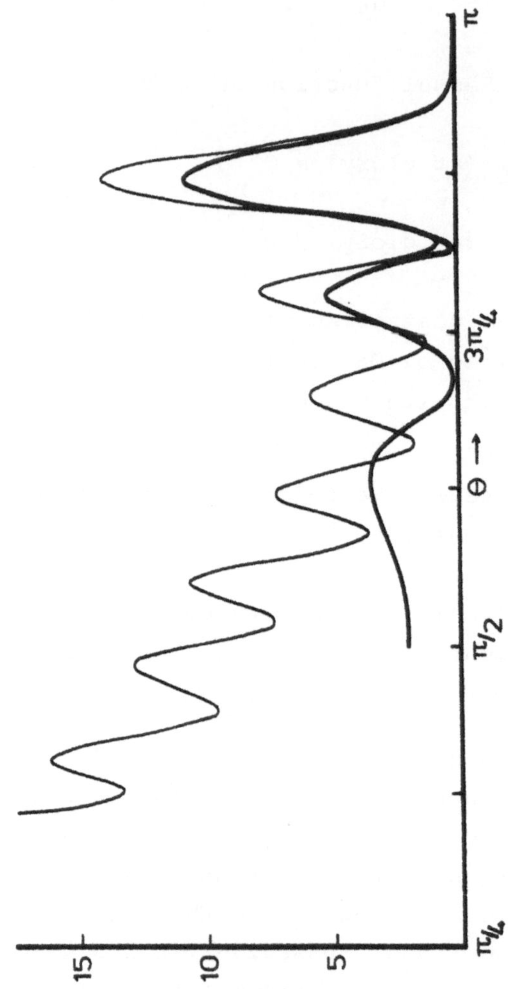

Fig. 1.

Fig.1: (Courtesy of Richard Matzner)
Cross sections of gravitational waves by Schwartzschild black holes of mass M, frequency ω, such that $M\omega = 2.5$. The thinner line is the cross section computed by Matzner and Handler using a partial wave decomposition of the solution of the wave equation. The thicker line is the glory cross section, valid when the scattering angle Θ is of order π, computed by Matzner from equation (3) with $B(\Theta)$ near glory [10] given by $B(\Theta) = M(3\sqrt{3}+3.48 \exp(-\Theta))$. The exponential dependence of B on Θ accounts for the differences in the maxima in using formula(3) beyond its validity, $\Theta \not\approx \pi$. In addition, glory scattering dominates the cross section only for Θ very close to π; other effects begin to be felt [6] for Θ close to π.

I shall now proceed from a definition of Feynman path integrals to the calculation of (3).

II. PRODISTRIBUTION

- How to compute cylindrical path integrals -

A number of years ago, at the suggestion of Yvonne Choquet-Bruhat, I found in the chapter of Bourbaki on integration on topological vector spaces [6] the means of defining an object, which could be used to build and compute some Feynman integrals. Whereas most texts on probability begin with (Ω, F, μ_F), a set Ω, a Borel σ-algebra F, a measure μ_F on F, Bourbaki begins with $(\Omega, \mathcal{Q}, \mu_\rho)$, a topological vector space Ω Hausdorff and locally convex, a projective family of finite dimensional spaces $\mathcal{Q}$ related to Ω and a projective family of bounded measures $\mu_\mathcal{Q}$ on $\mathcal{Q}$, called "promeasures".
A projective family of finite dimensional spaces is a mathe-

matical construction which accounts for Feynman's heuristic procedure. It states qualitatively the idea "Let us replace a path by an arbitrary finite set of its values."; it makes it simple to ensure that the result is independent of the chosen set. But is does more because the finite dimensional spaces are not restricted to spaces of paths defined by a finite number of their values. And it has computational power as we shall see shortly. It is defined as follows. Let $F(\Omega)$ be the set of closed subspaces V,W... of Ω of finite codimension partially ordered by the inclusion relation. The space V belongs to $F(\Omega)$ if and only if it consists of points $\omega \epsilon \Omega$ such that

$$\langle \omega_j', \omega \rangle = 0 \text{ for a finite set } V_o \stackrel{\text{def}}{=} \{\omega_j'\} , \tag{1}$$

where ω_j' Ω',the topological dual of Ω. If $V_o \subset W_o$, then $W \subset V$.

Let Ω/V be the equivalence class of points $[\omega]$ defined by the equivalence $\omega_1 \sim \omega_2 \Leftrightarrow (\omega_1 - \omega_2) \epsilon V$.
Let P_V be the canonical mapping from Ω into Ω/V. Let $W \subset V$ and let P_{VW} be defined by $P_V = P_{VW} P_W$. The quotient spaces Ω/V, Ω/W,... together with the canonical mappings P_{VW}: $\Omega/W \to \Omega/V$... form the projective system of finite dimensional quotient spaces of Ω indexed by $F(\Omega)$. A promeasure μ is a family of bounded measures $\{\mu_V\}$ on Ω/V such that

$\mu_V(\Omega/V)$ is independent of V ;

when $W \subset V$, μ_V is the image of μ_W under P_{VW} . (2)

It is an easy matter to restate the coherence conditions satisfied by the family μ_Ω of bounded measures as coherence conditions satisfied by the family $F\mu_\Omega$ of their Fourier transforms. Indeed, let $\tilde{P}_V$: $(\Omega/V)' \to \Omega'$ by $u' \to \omega'$ be the transposed mapping defined by

$$\langle u', P_V\omega \rangle_{\Omega/V} = \langle \tilde{P}_V u', \omega \rangle_\Omega \quad \text{for } \omega \in \Omega , \tag{3}$$

where $<\ ,\ >_{\Omega/V}$ is the duality in Ω/V, and $<\ ,\ >_{\Omega}$ is the duality in Ω. Then for $\omega' = \tilde{P}_V u'$

$$F\mu_{\Omega}(\omega') = F\mu_V(u') = \int_{\Omega/V} \exp(-i\langle u', u\rangle)\, d\mu_V(u) . \tag{4}$$

Since Ω' is the union of all V_o for all $V \in F(\Omega')$, equation (4) defines $F\mu_{\Omega}$ on Ω'.

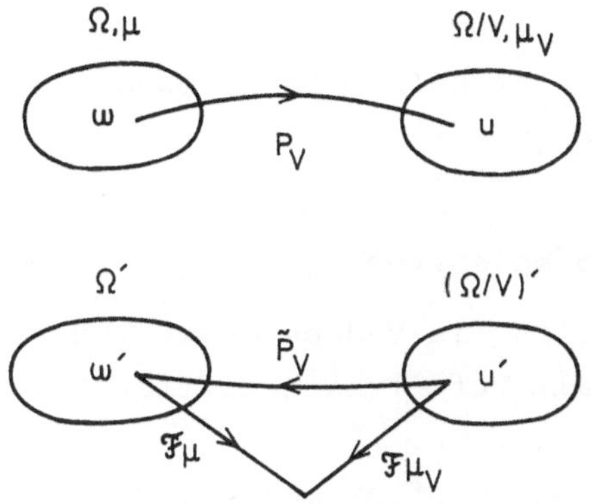

Fig. 2: P_V is a linear continuous mapping; $\tilde{P}_V$ its transposed. μ is a promeasure on Ω defined as a family of measures μ_V on Ω/V for all $V \in F(\Omega)$. The Fourier transform $F\mu_V = F\mu \circ \tilde{P}_V$.

The coherence conditions satisfied by $F\mu_\rho$ are

$F\mu_V(0)$ is independent of V ;

$$\text{when } W \subset V,\ F\mu_V = F\mu_W \circ \tilde{P}_{VW} . \tag{5}$$

We note that the coherence conditions (5) are simpler to implement than the coherence conditions (2) because they are not set statements.

Now, measures need not be defined as set functions, they can equivalently be defined as distributions of order zero. As set functions, the distinction between bounded and unbounded measures is formidable; as distributions, the distinction is interesting but not overwhelming. Given the fact that

(5) is easier to handle than (2) ,

unbounded measures belong more naturally to distribution theory than to set theory,

I propose to consider the projective family $\mathcal{F}\mu_{\Omega}$ of distributions $\{\mathcal{F}\mu_V; V \in \mathcal{F}(\Omega)\}$ satisfying the coherence conditions (5) and to call it the prodistribution $\mathcal{F}\mu_{\Omega}$. Originally called "pseudomeasure", it was shortly afterwards called "prodistribution" by Dieudonné. The Fourier transform $\mathcal{F}\mu_V$ is not restricted to be the Fourier transform of a bounded measure, thus prodistributions are generalizations of promeasures.

Important issues remained to be solved:

i) The space of test functions for $\mathcal{F}\mu_V$ is V-dependent. The set of all test functions for all $V \in \mathcal{F}(\Omega)$ is a difficult object to handle.

ii) The following theorem in the theory of promeasures has not been generalized to prodistributions:

"The mapping $\mu_{\Omega} \to \mathcal{F}\mu_{\Omega}$ of the set of promeasures on Ω to the set of functions on Ω' is injective."

However it is possible to bypass these issues if one develops a theory of integration based on $\mathcal{F}\mu_{\Omega}$ rather than on μ_{Ω}. This theory is still rudimentary but sufficient to compute many nontrivial physics problems.

Let $F = f \circ P_V$ for $V \in \mathcal{F}(\Omega)$ and $f \in \mathcal{S}(\Omega/V)$. Then we can define

$$\int_{\Omega} F(\omega)\, d\mu_{\Omega}(\omega) \text{ to be the number equal to } \int_{\Omega/V} f(u)\, d\mu_V(u) \tag{6}$$

with

$$\mathcal{F}\mu_V = \mathcal{F}\mu_{\Omega} \circ \tilde{P}_V \; . \tag{7}$$

This definition has computing power.

Example. Let Ω be the space of continuous functions defined on $T\subset\mathbb{R}$. Its dual Ω' is the space of bounded measures on T

$$\langle\omega',\omega\rangle = \int_T d\omega'(t)\omega(t) \quad \text{or} \quad \langle\delta_t,\omega\rangle = \omega(t) \text{ as the case may be.} \tag{8}$$

Let the functions which take the same values at a certain partition $\theta_V = \{t_1,\dots t_V\}$ be called equivalent, i.e.

$\omega_1 \sim \omega_2 \iff \omega_2 = \omega_1 + \chi$ with $\langle\delta_{t_i},\chi\rangle = 0$ for every $t_i \in \theta_V$,

$V_o = \{\delta_{t_i}; t_i \in \theta_V\}$, $\chi \in V$.

X/V is the space of equivalence classes $[\omega] \equiv \{\omega_i;\omega_i \sim \omega\}$.

A point $[\omega]\in X/V$ is defined by the values of $\{\omega(t_i);\delta_{t_i}\in V_o\}$,

$$P_V: \Omega \to X/V \text{ by } \omega \to \{u^1,\dots,u^V\} \text{ with } u_i = \omega(t_i), \tag{9}$$

$$\tilde{P}_V: (X/V)' \to \Omega' \text{ by } \{u_1',\dots,u_V'\} \to \sum_{i=1}^{V} u_i'\delta_{t_i}. \tag{10}$$

Proof of (10): $\langle\tilde{P}_V u',\omega\rangle = \langle u',P_V\omega\rangle = \sum_{i=1}^{V} u_i'\langle\delta_{t_i},\omega\rangle$. □

If $F\mu_Q$ is a complex gaussian prodistribution of variance W,

$$F\mu_Q(\omega') = \exp(-\tfrac{i}{2} W(\omega',\omega')) \tag{11}$$

and

$$F\mu_V(u') = \exp(-\tfrac{i}{2} \sum_{i=1}^{V}\sum_{j=1}^{V} u_i' u_j' W^{ij}, \text{ with } W^{ij} = W(\delta_{t_i},\delta_{t_j}), \tag{12}$$

$$d\mu_V(u) = \frac{du_1}{\sqrt{2\pi i}} \cdots \frac{du_V}{\sqrt{2\pi i}} (\det_{ij}(W^{-1})_{ij})^{1/2} \exp(\tfrac{i}{2} u^i u^j (W^{-1})_{ij}). \tag{13}$$

The work then consists in inverting W and computing its determinant. We shall give some practical methods to do so

in section III.
Of course the method is not limited to V_o a set of δ_{t_i}. The above expressions can be immediately rewritten when V_o^i is a set of arbitrary points in Ω'.

Do we, in practice, often encounter integrands F of the form $f \circ P_V$?
Yes, even in situations which seem far removed from linear mappings, such as path integrals for paths with values in a Riemannian space. There is a long list of problems which have been solved using (6) and (7). The use of (6) and (7) is not limited to mappings P into finite dimensional spaces. They can be used for linear continuous mappings $P: \Omega \to X$ where X is an infinite dimensional space with a countable basis; e.g. the basis made by a complete set of eigenvectors in a Sturm-Liouville problem [2e, 4c].

Since Fourier transforms can be defined on group manifolds, the generalization from (vector spaces, linear mappings) to (group manifolds, homomorphisms) is being investigated.

III. SEMICLASSICAL EXPANSIONS

- When the critical points of the action are degenerate -

Glory scattering is best characterized in terms of momentum-to-momentum transitions since it is a transition in which the final momentum is equal to plus or minus the initial momentum. Feynman-Kac formulae, on the other hand, give only position-to-position, position-to-momentum, momentum-to-position transitions. Indeed, let for instance, the Hamiltonian of the system be

$$H(p,q) = \frac{1}{2}||p||^2 + V(q) \quad , \tag{1}$$

where $||\cdot||$ is the norm defined by the Legendre metric tensor $g_{\alpha\beta} = \partial^2 L/\partial\dot{q}^\alpha \partial\dot{q}^\beta$. Then, the probability amplitude for a transition from a state $|\alpha\rangle$ to a state $|\beta\rangle$ where α and β are position or momentum states can be represented by the following Feynman-Kac formulae provided α and β are not both momentum states:

$$\langle\beta|\exp(-\frac{i}{\hbar}(t_b-t_a)\hat{H})|\alpha\rangle=\int_{\Omega_+} dw_+(\omega)\exp[-\frac{i}{\hbar}\int_{t_a}^{t_b} V(b+\sqrt{\hbar}\omega(s))ds]\phi_+(b+\sqrt{\hbar}\omega(t_a)) \quad (2)$$

or

$$\langle\beta|\exp(-\frac{i}{\hbar}(t_b-t_a)\hat{H}))|\alpha\rangle=\int_{\Omega_-} dw_-(\omega)\exp[-\frac{i}{\hbar}\int_{t_a}^{t_b} V(a+\sqrt{\hbar}\omega(s))ds]\phi_-(a+\sqrt{\hbar}\omega(t_b)). \quad (3)$$

Ω_+ is the space of continuous paths $\omega: [t_a,t_b] \to R^n$ such that $\omega(t_b) = 0$,
Ω_- is the space of continuous paths such that $\omega(t_a) = 0$,
ϕ_+ is the wave function of the initial state,
ϕ_- the wave function of the final state.

$$\mathcal{F}w_\pm(\omega') = \exp(-\frac{i}{2} W_\pm(\omega',\omega'))=\exp[-\frac{i}{2}\int_T d\omega'_\alpha(t)\int_T d\omega'_\beta(s) G_\pm^{\alpha\beta}(t,s)] \quad (4)$$

$$G_+^{\alpha\beta}(t,s) = (\theta(t-s)(t_b-t)+\theta(s-t)(t_b-s))g^{\alpha\beta} \quad (5)$$

$$G_-^{\alpha\beta}(t,s) = (\theta(t-s)(s-t_a)+\theta(s-t)(t-t_a))g^{\alpha\beta} \ . \quad (6)$$

T will be both $[t_a,t_b]$ and t_b-t_a.
For position-to-position transitions use either
(2) with $\phi_+(x) = \delta(x-a)$ or
(3) with $\phi_-(x) = \delta(x-b)$.
For momentum-to-position-transitions use (2) with
$\phi_+(x) = (2\pi\hbar)^{-n/2} \exp[\frac{i}{\hbar}\langle p_a,x\rangle]$.
For position-to-momentum transitions use (3) with
$\phi_-(x) = (2\pi\hbar)^{-n/2} \exp[-\frac{i}{\hbar}\langle p_b,x\rangle]$.

Equations (2),(3) can be derived in several ways, using a version of Itô calculus adapted to prodistribution, or the theory of product integrals[2e]. Alternatively one can use the method outlined below to derive another equivalent path integral representation of $<\beta|\exp[-\frac{i}{\hbar}(t_b-t_a)\hat{H}]|\alpha>$. Expansions in powers of $\sqrt{\hbar}$ of the path integrals (2) and (3) or their Wiener counterparts have been computed

-by Itô calculus for the diffusion equation [17b] when the classical flow has an inverse,

- by Stratonovich calculus for the Schrödinger equation [8],[2e]. These latter calculations are not restricted to classical flows which have an inverse.

1. Construction of path integrals for momentum-to-momentum transitions

To obtain a Feynman-Kac formula for momentum-to-momentum transitions, one can solve the following problem: Determine F and w such that

$$\int_{\Omega} F(\omega)dw(\omega) = <\beta|\exp[-\frac{i}{\hbar}(t_b - t_a)\hat{H}]|\alpha> \quad , \tag{7}$$

where $|\alpha>$ and $|\beta>$ can be linear combinations of position and momentum states. There is flexibility in assigning terms to the integrand F or to the integrator w. We assume for the time being that w can be chosen to be a Gaussian. Thus determining w means determining its covariance G. We shall consider the following cases, the general case is obtained by linearity. Let $\Omega = X \times Y$ be the space of continuous paths $\omega \equiv (x,y): T \to R^{2n}$.

(case a) Ω_0 the space of paths such that $x(t_a) = 0$, $x(t_b)=0$

(case b) Ω_+ the space of paths such that $y(t_a) = 0$, $x(t_b)=0$

(case c) Ω_- the space of paths such that $x(t_a) = 0$, $y(t_b)=0$

(case c) $\bar{\Omega}$ the space of paths such that $y(t_a) = 0$, $y(t_b)=0$

Recall that prodistributions are defined on vector spaces, hence the boundary conditions on the paths (if any) have to be vanishing conditions. Indeed let X be the space of paths x such that $x(t_a) = c$. It is a vector space if $x_1 \in X$, $x_2 \in X$ implies $(x_1 + x_2)(t_a) = c$. This is possible only if $c = 0$. The choice of space determines the boundary values of the covariances G_o, G_+, G_-, $\bar{G}$ of the integrators w_o, w_+, w_-, $\bar{w}$ defined on them, since

$$\int_\Omega \begin{bmatrix} x^\alpha(t) \\ \\ y_\alpha(t) \end{bmatrix} (x^\beta(s), y_\beta(s))\, dw(x,y) = i \begin{bmatrix} {}^1G^{\alpha\beta}(t,s) & {}^2G^\alpha{}_\beta(t,s) \\ \\ {}^3G_\alpha{}^\beta(t,s) & {}^4G_{\alpha\beta}(t,s) \end{bmatrix} . \tag{8}$$

The numbers 1-4 attached to each block will make it possible to drop the indices later on. For instance in case a, ${}^1G_o(t_a,s) = {}^1G_o(t_b,s) = 0$.
The four blocks are not independent (see Appendix A), and the boundary values of one block determine the boundary values of the full G.

The right hand side of (7) is defined formally as the limit $p \to \infty$ of

$$I_p \equiv \int_{R^{np}} \langle\beta| 1-\tfrac{i}{\hbar}\varepsilon\hat{H}|\gamma_p\rangle\langle\gamma_p| 1-\tfrac{i}{\hbar}\varepsilon\hat{H}|\gamma_{p-1}\rangle \ldots \langle\gamma_1| 1-\tfrac{i}{\hbar}\varepsilon\hat{H}|\alpha\rangle d\gamma_1 \ldots d\gamma_p ,$$

$$p\varepsilon = t_b - t_a . \tag{9}$$

Thus to determine the left hand side of (7) we must compare its appropriate lattice approximation J_p with I_p where

$$J_p \equiv \int_\Omega (f \circ P_p)(\omega)\, dw(\omega) = \int_{R^{np}} f(u,v)\, d(P_p w)(u,v) \tag{10}$$

with $P_p : \Omega \to R^{np}$ chosen to correspond to I_p. We shall set $n = 1$ but indicate at each key step the result for $n>1$.

Let $t_a = \theta_o < t_o < \theta_1 < t_1 \ldots < \theta_{p+1} < t_{p+1} = t_b$. (11)

Let P_p: $(x,y) \mapsto (u,v)$ by $x \mapsto u = \{u^j = \langle\delta_{t_j}, x\rangle\}$

$$y \mapsto v = \{v_j = \langle v_j = \delta_{\theta_j}, y\rangle\} \ . \qquad (12)$$

For $n > 1$, use the vector valued $\delta^{\alpha}_{t_j}$ whose β-component $\delta^{\alpha}_{t_j\beta} = \delta_{t_j}\delta^{\alpha}_{\beta}$ and the vector valued $\delta_{\theta_j\alpha}$ whose β-component $\delta_{\theta_j\alpha}{}^{\beta} = \delta_{\theta_j}\delta_{\alpha}{}^{\beta}$.

Whether to choose $\theta_k < t_k$ or $t_k < \theta_k$ depends on the way one has chosen to write I_p.
It follows from (II.12) that

$$\mathcal{F}(P_p w)(u', v') = \exp[-\tfrac{i}{2}(v'^{o}, u'_{o}, \ldots, v'^{p+1}, u'_{p+1}) \cdot$$

$$G(v'^{o}, u'_{o} \ldots v'^{p+1}, u'_{p+1})] \,\hat{} \qquad (13)$$

with

$G =$

$$\begin{bmatrix}
{}^4G(0,0) & {}^3G(0,0) & {}^4G(0,1) & {}^3G(0,1) & {}^4G(0,2) & \ldots & {}^3G(0,p+1) \\
{}^2G(0,0) & {}^1G(0,0) & {}^2G(0,1) & {}^1G(0,1) & {}^2G(0,2) & \ldots & {}^1G(0,p+1) \\
{}^4G(1,0) & {}^3G(1,0) & {}^4G(1,1) & {}^3G(1,1) & {}^4G(1,2) & \ldots & {}^3G(1,p+1) \\
{}^2G(1,0) & {}^1G(1,0) & {}^2G(1,1) & {}^1G(1,1) & {}^2G(1,2) & \ldots & {}^3G(1,p+1) \\
{}^4G(2,0) & {}^3G(2,0) & {}^4G(2,1) & {}^3G(2,1) & {}^4G(2,2) & \ldots & {}^3G(2,p+1) \\
\cdot & \cdot & \cdot & \cdot & \cdot & & \cdot \\
\cdot & \cdot & \cdot & \cdot & \cdot & & \cdot \\
\cdot & \cdot & \cdot & \cdot & \cdot & & \cdot \\
{}^2G(p+1,0) & {}^1G(p+1,0) & {}^2G(p+1,1) & {}^1G(p+1,1) & {}^2G(p+1,2) & \ldots & {}^1G(p+1,p+1)
\end{bmatrix}$$

(14)

where (for $n > 1$)

$$^4G(k,k+1) \equiv {}^4G_{\alpha\beta}(\theta_k,\theta_{k+1}) \qquad {}^3G(k,k+1) \equiv {}^3G_\alpha{}^\beta(\theta_k,t_{k+1})$$

$$^2G(k,k+1) \equiv {}^2G^\alpha{}_\beta(t_k,\theta_{k+1}) \qquad {}^1G(k,k+1) \equiv {}^1G^{\alpha\beta}(t_k,t_{k+1}) \ . \quad (15)$$

If we had chosen the partition such that $t_k < \theta_k$, the blocks which make G would have been positioned as in (8). The choice of space Ω determines P_p as follows:

$$\begin{array}{lll} \Omega_o & P_p:(x\to\{u^1,\dots,u^p\},\ y\to\{v_1,\dots v_{p+1}\}) & \text{since } u_o = u_{p+1} = 0 \\ \Omega_+ & P_p:(x\to\{u^o,\dots,u^p\},\ y\to\{v_1,\dots v_{p+1}\}) & \text{since } v_o = u_{p+1} = 0 \\ \Omega_- & P_p:(x\to\{u^1,\dots,u^p\},\ y\to\{v_1,\dots v_p\}) & \text{since } u_o = v_{p+1} = 0 \\ \bar{\Omega} & P_p:(x\to\{u^o,\dots,u^p\},\ y\to\{v_1,\dots v_p\}) & \text{since } v_o = v_{p+1} = 0 \ , \end{array} \quad (16)$$

where $\theta_o < t^o < \theta_1 < t^1 \dots \theta_{p+1} < t^{p+1}$, and

the choice of P_p determines the block of G, say G_p, which will be used. It is clear from this construction how the boundary conditions on G are related to the corners of the matrix G_p.

<u>A specific example</u>. At this point we choose a specific situation, say Ω_+, to simplify the presentation, and we choose p to be large enough to display the key issues but small enough to avoid unnecessary luggage, say p = 1. The relevant block G_p is marked on (14). For p = 1 and $\Omega = \Omega_+$,

$$J_p = \int_{R^4} f(u,v)\,d(P_p w)(u,v) \quad (17)$$

with

$$d(P_p w)(u,v) = du^o dv_1 du^1 dv_2 (2\pi i)^{-4/2} (\det G_p^{-1})^{1/2} \ .$$
$$\exp[\tfrac{i}{2}(u^o,v_1,u^1,v_2) G_p^{-1} (u^o,v_1,u^1,v_2)^\wedge] \ , \quad (18)$$

provided G_p can be inverted, which we shall assume for the time being.

J_p is to be set equal to

$$I_p = \int_{R^4} \langle b|p_2\rangle\langle p_2|1 - \frac{i}{\hbar}\varepsilon\hat{H}|q^1\rangle\langle q^1|p_1\rangle\langle p_1|1 - \frac{i}{\hbar}\varepsilon\hat{H}|q^o\rangle \cdot$$

$$\langle q^o|p_a\rangle dq^o dp_1 dq^1 dp_2 \quad , \tag{19}$$

where

$$\langle q^k|p_k\rangle = (2\pi\hbar)^{-1/2}\exp[\frac{i}{\hbar}\langle q^k,p_k\rangle] \text{ (if } t_k < \theta_k \text{ change the sign of the phase)} \ , \tag{20}$$

$$\langle p_{k+1}|1-\frac{i}{\hbar}\varepsilon\hat{H}|q^k\rangle = \langle p_{k+1}|q^k\rangle[1-\frac{i}{\hbar}\varepsilon H(p_{k+1},q^k)] \simeq$$

$$\simeq (2\pi\hbar)^{-1/2}\exp[\frac{i}{\hbar}(-p_{k+1}q^k-\varepsilon H(p_{k+1},q_k))] \ . \tag{21}$$

Set

$$I_p = \frac{1}{\sqrt{2\pi\hbar}}\int \exp[\frac{i}{\hbar}S_p(b,p_2,q^1,p_1,q^o,p_a)]dq^o dp_1 dq^1 dp_2 (2\pi\hbar)^{-4/2} \tag{22}$$

with

$$S_p(b,p_2,q^1,p_1,q^o,p_a) = S(b,p_2)+S(p_2,q^1)+S(q^1,p_1)+S(p_1,q^o)+ + S(q^o,p_a) \tag{23}$$

$$S(p_{k+1},q^k) = p_{k+1}(q^{k+1}-q^k)-\varepsilon H(p_{k+1},q^k)-p_{k+1}q^{k+1} \tag{24}$$

$$S(q^k,p_k) = q^k p_k \ . \tag{25}$$

In the limit $p \to \infty$ one thinks of p_k as the value of a path p at time θ_k and q^k the value of a path q at time t_k. The paths (q,p) cannot be identified with the paths (x,y) because the paths (x,y) have to have vanishing boundary values $y(t_a) = 0$, $x(t_b) = 0$, whereas $p(t_a) = p_a$, and $q(t_b) = b$.

It is convenient to set

$$q(t) = \bar{q}(t) + \sqrt{\hbar}\, x(t),\; p(t) = \bar{p}(t) + \sqrt{\hbar}\, y(t) \;,$$

i.e.

$$q^k = \bar{q}^k + \sqrt{\hbar}\, u^k \;, \qquad p_k = \bar{p}_k + \sqrt{\hbar}\, v_k \;, \tag{26}$$

where $(\bar{p},\bar{q})$ is an arbitrary fixed path such that $\bar{p}(t_a) = p_a,\ \bar{q}(t_b) = b$, assuming (p_a,b) defines a unique classical path. It is considerably easier but not necessary in principle to choose $(\bar{p},\bar{q})$ to be a classical path. If, in the limit $p \to \infty$, we <u>define</u>

$$\int_{t_k}^{\theta_{k+1}} p dq - H(p,q)dt \overset{\text{def}}{=} p_{k+1}(q^{k+1} - q^k) - \varepsilon H(p_{k+1},q^k) \;, \tag{27}$$

we can interpret $S(p_{k+1},q_k)$ as the action function along the classical path defined by p_{k+1} at θ_{k+1} and q_k at t_k. (See (A2c) for the relationship between the action function and the action functional.)

<u>Remark</u>: dq "sticks out in the past" of the integrand because we chose to work with $\langle p_{k+1}|1-i\varepsilon\hat{H}|q_k\rangle$. We could have chosen $\langle q_{k+1}|1-i\varepsilon\hat{H}|p_k\rangle$ and be led to defining $\int pdq$ with "dq sticking out in the future". We would then have chosen the partition $t_k < \theta_k$ in evaluation J_p. Either choice is acceptable; it determines the version of stochastic calculus one must use henceforth.

Since I_p is any approximation of I such that $\lim\limits_{p=\infty} I_p = I$, we shall set $S_p(b,p_2,q^1,p_1,q^0,p_a)$ = p-broken action defined by the arguments of S_p.

Inserting (26) into (23) and expanding in powers of (u,v) gives

$$S_p(b,p_2,q^1,p_1,p^0,p_a) = S(b,p_a) + \tfrac{1}{2}(u^0,\ldots v_2)\hbar S_p''(u^0,\ldots,v_2)^\sim + \\ + \Sigma_p(\sqrt{\hbar}u) \quad , \tag{28}$$

where S''_{p_k} is a 4×4 matrix consisting of second dervatives of $S(p_{k+1},q^k)$ and $S(q^k,p_k)$, and $\Sigma_p(\sqrt{\hbar}u)$ is a term in powers of u higher than 2. There are no linear terms in u because we have chosen $(\bar{q},\bar{p})$ to be a classical path. Inserting (28) into (22) gives

$$I_p = \frac{1}{\sqrt{2\pi\hbar}} \exp\left(\frac{i}{\hbar} S(b,p_a)\right) \int \exp\left[\frac{i}{2}(u^0,\ldots,v_2) S''_p (u^0,\ldots,v_2)^{\sim} + \right.$$

$$\left. + \frac{i}{\hbar}\Sigma_p(\sqrt{\hbar}u)\right] du^0 dv_1 du^1 dv_2 (2\pi)^{-4/2} \; . \qquad (29)$$

For J_p (given by (17), (18), and (14)) to be equal to I_p, G_p^{-1} must be equal to S''_p. We shall determine the covariance G of the prodistribution w in (7) by solving

$$S''_p G_p = 1 \; . \qquad (30)$$

This is a straightforward equation to solve, even for $n > 1$, when we express S''_p in terms of the Jacobi fields of the system. The procedure to compute the second derivatives of the action function in terms of Jacobi fields is outlined in Appendix A. Some of them can be found in the following table. The complete set can be found in the appendix of [4b]. The detailed solution of (30) for arbitrary p can be found in [4a].

Table 1

Jacobi matrices and their inverses

Let ϕ_t be a classical flow in phase space mapping (a,p_a) into $(\bar{q}(t,a,p_a),\ \bar{p}(t,a,p_a))$. The derivative mapping $D\phi_t(a,p_a)$ maps the tangent space to the phase space at (a,p_a) into its tangent space at $(\bar{q}(t,a,p_a),\ \bar{p}(t,a,p_a),\ \bar{p}(t,a,p_a))$. It is a $2n \times 2n$ matrix which can be split into 4 blocks which we call Jacobi matrices:

$$\partial\bar{q}^{\alpha}(t,a,p_a)/\partial a^{\beta} \equiv K^{\alpha}{}_{\beta}(t,t_a)\ ,\quad \partial\bar{q}^{\alpha}(t,a,p_a)/\partial p_{a\beta} \equiv J^{\alpha\beta}(t,t_a),$$
$$\partial\bar{p}_{\alpha}(t,a,p_a)/\partial a^{\beta} \equiv L_{\alpha\beta}(t,t_a)\ ,\quad \partial\bar{p}_{\alpha}(t,a,p_a)/\partial p_{a\beta} \equiv \tilde{K}_{\alpha}{}^{\beta}(t,t_a)\ .$$

Their inverses are components of the Hessian of the action functions known as Van Vleck determinants. [See apppendix A]

Let $M_{\alpha\beta}(t_a,t)J^{\beta\gamma}(t,t_a)=\delta^{\gamma}_{\alpha}$, then

$$M_{\alpha\beta}(t_a,t_b) = -\partial^2 S(b,t_b;a,t_a)/\partial b^{\beta}\partial a^{\alpha}\ ;$$

let $N^{\alpha}{}_{\beta}(t_a,t)K^{\beta}{}_{\gamma}(t,t_a) = \delta^{\alpha}_{\gamma}$, then

$$N^{\alpha}{}_{\beta}(t_a,t_b) = \partial^2 S(b,t_b;p_a,t_a)/\partial b^{\beta}\partial p_{a\alpha}\ ;$$

let $\tilde{N}_{\alpha}{}^{\beta}(t_a,t)\tilde{K}_{\beta}{}^{\gamma}(t,t_a) = \delta^{\gamma}_{\alpha}$, then

$$\tilde{N}_{\alpha}{}^{\beta}(t_a,t_b) = -\partial^2 S(p_b,t_b;a,t_a)/\partial p_{b\beta}\partial a^{\alpha}\ ;$$

let $P^{\alpha\beta}(t_a,t)L_{\beta\gamma}(t,t_a) = \delta^{\alpha}_{\gamma}$, then

$$P^{\alpha\beta}(t_a,t_b) = \partial^2 S(p_b,t_b;p_a,t_a)/\partial p_{b\beta}\partial p_{a\alpha}\ ;$$

here the action functions are, respectively,

$$S(b,t_b;a,t_a) \equiv S(b;a) = \mathcal{S}(\bar{q},\bar{p})\ ,$$

$(\bar{q},\bar{p})$ such that $\bar{q}(t_a) = a,\quad \bar{q}(t_b) = b$;

$$S(b,t_b;p_a,t_a) \equiv S(b;p_a) = \mathcal{S}(\bar{q},\bar{p})+p_a q(t_a)\ ,$$

$(\bar{q},\bar{p})$ such that $\bar{p}(t_a) = p_a,\quad \bar{q}(t_b) = b$;

$$S(p_b,t_b;a,t_a) \equiv S(p_b;a) = \mathcal{S}(\bar{q},\bar{p})-p_b q(t_b) ,$$
$$(\bar{q},\bar{p}) \quad \text{such that } \bar{q}(t_a) = a, \quad \bar{p}(t_b) = p_b ;$$

$$S(p_b,t_b;p_a,t_a) \equiv S(p_b;p_a) = \mathcal{S}(\bar{q},\bar{p})+p_a q(t_a)-p_b q(t_b) ,$$
$$(\bar{q},\bar{p}) \quad \text{such that } \bar{p}(t_a) = p_a, \quad \bar{p}(t_b) = p_b ,$$

with the actional functional $\mathcal{S}$ given by

$$\mathcal{S}(\bar{q},\bar{p}) = \int_{t_a}^{t_b} (<\bar{p}(t),\dot{\bar{q}}(t)>-H(\bar{p}(t),\bar{q}(t)))dt \quad .$$

Note that the action functions are not the same function of their arguments and should be labelled differently. I did not do so in order to avoid unwieldy notation and I hope that the reader will remember that the arguments determine the form of the function.

The solution G_p of (30) is the matrix constructed from the elementary kernel G_+ of the Jacobi operator of the system with Hamiltonian H and boundary conditions determined by Ω_+, namely

$$^1G_+(t,s)=\theta(s-t)K(t,t_a)N(t_a,t_b)J(t_b,s)-$$
$$- \theta(t-s)J(t,t_b)\tilde{N}(t_b,t_a)\tilde{K}(t_a,s)$$
$$^2G_+(t,s)=\partial_s{}^1G_+(t,s), \quad ^3G_+(t,s)=\partial_t{}^1G(t,s),$$
$$^4G_+(t,s)=\partial_t\partial_s{}^1G_+(t,s)-\delta(t-s) \quad , \tag{31}$$

where θ is the step function equal to 1 for positive arguments and 0 otherwise.

Example. For a free particle

$$^1G_+(t,s) = \theta(s-t)(t_b-s) + \theta(t-s)(t_b-t) \; ,$$

$$^2G_+(t,s) = -\theta(s-t) \; , \quad ^3G_+(t,s) = -\theta(t-s) \; , \quad ^4G_+(t,s) = 0 \; . \tag{32}$$

We note that $^1G_+(t,s)$ is the Wiener covariance on the space of paths with fixed end point.
Although the equation $S''_p G_p = 1$ is only a discretized approximation of the Jacobi equation, it yields via (14) the elementary kernels of the Jacobi operator without limiting procedure. We have established an important connection between path integration and the calculus of variation: The covariance G is a kernel of a Jacobi operator.

Without any formal step in the derivation we have expressed the p-th approximation of $\langle b|\exp(-\frac{i}{\hbar}(t_b-t_a)\hat{H})|p_a\rangle$ as a path integral. Indeed

$$I_p = (2\pi\hbar)^{-1/2}(i)^{4/2}(\det G_p^{-1})^{-1/2}\exp(\tfrac{i}{\hbar}S(b,p_a))\int_{\Omega_+} dw_+(\omega)\cdot \cdot\exp(\tfrac{i}{\hbar}\,\Sigma_p(\sqrt{\hbar}u)) \; , \tag{33a}$$

or for arbitrary p and arbitrary n

$$I_p = (2\pi\hbar)^{-n/2}(i)^{n(p+1)}(\det G_p^{-1})^{-1/2}\exp(\tfrac{i}{\hbar}\,S(b,p_a))\cdot \cdot\int_{\Omega_+} dw_+(\omega)\exp(\tfrac{i}{\hbar}\,\Sigma_p(\sqrt{\hbar}u)) \; . \tag{33b}$$

It now remains to compute the limit of I_p when $p = \infty$. Since I_p is expressed already as an integral over an infinite dimensional space, we are in a much better position to do so than when I_p was given by an integral over R^{np}.

The two limits which have to be computed are:

i) $\underset{p=\infty}{\text{limit}}(i)^{n(p+1)}(\det G_p^{-1})^{-1/2}$.

The properties of the Jacobi matrices make it possible [4a] to compute

$$\det G^{-1} = (-1)^{n(p+1)} \det N(t_{p+1},t_p) \ldots \det N(t_1,t_0)/\det N(t_b,t_a)$$

and

$$\lim_{p=\infty} (-1)^{n(p+1)} \det G^{-1} = \prod_{k=0}^{p} (1+O((t_{k+1}-t_k)^2)/$$

$$/ \det_{\alpha\beta} \partial^2 S(b;p_a)/\partial b^\alpha \partial p_{a\beta}$$

$$= (\det \partial^2 S(b;p_a)/\partial b^\alpha \partial p_{a\beta})^{-1} . \qquad (34)$$

ii) $\underset{p=\infty}{\text{limit}} \int_\Omega \exp(\frac{i}{\hbar} \Sigma_p(u))dw_+(\omega)$.

It is clear from (28), (23), (24), (27) that

$$\lim_{p=\infty} \Sigma_p(\sqrt{\hbar}u) = \int_T V(\sqrt{\hbar}\, x(t))\, dt), \qquad (35)$$

where

$$\int_T V(\sqrt{\hbar}\, x(t))dt = \int_T (\langle p(t),\dot{q}(t)\rangle - H(p(t),q(t)))dt -$$

$$- \int_T (\langle \bar{p}(t),\dot{\bar{q}}(t)\rangle - H(\bar{p}(t),\bar{q}(t)))dt -$$

$$-\frac{1}{2}\hbar \int_T (x(t),y(t)(J(\bar{q}(t),\bar{p}(t)))(x(t),y(t))^{\sim} dt .$$

$$(36)$$

One expects that it is possible to define

$$\lim_{p=\infty} \int_{\Omega} \exp(\frac{i}{\hbar} \Sigma_p) dw = \int_{\Omega} \lim_{p=\infty} \exp(\frac{i}{\hbar} \Sigma_p) dw . \qquad (37)$$

If this is so, then the limit of I_p is

$$\langle b|\exp(-\frac{i}{\hbar}(t_b-t_a)\hat{H})|p_a\rangle = K_{WKB}(b,t_b;p_a,t_a) \cdot$$
$$\cdot \int_{\Omega_+} dw_+(\omega)\exp(\frac{i}{\hbar}\int_T V(x(t))dt) , \qquad (38)$$

where w_+ is the Gaussian of covariance iG_+ given by (31). However until (37) has been proven, (38) is only formal. Formal! Yes, but more meaningful and certainly more practical than our starting point $\langle b|\exp(-\frac{i}{\hbar}(t_b-t_a)\hat{H})|p\rangle = \lim_{p=\infty} I_p$ given by (9).

A similar calculation can be done for the three other cases. The final result is

$$\langle \beta|\exp(-\frac{i}{\hbar}(t_b-t_a)\hat{H})|\alpha\rangle = K_{WKB}(\beta,t_b;\alpha,t_a)\cdot$$
$$\cdot\int_{\Omega_{\alpha\beta}} dw_{\alpha\beta}(\omega)\exp(\frac{i}{\hbar}\int_T V(x(t))dt) , \qquad (39)$$

$$K_{WKB}(\beta,t_b;\alpha,t_a) = C(\beta,\alpha)\,|\det_{\mu\nu} \partial^2 S(\beta,t_b;\alpha,t_a)/\partial\beta^\mu\partial\alpha^\nu|^{1/2} \cdot$$
$$\cdot \exp(\frac{i}{\hbar} S(\beta,t_b;\alpha,t_a)) , \qquad (40)$$

where $C(\beta,\alpha)$ is determined by the chosen normalisation for $\langle\beta|\alpha\rangle$. For β and α in R^n, and $\langle b|a\rangle = \delta(b-a)$, $C(\beta,\alpha)$ is given by the following table:

Table 2	$\langle p_b\|p_a\rangle=\delta(p_b-p_a)$	$\langle p_b\|p_a\rangle=(2\pi\hbar)^n\delta(p_b-p_a)$
$C(b,a)$	$(2\pi i\hbar)^{-n/2}$	$(2\pi i\hbar)^{-n/2}$
$C(p_b,a) = C(b,p_a)$	$(2\pi\hbar)^{-n/2}$	1
$C(p_b,p_a)$	$(2\pi\hbar)^{-n/2}$	$(2\pi\hbar)^{n/2}$

The action functions and the covariances of the pro-distributions for the 4 cases position-to-position, position-to-momentum, momentum-to-position, momentum-to-momentum can be found in Appendix A.

So far we have assumed that the initial and final state define a unique classical path and that all matrices were invertible. In fact this is rarely the case. Nevertheless (39) is still valid, or can easily be generalized, when the strict WKB approximation $K_{WKB}(\beta,t_b;\alpha,t_a)$ breaks down. K_{WKB} becomes infinite in a variety of situations. In paragraph 2 we show that under certain conditions the path integral in (39) combines with K_{WKB} to give a finite probability amplitude. In section IV we show how, in some situations, K_{WKB} needs to be generalized.

2. The critical points of the action functional are degenerate

To speak of critical points implies that one has defined a variational problem, for instance a function has to be extremized. To extremize the action functional

$$S : (q,p) \to \int_T (\langle p(t),\dot{q}(t)\rangle - H(p(t),q(t)))dt \qquad (41)$$

one must specify n initial and n final conditions for (q,p). Let $Z_{\alpha\beta}$ be the space of paths (q,p) which satisfies n initial α-conditions and n final β-conditions. The <u>critical points</u> of

$$S : Z_{\alpha\beta} \to R$$

are the points $(\bar{q},\bar{p}) \in Z_{\alpha\beta}$ such that

$$S'(\bar{q},\bar{p}) = 0,$$

where $S'(q,p): T_{(q,p)}Z_{\alpha\beta} \to T_{S(q,p)}R$, and $T_{(q,p)}Z_{\alpha\beta}$ is the space of vector fields along the path (q,p).

To speak of degenerate critical points of the action functional it is convenient to investigate a family of critical points obtained, for instance, by varying the n final β-conditions. Consider n one-parameter families of classical paths $\{(\bar{q}(u), \bar{p}(u)); u = \{u^1,\dots,u^n\}\}$ satisfying n fixed initial α-conditions and n varying final β(u)-conditions; or rather consider only one of these families at a time, i.e. $\{\bar{q}(u), \bar{p}(u)\}$ for $u \in [-1,1]\}$. Set $\beta(0) = \beta$, $\bar{q}(0) = \bar{q}_0$ and $\bar{p}(0) = p_0$. In general the second variation (Appendix A)

$$S''(\bar{q}_0,\bar{p}_0): T_{(\bar{q}_0,\bar{p}_0)}Z_{\alpha\beta} \times T_{(\bar{q}_0,\bar{p}_0)}Z_{\alpha\beta} \to T_{S(\bar{q}_0,\bar{p}_0)}R \tag{42}$$

is a non degenerate quadratic form: The Jacobi equation (Appendix A9)

$$S''(\bar{q}_0,\bar{p}_0)(h,j) = 0 \ , \ h(t) = \partial\bar{q}(u,t)/\partial u\Big|_{u=0} \ ,$$
$$j(t) = \partial\bar{p}(u,t)/\partial u\Big|_{u=0} \ , \tag{43}$$

has a solution with boundary conditions determined by the chosen one parameter family of classical paths $\{\bar{q}(u),\bar{p}(u)\}$. The solution (h,j) is a <u>Jacobi field along $(\bar{q}(0),\bar{p}(0))$ defined by the family $\{\bar{q}(u),\bar{p}(u)\}$</u>.

The n families of classical paths with n fixed α-conditions determine n Jacobi fields along $(\bar{q}_0,\bar{p}_0)$. Choose now n different initial conditions, the same procedure determines another set of n Jacobi fields. There are at most 2n linearly independent Jacobi fields. It can harpen that there are less than 2n linearly independent Jacobi fields along a classical path. This path is then called a <u>degenerate critical point of the action functional</u>. An equivalent definition of degenerate critical points is the following. The critical point $(\bar{q}_0,\bar{p}_0)$ is said to be degenerate if there is at least one (non zero) Jacobi field $(\overset{o}{h},\overset{o}{j})$ along $(\bar{q}_0,\bar{p}_0)$ with n vanishing initial conditions and n vanishing final conditions.

All the possible fields can be constructed from the derivative of the flow on phase space defined by classical paths $\{\bar{q}(u),\bar{p}(u);\ u = (u^1,\ldots u^{2n})\}$. In a given variational problem, one must specify n initial and n final conditions (for instance fixed initial momentum and varying final position (case b in the Appendix)) and thereby determine the relevant set of Jacobi fields.

Since we are interested in position, or momentum, to position, or momentum transitions we select the Jacobi fields defined by the $n \times n$ matrices $J,K,\tilde{K},L$ defined in Table 1. Given these choices of boundary conditions a path $(\bar{q}_o,\bar{p}_o)$ is degenerate if and only if the determinant of at least one of these matrices vanish. The problem is to analyze what happens to the path integrals (30) when the determinant of the Jacobi matrix associated to $Z_{\alpha\beta}$ vanishes. When it vanishes the corresponding Van Vleck determinant is infinite; however, it is easier to analyze the vanishing of the determinant of a Jacobi matrix than the blowing up of a Van Vleck determinant.

We can identify 3 situations when the determinant of a Jacobi matrix vanishes:

1) The Hamilton-Jacobi equation $S'(\bar{q},\bar{p}) = 0$ has a multiple root in $Z_{\alpha\beta}$.
2) The Hamilton-Jacobi equation has a one parameter family of solutions in $Z_{\alpha\beta}$ (or several). The action function does not depend on one of the variables.
3) The action functional is invariant under a continuous group of transformations. The initial and final boundary conditions do not define a classical path, unless they satisfy some conservation laws.

Case 1). We give in the appendix examples of double roots of $S'(\bar{q},\bar{p}) = 0$ for the 4 spaces $Z_{\alpha\beta}$ in which we are interested. We have shown [4c] that in all cases equation (39) remains finite because of the contribution of the path integral

$\int_{Z_{\alpha\beta}} dw_{\alpha\beta}(\omega) \exp(\frac{i}{\hbar}\int_T V(x(t))dt)$. The calculation is fairly involved and we refer the reader to [4c] where the full proof, starting with equation (2), is given.

We shall give the key elements of the proof on a specific example; namely, the scattering of charged particles with fixed initial momentum p_a by a repulsive Coulomb potential (case b in the appendix).

i) Zero eigenvalues of Jacobi operators.

Consider a path $(\bar{q}_o, \bar{p}_o)$ such that $\bar{q}_o(t_b) = b$ be on the caustic. The path $(\bar{q}_o, \bar{p}_o)$ is a degenerate critical point of $S(q,p)$ in the space $Z_{\alpha\beta}$ of paths (q,p) such that $p(t_a) = p_a$, $q(t_b) = b$. Consider now an n-parameter family of classical paths $\{\bar{q}(u), \bar{p}(u);\ u=(u^1,\dots u^n)\}$ such that

$$\bar{q}(0) = q_o\ , \qquad \bar{p}(0) = p_o\ ,$$

$$\bar{p}(u,t_a) = p_a \quad \text{and} \quad \bar{q}(u,t_b) = b(u)\ ,$$

$$\partial\bar{q}(u,t_b)/\partial u^1\Big|_{u^1=0} \equiv \mathring{h}_1(t_b) = 0\ . \tag{44}$$

Figure (Ab) shows the family $\{\bar{q}(u), \bar{p}(u)\}$ as u^1 varies.

The Jacobi matrix

$$K^\alpha{}_\beta(t_b,t_a) = \partial\bar{q}^\alpha(u,t_b)/\partial u^\beta\Big|_{u=0} \tag{45}$$

has a vanishing determinant.

The Jacobi field $(\mathring{h}_1, \mathring{j}_1)$ along $(\bar{q}_o, \bar{p}_o)$ has vanishing boundary conditions $\mathring{j}_1(t_a) = 0$, $\mathring{h}_1(t_b) = 0$; it is also an eigenvector of the Jacobi operator $J(\bar{q}_o(t), \bar{p}_o(t)) \equiv J_t^o$ with zero eigenvalues in the space $\Omega_{\alpha\beta}$ of paths with vanishing boundary condition

$$J_t^o(\mathring{h}_k(t),\ \mathring{j}_k(t)) = \lambda_k(\mathring{h}_k(t),\ \mathring{j}_k(t))\ , \qquad \lambda_1 = 0\ .$$

Let $K^u(t_b,t_a)$, J_t^u, λ_k^u be the similar quantities in the space

$Z_{\alpha\beta(u)}$, and assume that $(\bar{q}(u),\bar{p}(u))$ is not degenerate. It can be shown that

$$\lim_{u=0} (\lambda_1^u)^{-1} \det K^u(t_b,t_a) = \det \hat{K}(t_b,t_a)/\overset{\circ}{h}{}_1^{\alpha}(t_a)\overset{\circ}{j}_{1\alpha}(t_b) \quad , \quad (46)$$

(no summation over α)

where $\det \hat{K}(t_b,t_a)$ is the truncated determinant of K equal to the product of its non zero eigenvalues, and $\overset{\circ}{h}{}_1^{\alpha}$ and $\overset{\circ}{j}_{1\alpha}$ are the α coordinate of $(\overset{\circ}{h}_1,\overset{\circ}{j}_1)$ such that $\alpha = 1$ in the coordinate system which block diagonalize $K^u(t_b,t_a)$ into a 1-1 term which tends to zero with u and an $(n-1)\times(n-1)$ matrix with non vanishing determinant.

ii) Probability amplitude in the Airy regime.

We compute the probability amplitude $\langle b(u)|\exp(-\frac{i}{\hbar}(t_b-t_a)\hat{H})|p_a\rangle$ by equations (19) - (25). However, in (26) we do not expand around the classical path such that $\bar{p}(u,t_a) = p_a$, $\bar{q}(u,t_b) = b(u)$ but around the classical path such that $\bar{p}_o(t_a) = p_a$, $\bar{q}_o(t_b) = b$. One of the reasons is that if b(u) is on the dark side of the caustic, there is no classical path in the family ending at b(u), and on the other hand we want a probability amplitude valid on and near the caustic on either side.

We make a change of variable of integration in $\int_\Omega F(\omega)dw(\omega)$ by expanding $(x(t),y(t))$ in a set of eigenfunctions of the Jacobi operator J_t^u. This change of variable diagonalizes the covariances. It serves two purposes: it brings out explicitly the eigenvalues of J_t^u which makes it possible to use (46); it makes the path integral easier to compute to the order in $\sqrt{\hbar}$ required. The final result is, for $|b(u)-b|$ of order $\hbar/mc$,

$$\langle b(u)|\exp(-\tfrac{i}{\hbar}(t_b-t_a)\hat{H})|p_a\rangle =$$

$$= -i(2\pi\hbar)^{-n/2}(2\pi i\hbar)^{-1/2}\frac{\widehat{\det} K^{\alpha}{}_{\beta}(t_b,t_a)}{\overset{\circ}{j}_{1\alpha}(t_a)\,\overset{\circ}{h}{}_1^{\alpha}(t_b)}\exp(\tfrac{i}{\hbar}S(b(u),t_b;p_a,t_a))\,I(\nu,c \quad (47)$$

where $I(\nu,c)$ is the Airy function

$$I(\nu,c) = \int_R ds \exp (i(cs - \frac{1}{3} \nu s^3))$$

with

$$\nu = -(2\hbar)^{-1} \frac{\partial p_\alpha}{\partial u^1} (t_b, u) \frac{\partial^2 q^\alpha}{(\partial u^1)^2} (t_b) ,$$

$$c = \hbar^{-1} (b(u) - b)^\alpha \overset{\circ}{j}_{1\alpha}(t_b) .$$

Equation (47) is finite on and near the caustic; the probability amplitude oscillates rapidly on the bright side; it decays exponentially on the dark side.

The conclusions of this calculation are:

- Quantum Mechanics softens up the caustics.
- Path integration has computing power. Its expansion in powers+) of $\sqrt{\hbar}$ gives term by term the semi-classical expansion.
- Terms of order $\hbar^{-1}$ give the phase of the WKB approximation, namely the action function for a classical path determined by the initial and final state.
- Terms of order $\hbar^{-1/2}$ give no contribution by virtue of the Hamiltonian-Jacobi equation.
- Terms of order $\hbar^0$ give the Van Vleck determinants, namely properties of the flow of neighboring classical paths.
- Terms of order $\hbar^{1/2}$ give the Airy regime, if any.
- Terms of higher order give the semi-classical expansion, in principle to any order described.

Cases 2 and 3 will be discussed in the context of glory scattering.
Case 2 is analyzed in [4d] and also briefly in [4f]. Case 3 is analyzed in [4b].

+) Disregarding the factor $(2\pi\hbar)^{-n/2}$ which comes from normalizing the states $\langle p_a | p_b \rangle = \delta(p_a - p_b)$.

IV GLORY SCATTERING

- In quantum mechanics and in classical wave scattering -

1. WKB glory scattering cross section in quantum mechanics

The set up: A flow of incoming particles with momentum p_a parallel to the axis of symmetry of the potential (not assumed to have spherical symmetry). The surface of impact is the annulus centered at the axis of symmetry of area $2\pi B dB$, where B is called the impact parameter. The exit area at a distance R from the scatterer is $R^2 d\Omega = 2\pi R^2 \sin\Theta d\Theta$. The potential is assumed to be of compact support. (This restriction is removed in paragraph 2.) The incoming and outgoing particles propagate freely in the distant past $t \leq t_a$ and in the distant future $t \geq t_b$. For a point $x \in R^3$ we shall use both cartesian coordinates (x^α) and cylindrical coordinates (z,ρ,ϕ) such that

$$x^1 = \rho\cos\phi \; , \quad x^2 = \rho\sin\phi \quad , \qquad x^3 = z \quad , \tag{1}$$

with the z-axis being the axis of symmetry of the potential.

The first question is "What is a scattering cross section?". The answer in classical physics is: " A cross section is, by definition, the ratio of the number of particles hitting the area $R^2 d\Omega$ per unit time to the numbers of incident particles per unit area per unit time.", namely eq. (I.1). In quantum mechanics, there are basically three formulae for scattering cross sections:

$$d\sigma(\Omega) = \frac{|j(x(R,\Omega),t_b)|^2}{|j(a,t_a)|^2} R^2 d\Omega \quad , \tag{2}$$

where the current density j is a function of position and time given by

$$j_\mu(x,t) = (-i\hbar/2m)(\phi^*(x,t)\partial_\mu\phi(x,t)-\phi(x,t)\partial_\mu\phi^*(x,t)) \tag{3}$$

for a particle of mass m and wave function $\phi(x,t)$. Another formula in terms of the probability amplitude $K(p_b,t_b;p_a,t_a) \equiv K(p_b;p_a)$ that the particle with momentum p_a at t_a be found with momentum p_b at t_b. Indeed the cross section is the transition probability into the direction Ω, per unit time per unit incident flux, and is given by

$$d\sigma(\Omega) = \frac{d\Omega}{t_b-t_a}(|p_a|/m)^{-1}\int_0^\infty (2\pi\hbar)^3|p_b|^2 dp_b|K(p_b,t_b;p_a,t_a|^2 \tag{4}$$

for states normalized by $\langle p_b|p_a\rangle = \delta(p_b-p_a)$.
This formula suffices for our purpose, although a third formula from the S-matrix formulation [4c] is more elegant. One can check [4a,4d] that if the classical flow has an inverse, the WKB approximation σ_{WKB} of either (2) or (4) is equal to the classical or geometrical σ_{cl} given by (I.1). If the classical flow does not have an inverse, but if the initial and final conditions define a finite number of classical paths "sufficiently distinct", the total WKB probability is the sum of the probabilities contributed by each classical path, and σ_{WKB} is still equal to σ_{cl}. However, this is not so if the classical paths are "sufficiently close to each other" for the wave packets representing the classical particles to interfere with each other. Such quantum interferences "soften up" the classical cross sections. This is the case in glory scattering: the family of classical paths entering an annulus of radius equal to the glory impact parameter B_g satisfying (I.2) does not scatter but remains bound together.

Glory scattering is a momentum-to-momentum transition. The WKB approximation for the momentum-to-momentum transition is (III.40):

$$K_{WKB}(p_b,t_b;p_a,t_a) = (2\pi\hbar)^{-n/2}(\det L_{\mu\nu}(t_b,t_a))^{-1/2} \cdot$$

$$\cdot \exp(\frac{i}{\hbar} S(p_b,t_b;p_a,t_a)) \quad , \tag{5}$$

provided

$$\det L_{\mu\nu}(t_b,t_a) = \det \partial p_\mu(t_b,a,p_a)/\partial a^\nu \qquad \text{for } a = q(t_a)$$

is not vanishing.

In glory scattering $\det L_{\mu\nu}$ vanishes for t_a and t_b in the distant past and future, respectively, on two accounts:

i) $\hat{e}_z^\nu \, \partial p/\partial a^\nu = 0$ where $\hat{e}_z$ is a vector in the direction of motion;

i)) $\hat{e}_\phi^\nu \, \partial p/\partial a^\nu = 0$ where $\hat{e}_\phi$ is a vector tangent to the circle of radius B_g perpendicular to the flow.

Case i) is a consequence of the invariance of the system under time translation;case ii) is due to the fact that in glory scattering $(p_b = -p_a, t_b)$ and (p_a, t_a) determine a one parameter family of classical paths.
Since $\det L_{\mu\nu} = 0$, we cannot use (4) and we have to derive $K_{WKB}(p_b,t_b;p_a,t_a)$ again. Since $K_{WKB}(b,t_b;p_a,t_a)$ is not infinite (eq. III.40) for glory scattering,one can, in flat space, simply compute

$$K_{WKB}(p_b,t_b;p_a,t_a) = \text{s.p.a.}\ (2\pi\hbar)^{-n/2}\int_{R^3} dx$$

$$\cdot \exp(-\frac{i}{\hbar} p_{b\alpha} x^\alpha) K_{WKB}(x,t_b;p_a,t_a) \quad , \tag{6}$$

where s.p.a. stands for stationary phase approximation.
The problem is now shifted to the calculation by stationary phase approximation of an integral in R^3 where the critical points are degenerate. We shall do the calculation in cylindrical coordinates (z,ρ,ϕ) given by (1).

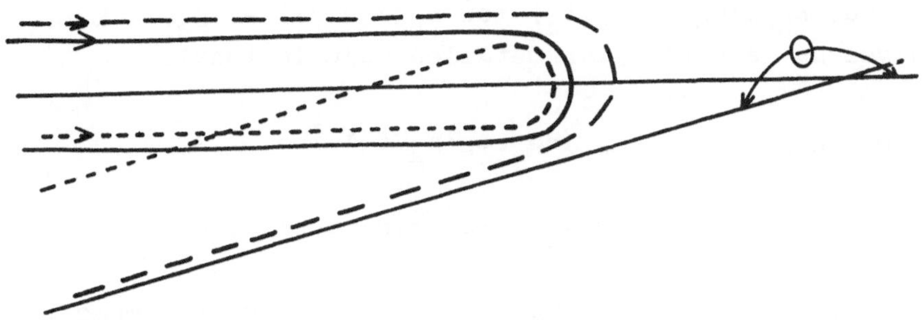

Fig.3: Interferences of two paths with impact parameter close to the glory impact parameter. The solid line is a glory path. The dashed and dotted lines are two paths with same initial momentum and final momenta equal "to first order".

Let (p_z, p_ρ, p_ϕ) be the corresponding momenta

$$p_1 = p_\rho \cos\phi - \rho^{-1} p_\phi \sin\phi$$
$$p_2 = p_\rho \sin\phi + \rho^{-1} p_\phi \cos\phi$$
$$p_3 = p_z \quad . \tag{7}$$

Then, at glory angle, $p_b = -p_a$,

$$K_{WKB}(-p_a, t_b; p_a, t_a) = \text{s.p.a.}\,(2\pi\hbar)^{-n/2} \int g(z,\rho) \cdot \exp\left(\frac{i}{\hbar} f(z,\rho)\,\rho d\rho dz \int d\phi\right) , \tag{8}$$

where

$$f(z,\rho) = z(p_a)_z + \rho(p_a)_p + S(z,\rho,t_b;p_a,t_a) \ , \tag{9a}$$

$$g(z,\rho) = |\tau(q(t_b))/\tau(q(t_a))|^{-1/2} \quad , \tag{9b}$$

with $\tau(x)$ the volume element at x defined by the flow of classical paths with fixed initial momenta. As expected f and g are independent of ϕ. If, however, we want the cross section not only at glory but also near glory, we note that $K_{WKB}(p_b,t_b;p_a,t_a)$ for $p_b(1+\sin\Theta) = -p_a$, $\Theta \approx \pi$, is given by (see [4d] for detailed calculations)

$$K_{WKB}(p_b,t_b;p_a,t_a) = K_{WKB}(-p_a,t_b;p_a,t_a) \cdot \int \exp(\tfrac{i}{\hbar}|p_b|B_g \sin\Theta \cos\phi)d\phi(\int d\phi)^{-1} . \quad (10a)$$

The new term comes from the fact that near glory impact there are for each value of ϕ two paths which exit at the same scattering angle, one with impact parameter $B_g+\delta B$ and one with impact parameter $B_g-\delta B$. Its contribution is

$$I \equiv \int_0^{2\pi} \exp(\tfrac{i}{\hbar}|p_b|B_g \sin\Theta \cos\phi)\, d\phi = 2\pi\, J_0(\frac{|p_b|}{\hbar} B_g \sin\Theta) , \quad (10b)$$

where J_0 is the Bessel function of order zero.

The critical points of f are the solutions of $\nabla f(z_0,\rho_0) = 0$, namely

$$\begin{cases} 0 = \partial f/\partial z_0 = -(p_b)_z + \partial S(z_0,\rho_0,t_b;p_a,t_a)/\partial z_0 , \\ 0 = \partial f/\partial \rho_0 = -(p_b)_\rho + \partial S(z_0,\rho_0,t_b;p_a,t_a)/\partial \rho_0 . \end{cases} \quad (11)$$

Since we do not spell out the nature of the potential responsible for glory scattering, we can solve these equations only in terms of the properties of glory scattering. Any point (z_0,ρ_0) such that

$$\begin{cases} z_0 \quad \text{sufficiently far from the scatterer} , \\ \rho_0 = B_g , \end{cases}$$

is a critical point. Hence the sum of the contributions of the critical points is an integral over z_o.

For $t \geq t_b$,

$$S(z,\rho,t;p_a,t_a) = zp_z(t) + \hat{S}(\rho,t;p_a,t_a) \tag{12}$$

is linear in z and $p_z(t)$ is constant. The phase $f(z,\rho)$ is linear in z <u>for all z_o which are critical points</u>, so that the integral <u>over z_o</u> contributes only

$$2\pi\hbar\delta((p_a)_z + (p_b)_z) = 2\pi\hbar\delta(-|p_a|+|p_b|) = 2\pi\hbar|p_a/m|\delta(E_b-E_a) . \tag{13}$$

The integral over ρ is a regular stationary phase integral (see details in [4d]), and the final result is

$$d\sigma_{WKB}(\Omega) = 2\pi\hbar^{-1}|p_a|B_g^2 \frac{dB}{d\Theta} J_o^2(\hbar^{-1}|p_b|B_g \sin\Theta)d\Omega \text{ for } \Theta \approx \pi . \tag{14}$$

This formula is identical with the formula derived by Ford and Wheeler [1e] via partial wave decomposition. Arriving at the same result by two entirely different methods strengthens both methods: it justifies the several unrelated approximations made using partial wave decomposition; it shows that path integration is a practical tool for scattering theory. It confirms and completes Pechukas' result [9]. In addition to being a more direct approach requiring only one expansion in power of $\sqrt{\hbar}$, the path integration method underscores and clarifies the roles of the degenerate critical points: In glory scattering the degeneracies which made det $L_{\mu\nu} = 0$ are of two kinds.

i) $\partial p/\partial z = 0$ for z far from the scatterer is a consequence of the invariance of the system under time translation and yields the conservation of energy in the WKB limit. The general problem of quantum systems whose classical limits are constrained by conservation laws has been analyzed in

[4b].

ii) $\partial p/\partial\phi = 0$ for any ϕ because the action function $S(z,\rho,t_b;p_a,t_a)$ is independent of ϕ.
The problem was shifted to the calculation of a stationary phase integral in R^3 where the critical points are degenerate on two acccounts corresponding respectively to these two situations; case i) the phase $f(z_o.\rho_o)$ is linear in z_o for all critical points (z_o,ρ_o,ϕ_o); case ii) the phase $f(z_o,\rho_o)$ is independent of ϕ_o.

To obtain the glory cross section of particles with arbitrary spin, one has to give some information on the interaction of the spin with the scatterer. On the other hand, the WKB glory cross section of polarized waves propagating in curved space time can be computed without specifying the mechanism responsible for glory scattering.

2. WKB glory scattering cross section in classical wave scattering

i) Introduction. We shall compute the glory cross section of polarized waves propagating in curved space time. This calculation is considerably simplified at the WKB approximation by the two following facts, valid only at the WKB approximation.

a) The polarization vector is parallel transported along the rays [12].

b) Path integrals for paths with values in Riemannian manifolds [2e,17] give WKB propagators similar to their flat counterparts. Equation (III.40) can readily be transcribed in terms of geodesics and geodesic flows. This result is so intuitively simple that it could be accepted without proof - if it were not for the controversies which have surrounded the subject of path integrals in curved spaces for more than 40 years (see a brief historical account in [17a]). I wish I had the time to present the proof because it is one of the very beautiful chapters in the theory of path integration. The proof, rigorous for Wiener integrals and Riemannian spaces, formal for Feynman integrals and pseudo-Riemannian spaces, starts with a stochastic differential equation on the frame bundle of the configuration space. It has recently been simplified [17b] by using the properties of the Hamilton-Jacobi equation rightaway in the stochastic differential equation - rather than at a later stage of the calculation. It displays path integration in its full setting at the crossroad of stochastic and differential calculus, and even suggests a different approach to quantization [18]. But this is another story.

ii) General procedure (not limited to glory scattering). You may have questioned in the previous remarks the relevance of path integrals as a tool for computing wave scattering since it is not governed by a diffusion or Schrödinger equation but by a hyperbolic one. There exists [16,18] a beautiful path integral construction of the Feynman-Green function of hyperbolic operators, and all Green functions can be extracted from the Feynman-Green function, provided one can identify its positive and negative frequency parts. So, if a space-time is asymptotically flat, path integral techniques can indeed be used in classical wave scattering problems, in principle. However, it is not necessary to follow this circuitous route. We shall recall briefly the path integral construction of the Feynman-Green function, contrast classical and quantum field theory, and present a path integral solution of classical wave scattering which bypasses the construction of the Feynman-Green function.

Let the wave equation in space-time (V^4,g) be written in the form

$$(H(x,-i\partial_\mu)-m^2)\Phi(x) = 0, \qquad \mu = 0,1,2,3 \quad , \tag{15}$$

with m not necessarily 0. We choose to write the operator $H(x,-i\partial_\mu)$ rather than $H(x,\partial_\mu)$ for easier comparison of (15) and (21). As a first example, we can consider a scalar wave in a Schwarzschild space ,

$$H(x,-i\partial_\mu) = g^{\mu\nu}(x)\nabla_\mu\nabla_\nu = (-g)^{-1/2}\partial_\mu(-g)^{1/2}g^{\mu\nu}\partial_\nu \quad ,$$
$$g = \det g_{\mu\nu} \quad , \tag{16}$$

where the Schwarzschild line element

$$g_{\mu\nu}dx^\mu dx^\nu = (1-\frac{2M}{r})dt^2 - (1-\frac{2M}{r})^{-1}dr^2 - r^2 d\Omega^2 \quad , \tag{17}$$

r is the distance to the black hole, M its mass, $d\Omega^2 = \sin^2\theta d\theta^2 + d\phi^2$. We shall consider monoenergetic solutions of the wave equation (15) which for r very large are asymptotic to distorted plane waves [13],

$$\Phi(x) = \exp(-i\omega t)\phi(\vec{x}) \quad . \tag{18}$$

For large r, $\phi(\vec{x})$ is asymptotic to a distorted plane wave of momentum $\vec{k}'$,

$$\Phi_{(k')}(x) = (2\pi\hbar)^{-n/2}\frac{r^*}{r(1-2M/r)^{1/2}}\exp(ik'\cdot x) \quad ,$$
$$x^* \equiv (x^0, \frac{r^*}{r}\vec{x}) \, , \tag{19}$$

where r^* is the "tortoise" [14] distance to the black hole ,

$$dr^* = (1-\frac{2M}{r})^{-1} dr \; , \tag{20a}$$

$$r^* = r + 2M \log(\frac{r}{2M} - 1) + \text{const.} \tag{20b}$$

The constant of integration is chosen for convenience [13], and need not to be specified in our case.
The set of distorted plane waves $\{\Phi_{(k)}\}$ is a convenient set to discuss scattering problems for two reasons:
1. It is a complete orthogonal set normalized to

$$\int \Phi_{(k')}(x)\overline{\Phi_{(k)}(x)}\ \sqrt{g(x)}\ dx = \delta(k'-k)\ .$$

2. As r from $+\infty$ towards 2M, the tortoise distance r^* goes from $+\infty$ to $-\infty$. The description of motion in terms of t and r^* leaves out the range of r values from r = 2M to r = 0.

Given a solution Φ asymptotic to a distorted plane wave of momentum k', we can define the scattering cross section $d\sigma(\Omega)/d\Omega$ as the normalized intensity of the "fraction of Φ" which is asymptotic to a distorted plane wave of momentum $\vec{k}$, for $\vec{k}$ in the solid angle $d\Omega$ defined by $(\vec{k},\vec{k}')$.

Rather than solve the WKB approximation of (15) by making the ansatz $\Phi = A \exp(i\ S/\hbar)$ and expanding in powers of a parameter $\hbar$ (no physical meaning being given to $\hbar$) we shall solve a "scaled" equation

$$(H(x,-i\hbar\partial_\mu)\ -m^2)\Psi(x) = 0 \tag{21}$$

and extract solutions Φ of the wave equation (15) from solutions Ψ of the scaled equation (21).

The Feynman-Green function $G^F(x;x')$ of (21) can be obtained [16,15] from the propagator $K(x,s;x',0)$ defined by the following equations:

$$i\hbar\partial_s K(x,s;x',0) = H(x,-i\hbar\partial_\mu)K(x,s;x',0)\quad ,\quad s > 0\ , \tag{22}$$

$$\lim_{s=0} K(x,s;x',0) = \delta(x-x')\quad . \tag{23}$$

Indeed,

$$G^F(x;x') = \lim_{\varepsilon=0} \frac{i}{\hbar}\int_0^\infty \exp(\frac{i}{\hbar}\ m^2\ s - \frac{\varepsilon}{\hbar}\ s)K(x,s;x',0)ds. \tag{24}$$

The convergence factor $\exp(-\varepsilon s/\hbar)$ is precisely the term which makes G^F a Green function and gives it the boundary values characterizing the Feynman-Green function.

All Green functions and propagators of a wave equation can be obtained from its Feynman-Green function [15] provided one can identify the positive and negative frequency terms in the Feynman-Green function G^F. Indeed, let

$$G^F(x;x') = \theta(x^o - x'^o)G^{(+)}(x;x') - \theta(x'^o - x^o)G^{(-)}(x;x') \quad ,$$

where $G^{(\pm)}$ are the positive and negative frequency contributions, respectively; then the commutator function which solves the Cauchy problem is

$$\tilde{G}(x;x') = G^{(+)}(x;x') + G^{(-)}(x;x') \quad . \tag{25}$$

The advanced and retarded Green functions are

$$G^+(x;x') = \theta(x'^o - x^o)\tilde{G}(x;x') \quad , \tag{26}$$

$$G^-(x;x') = -\theta(x^o - x'^o)\tilde{G}(x;x') \quad . \tag{27}$$

Note that G^F, $G^{\pm}$ are Green functions and that $\tilde{G}$, $G^{(\pm)}$ are solutions of the wave equation.

The fact that the commutator function (25) governs classical wave scattering,while the Feynman-Green functions governs quantum field theory,can be seen from the following example in flat space time. A wave Φ in space time can be computed from its Cauchy data at time t_o. Let the wave at time t_o be a real plane wave $(\exp(ik'x') + \exp(-ik'x'))\big|_{x'^o = t_o}$, then the wave at any time is given by

$$\Phi(x) = \int d\vec{x}'\,(\tilde{G}(x;x')\frac{\partial}{\partial x'^o}(\exp(ik'x') + \exp(-ik'x'))$$

$$- (\frac{\partial}{\partial x'^o} G(x;x'))(\exp(ik'x') + \exp(-ik'x')) \quad , \tag{28}$$

where $\tilde{G}$ is the commutator function solution of the wave equation. If the wave operator is $\partial_o^2 - \partial_i^2 + m^2$, then as expected equation (22) yields

$$\Phi(x) = \exp(ik'x) + \exp(-ik'x) \quad . \tag{29}$$

If, however, we insert G^F instead of $\tilde{G}$ in (28) we obtain

$$\Phi^F(x) = \theta(x'^o - x^o)\exp(ik'x) - \theta(x^o - x'^o)\exp(-ik'x) \; . \tag{30}$$

The particle interpretation of quantum field theory and the vacuum definition are linked to the Feynman-Green function which "propagates positive frequencies in the future and negative frequenciesin the past". On the other hand, in classical field theory a real wave remains a real wave. It seems unnecessary, and indeed it is, to use G^F to construct $\tilde{G}$. Why identify positive and negative frequency parts in a problem which keeps them together?

Let us return to the starting point, namely the propagator K, and investigate how it can be used to solve classical wave scattering problems. Since we want to solve (21) with solutions asymptotic to distorted plane waves

$$\Psi_{p'}(x) = (2\pi\hbar)^{-n/2} \frac{r^*}{r(1-2M/r)^{1/2}} \exp(\frac{i}{\hbar} p'x^*) , \; x^* = (x^o, \frac{r^*}{r}\vec{x}) \; , \tag{31}$$

we shall solve the following associated problem ,

$$i\hbar\, \partial_s K(x,s;p',0) = H(x,-i\hbar\partial_\mu)K(x,s;p',0), \quad s > 0 \; ,$$

$$K(x,0;p',0) = (2\pi\hbar)^{-n/2} \frac{r^*}{r(1-2M/r)^{1/2}} \exp(\frac{i}{\hbar} p'x^*) \; . \tag{32}$$

The WKB solution of (32) is

$$K_{WKB}(x,s;p',0) = (2\pi\hbar)^{-n/2}|\det\partial^2(x,s;p',0)/\partial x^\mu \partial p'_\nu|^{1/2} \cdot \exp(\frac{i}{\hbar}\, S(x,s;p',0)) \quad , \tag{33}$$

where $S(x,s;p',0)$ is the solution of the Hamilton-Jacobi equation

$$\partial_s S + H(x,\partial_\mu S) = 0 \quad , \tag{34a}$$

$$S_0(x,0;p',0) \equiv S_0(x) = p'x^* \ . \tag{34b}$$

The characteristic system for this Hamilton-Jacobi equation is [19]

$$ds = \frac{d\bar{q}^\mu}{\partial H/\partial \bar{p}_\mu} = \frac{-d\bar{p}_s}{\partial H/\partial s} = \frac{-d\bar{p}_\mu}{\partial H/\partial \bar{q}^\mu} = \frac{dS}{\Sigma \bar{p}_\mu \partial H/\partial \bar{p}_\mu - H} \quad , \tag{35}$$

$$\bar{p}_\mu(0,\bar{q}(0)) = \partial_\mu S_0(\bar{q}(0)) \ . \tag{36}$$

For $\bar{q}(0)$ far from the black hole, $\bar{p}(0,\bar{q}(0)) = p'$.
The Hamilton-Jacobi equation (34) generalizes the more conventional [20] relativistic Hamilton-Jacobi equation

$$H(x,\partial_\mu S_m) - m^2 = 0 \tag{37}$$

in the same sense that the action generalizes the abbreviated action in classical dynamics. The characteristic system of the relativistic Hamilton-Jacobi equation (37) is

$$\frac{d\bar{q}^\mu}{\partial H/\partial \bar{p}_\mu} = \frac{-d\bar{p}_\mu}{\partial H/\partial \bar{q}^\mu} = \frac{dS_m}{\Sigma \bar{p}_\mu \partial H/\partial \bar{p}_\mu - H} \ . \tag{38}$$

In contrast to the characteristic system of a <u>non relativistic</u>

Hamilton-Jacobi equation, $\bar{q}^0$ cannot be used to parametrize the path since now $\partial H/\partial \bar{p}_0 \neq 1$. Hence the need to introduce a parameter, say s, if one wishes to parametrize the rays of geometrical optics. Such a parametrization is necessary, for instance, for analyzing a flow of geodesics in terms of Jacobi fields. Equation (35) does more than introduce a parameter s to solve (38), it introduces a pair of conjugate variables, s and p_s.

We shall now exploit the fact that H does not depend explicitly on s. It follows that p_s is constant, and with (21) in mind we choose this constant equal to m^2,

$$H(\bar{q}(s), \bar{p}(s)) = -p_s = m^2 \quad . \tag{39}$$

The characteristic system (35) can be used to compute the proper time τ,

$$d\tau^2 \equiv g_{\mu\nu} d\bar{q}^\mu d\bar{q}^\nu = g_{\mu\nu} (\partial H/\partial \bar{p}_\mu)(\partial H/\partial \bar{p}_\nu) ds^2 \quad . \tag{40}$$

If, for instance,

$$H = g^{\mu\nu}(p_\mu - eA_\mu)(p_\nu - eA_\nu) + V(q) \quad , \tag{41}$$

then

$$d\tau^2 = 4H ds^2 = 4m^2 ds^2 \quad . \tag{42}$$

The parameter s is positive by definition, but the proper time can be positive, negative or null. The parameter s is related to the proper time only along a solution of the characteristic system, but not when it parametrizes an arbitrary path in a path integral.

In clonclusion, $H(x, -i\hbar\partial_\mu)$ is a mass operator, the propagators $K(x,s;x',0)$ and $K(x,s;p',0)$ defined by (22) and (32) propagate particles of all masses, and the integral (24) selects a mass mode.

Equation (24) is not the only procedure which selects a mass mode. In classical relativistic mechanics one can select a mass mode when $\partial H/\partial s = 0$ by shifting from the action to the abbreviated action. Indeed, when $\partial H/\partial s = 0$, the action functional $S(\bar{q},\bar{p})$ for the classical path

$$\bar{q} : [0,s] \to R^n , \quad \bar{p} : [0,s] \to R^n , \quad H(\bar{p}(r), \bar{q}(r)) = m^2, \tag{43}$$

can be split into an abbreviated action functional S and a term linear in s ,

$$S(\bar{q},\bar{p}) = \int_{x_0}^{x} \bar{p}(r)\, d\bar{q}(r) - m^2 s , \quad \text{with } \bar{q}(0) = x_0 , \ \bar{q}(s) = x. \tag{44}$$

The abbreviated action functional is

$$S_m(\bar{q}) = \int_{x_0}^{x} \bar{p}(\bar{q},d\bar{q})\, d\bar{q} , \tag{45}$$

where $\bar{p}$ is the function of $\bar{q}$ and $d\bar{q}$ obtained by eliminating ds from the system of equations

$$\bar{p}_\mu(\bar{q},\dot{\bar{q}}) = \partial L(\bar{q},\dot{\bar{q}})/\partial\dot{\bar{q}}^\mu , \qquad \dot{\bar{q}} = d\bar{q}/ds , \tag{46}$$

$$H(\bar{q},\bar{p}(\bar{q},\dot{\bar{q}})) = m^2 , \tag{47}$$

where L is the "Lagrangian" which yields the "Hamiltonian" H.

Example. If $L(q,\dot{q}) = g_{\mu\nu}(q)\dot{q}^\mu\dot{q}^\nu - V(q)$, then (46) gives

$$p_\mu(q,\dot{q}) = \frac{1}{2} g_{\mu\nu}(q)\, dq^\nu/ds ; \tag{48}$$

equation (47) gives

$$m^2 = \frac{1}{4} g_{\mu\nu}\, d\bar{q}^\mu d\bar{q}^\nu/ds^2 + V(\bar{q}) = \frac{1}{4}\, d\tau^2/ds^2 + V(\bar{q}) . \tag{49}$$

Solving this equation for ds and inserting in (48) gives for the abbreviated action

$$S_m(\bar{q}) = \int_{x_o}^{x} (m^2 - V(\bar{q}(\tau)))^{1/2} d\tau \ , \quad d\tau^2 = g_{\mu\nu}(\bar{q})\, d\bar{q}^{\mu} d\bar{q}^{\nu} \ . \tag{50}$$

The parameter s is, in that case, given by (49). Note, incidentally, that if we had chosen

$$L(q,\dot{q}) = g^{\mu\nu}(\tfrac{1}{2}\, g_{\mu\rho}\dot{q}^{\rho} - e\, A_{\mu})(\tfrac{1}{2}\, g_{\nu\sigma}\dot{q}^{\sigma} - e\, A_{\nu}) - V(q) \ , \tag{51}$$

ds would be a complicated function of $d\tau = (g_{\mu\nu} d\bar{q}^{\mu} d\bar{q}^{\nu})^{1/2}$.

Let r_m be the parameter such that if the geodesic is parametrized by r_m, then

$$H(\partial L/\partial \dot{\bar{q}}(r_m) \ , \quad \bar{q}(r_m)) = m^2 \ , \qquad 0 \leq r_m \leq s_m \ . \tag{52}$$

The abbreviated action functional can be written

$$S_m(\bar{q},\bar{p}) = \int_{x_o}^{x} \bar{p}(r_m)\, d\bar{q}(r_m) \ , \ \bar{q}(0) = x_o \ , \quad \bar{q}(s_m) = x. \tag{53}$$

We can construct abbreviated actions S_m, function of the endpoints, corresponding to given geodesic flows. For a flow such that all the geodesics start at x_o, i.e. $\bar{q}(0) = x_o$ for all geodesics, set

$$S_m(x;x_o) = S_m(\bar{q},\bar{p}) \quad \text{with} \quad \bar{q}(s_m) = x \ . \tag{54}$$

For a flow defined by

$$\bar{p}(\bar{q}(0)) = \nabla S_o(\bar{q}(0)) \quad \text{with the constraint } H(\bar{p}(\bar{q}(0)),\, \bar{q}(0)) = m^2, \tag{55}$$

set

$$S_m(x;S_o) = S_m(\bar{q},\bar{p}) + S_o(\bar{q}(0)) \ , \tag{56}$$

where $(\bar{q},\bar{p})$ is the geodesic in the flow such that $\bar{q}(s_m) = x$.

Remark. The condition (39) together with either set of boundary conditions, $\{\bar{q}(0) = x_0, \bar{q}(s_m) = x\}$ or $\{\bar{p}(\bar{q}(0)) = \nabla S_0(\bar{q}(0)), \bar{q}(s_m) = x\}$ determines s_m only when $m \neq 0$.

This analysis of the abbreviated relativistic action function S_m suggests that we compute

$$K^m_{WKB}(x;\alpha) \equiv \text{s.p.a.} \int_0^\infty \exp(\frac{i}{\hbar} m^2 s) K_{WKB}(x,s;\alpha)\,ds \quad , \tag{57}$$

where α refers to the initial conditions, either $(x_0,0)$ or an initial manifold S_0. If $m\neq 0$ the critical point of the phase is the value s_m defined by (39) together with the boundary conditions. If $m=0$, all values of $s\in[0,\infty)$ are critical and one must sum over all the critical points (see [4b]).

Consider now the space $\{[0,\infty) \times$ extended phase space$\}$ parametrized by (s,q,p). Consider its subspace defined by all the paths $\{q(s),p(s)\}$ which satisfy the initial conditions. Consider the "mass shell" Σ_m of this subspace, i.e. the space defined by the classical paths $\{\bar{q}(s),\bar{p}(s)\}$ which satisfies the appropriate initial conditions and the conservation law (39). One can construct path integral representations for the propagators defined on the mass shell Σ_m.

The composition law of WKB approximations (A.19) holds in Riemannian manifolds of arbitrary dimensions, whether or not the critical points which contribute to the integral are degenerate, and the technique developed in the non relativistic case can be used with minor modifications.

iii) Polarized glory scattering

In the case of glory scattering of scalar classical waves, one obtains for $K^m_{WKB}(p;p')$ an expression similar to (10) or rather similar to its double Fourier transform $K_{WKB}(p_b,E_b;p_a,E_a)$. In the case of glory scattering of polarized classical waves, equation (10b) becomes

$$I \equiv \int_0^{2\pi} \hat{e}_{out}(\phi) \exp(\frac{i}{\hbar}|\vec{p}|B_g \sin\theta \cos\phi)d\phi \ , \tag{58}$$

where $\hat{e}_{out}$ is the projection on space of the polarization vector obtained by parallel transporting along a space time glory trajectory the initial polarization. Let $\hat{e}_{in}$ be the projection of the initial polarization at $\bar{q}(0)$.
Since the polarization vector, tensor, or spinor of any massless wave has only two independent components [24,14,15] in the plane perpendicular to the propagation, we write

$$\hat{e}_{out}(\phi) = h_1(\phi)\hat{e}_1 + h_2(\phi)\hat{e}_2 \ . \tag{59}$$

Let $\hat{e}_{in}$ be a linearly polarized state, $\hat{e}_{in} = \hat{e}_1$, and let the origin in the ϕ coordinate be chosen so that $\hat{e}_1$ makes an angle ϕ with the tangent to the circle of radius B_g perpendicular to the incoming rays. Decompose $\hat{e}_{in}$ into a component $\hat{e}_{in||}$ in the plane of the glory ray and a component $\hat{e}_{in\perp}$ perpendicular to the plane of the glory ray. Parallel propagate $\hat{e}_{in||}$ and $\hat{e}_{in\perp}$ along the space time geodesic and project them in three dimensional space. As expected from the symmetry of the problem, one finds that

$$\hat{e}_{out\perp} = \hat{e}_{in\perp}, \quad \hat{e}_{out||} = -\hat{e}_{in||} \ . \tag{60}$$

It follows that if $\hat{e}_{in}$ makes an angle ϕ with $\hat{e}_{in\perp}$, $\hat{e}_{out}$ makes an angle 2ϕ with $\hat{e}_{in}$. Recall [24,14] that circular polarizations $\hat{e}_{\pm} = \hat{e}_1 \pm i\hat{e}_2$ transform under a rotation α of $\hat{e}_1$ and $\hat{e}_2$ about the propagation as follows,

$$\hat{e}'_{\pm} = \exp(\pm i \ s \ \alpha)\hat{e}_{\pm} \ , \tag{61}$$

where s is the spin of the wave.

Hence under the rotation (61) the state $\hat{e}_{out}$ is given

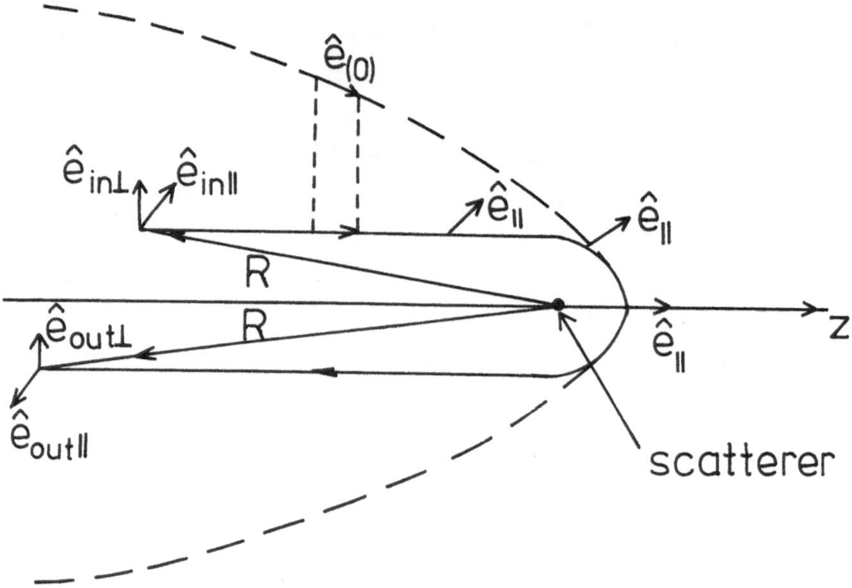

Fig.4: (Courtesy of Tian-Rong Zhang) [22]
Sketch for polarized glories. The projection on space of the space time trajectory is a planar curve. The projection $\hat{e}_{in}$ of the initial polarization is decomposed into a component in the plane and a component perpendicular to the plane. $\hat{e}_{in}$ is parallel transported along the space time geodesic (dashed line) and projected on three dimensional space. $\hat{e}_{out\perp}$ is parallel to $\hat{e}_{in\perp}$, but $\hat{e}_{out||}$ is antiparallel to $\hat{e}_{in||}$.

by

$$\hat{e}_{out} = \hat{e}_1 \cos(2s\phi) - \hat{e}_2 \sin(2s\phi) \quad . \tag{62}$$

A rotation of the state by 2ϕ is equivalent to a rotation of the coordinate system by -2ϕ. Inserting

$$\hat{e}_{out}(\phi) = \hat{e}_1 \cos(2s\phi) + \hat{e}_2 \sin(2s\phi) \tag{63}$$

in (59) gives [23]

$$I = J_{2s}(\hbar^{-1}|\vec{p}|B_g \sin\Theta) \ , \tag{64}$$

where J_{2s} is the Bessel function of order 2s.

Finally we return from the solution of (21) to the solution of (15), i.e. we replace $|\vec{p}|/\hbar$ by $2\pi\lambda^{-1}$. It is clear from (14) that the WKB cross section for backward polarized glories is

$$d\sigma_{WKB}(\Omega)/d\Omega = 4\pi^2\lambda^{-1}|\vec{p}'|B_g^2 \frac{dB}{d\Theta} J^2{}_{2s}(2\pi\lambda^{-1}B_g \sin\Theta) \quad . \tag{65}$$

ACKNOWLEDGEMENT

The glory cross sections have been computed jointly with Bruce Nelson who derived the non relativistic particle cross section (IV 14) and Tian-Rong Zhang who computed the polarized cross section (IV 67-73). Over the years, Richard Matzner has shared with us his expertise on the scattering of gravitational waves by black holes, computed the glory cross section for the scattering of a gravitational wave by a Schwarzschild black hole and compared it with his earlier results obtained by partial wave scattering (Figure 1). A full account of glory cross sections will be published jointly with Matzner, Nelson and Zhang. I am very grateful to them for allowing me to present the key results in these lectures before our joint publication.

Much of the preparation of these notes has been done at the Zentrum für Interdisziplinäre Forschung (Universität Bielefeld) where I have enjoyed the leadership and hospitality of L.Streit, excellent working conditions, and in particular discussions with G.F. Dell'Antonio, and S. Low.

Lecturing at the Schladming Winter School has been a real pleasure.

<u>Appendix A:</u> Jacobi fields in phase space+)

Let $\mathcal{Q} \times \mathcal{P}$ be the phase space of a given system, $\mathcal{Q}$ its position (or configuration) space, $\mathcal{P}$ its momentum space. We shall assume here that the position space $\mathcal{Q} = R^n$ with the Legendre metric tensor $g_{\alpha\beta} = \partial^2 L/\partial\dot{x}^\alpha \partial\dot{x}^\beta$. We use $||\cdot||$ for the corresponding norm and $|\cdot|$ for the Euclidean norm. Thus, for $L = \frac{1}{2}m|\dot{q}|^2$, $||\dot{q}||^2 = m|\dot{q}|^2$, $||p||^2 = |p^2|/m$ and $p_\alpha = g_{\alpha\beta}\dot{q}^\beta$. For $\mathcal{Q}$ a Riemannian manifold see [2e]. A path $\Omega \equiv (q,p)$ maps $T \equiv [t_a, t_b]$ into $\mathcal{Q}\times\mathcal{P}$. A path satisfying Hamilton-Jacobi equation is labelled $\bar{Q} = (\bar{q},\bar{p})$. In a variational problem a path $\bar{Q}$ is defined by n initial conditions at $t = t_a$ and n final conditions at $t = t_b$. For simplicity we shall consider only the following 4 cases. But it should be remembered that any n initial and n final conditions would define a similar variational problem.

Case a: paths $\bar{Q}$ such that $\bar{q}(t_a) = a$, $\bar{q}(t_b) = b$;

case b: paths $\bar{Q}$ such that $\bar{p}(t_a) = p_a$, $\bar{q}(t_b) = b$;

case c: paths $\bar{Q}$ such that $\bar{q}(t_a) = a$, $\bar{p}(t_b) = p_b$;

case d: paths $\bar{Q}$ such that $\bar{p}(t_a) = p_a$, $\bar{p}(t_b) = p_b$.

Equations referring to a particular case will henceforth be given the same letter.

+) For more properties of Jacobi fields and for the proofs of the results given in this appendix see [2c,2e,4b].

The <u>action functional</u> is

$$S(q,p) = \int_T (\langle p(t),\dot{q}(t)\rangle - H(p(t),q(t)))\,dt \ . \qquad (A.1)$$

The <u>action function</u> S is a generating function of a classical path.

Case a: $S(b;a) = S(\bar{q},\bar{p})$,

$$\partial S/\partial b = \bar{p}(t_b), \quad \partial S/\partial a = -\bar{p}(t_a) \ ; \qquad (A.2a)$$

case b: $S(b;p_a) = S(\bar{q},\bar{p}) + \langle p_a,\bar{q}(t_a)\rangle$,

$$\partial S/\partial b = \bar{p}(t_b), \quad \partial S/\partial p_a = \bar{q}(t_a) \ ; \qquad (A.2b)$$

case c: $S(p_b;a) = S(\bar{q},\bar{p}) - \langle p_b,\bar{q}(t_b)\rangle$,

$$\partial S/\partial p_b = -\bar{q}(t_b), \ \partial S/\partial a = -\bar{p}(t_a) \ ; \qquad (A.2c)$$

case d: $S(p_b;p_a) = S(\bar{q},\bar{p}) + \langle p_a,\bar{q}(t_a)\rangle - \langle p_b,\bar{q}(t_b)\rangle$,

$$\partial S/\partial p_b = -\bar{q}(t_b), \ \partial S/\partial p_a = \bar{q}(t_a) \ . \qquad (A.2d)$$

The action functions are different functions of their arguments.
Let $\bar{Q}(t,Q_a)$ be a classical path such that $\bar{Q}(t_a) = Q_a$. A flow in phase space is a mapping

$$\Phi_t : \mathcal{Q} \times \mathcal{P} \to \mathcal{Q} \times \mathcal{P} \quad \text{by} \quad \bar{Q}_a \mapsto \bar{Q}(t,\bar{Q}_a) \ . \qquad (A.3)$$

Its derivative mapping

$$D\Phi_t(\bar{Q}_a) : T_{Q_a}(\mathcal{Q}\times\mathcal{P}) \to T_{\bar{Q}(t,Q_a)}(\mathcal{Q}\times\mathcal{P}) \text{ by } V_a \mapsto V_t = \frac{\partial\bar{Q}(t,Q_a)}{\partial Q_a} V_a \ . \qquad (A.4)$$

1. <u>V_t is the Jacobi field along $\bar{Q}(t,Q_a)$ defined by V_a</u>

The mapping $D\Phi_t(Q_a)$ is a 2n×2n time dependent matrix, but in any variational problem characterized by n initial and n final conditions, one deals only with an n×n submatrix of $D\Phi_t(Q_a)$.
In the 4 cases we have chosen they are

$$\partial\bar{q}^{\alpha}(t,a,p_a) \;/\; \partial p_{a\beta} \equiv J^{\alpha\beta}(t,t_a) \quad , \tag{A.5a}$$

$$\partial\bar{q}^{\alpha}(t,a,p_a) \;/\; \partial a^{\beta} \equiv K^{\alpha}{}_{\beta}(t,t_a) \quad , \tag{A.5b}$$

$$\partial\bar{p}_{\alpha}(t,a,p_a) \;/\; \partial p_{a\beta} \equiv \tilde{K}_{\alpha}{}^{\beta}(t,t_a) \quad , \tag{A.5c}$$

$$\partial\bar{p}_{\alpha}(t,a,p_a) \;/\; \partial a^{\beta} \equiv L_{\alpha\beta}(t,t_a) \quad . \tag{A.5d}$$

V satisfies the Jacobi equation obtained from the second variation of the action functional. Let (q(u), p(u)) be a one parameter family of paths such that

$q(0) = \bar{q}$, $p(0) = \bar{p}$ is a classical path.

Set $(q(u))(t) = q(u,t)$ and $(p(u))(t) = p(u,t)$. Set

$$\partial q(u,t)/\partial t = \dot{q}(u,t) \quad , \quad \partial p(u,t)/\partial t = \dot{p}(u,t)$$

$$\partial q(u,t)/\partial u\Big|_{u=0} = h(t) \quad , \quad \partial p(u,t)/\partial u\Big|_{u=0} = j(t) \quad . \tag{A.6}$$

Then

$$\partial S(q(u),p(u))/\partial u\Big|_{u=0} = \int_T (h(t)(-\dot{\bar{p}}(t)-\partial H/\partial\bar{q}(t)) +$$

$$+ j(t)(\dot{\bar{q}}(t)-\partial H/\partial\bar{p}(t)))dt+\bar{p}(t_b)h(t_b)-\bar{p}(t_a)h(t_a) \quad . \tag{A.7}$$

The path $(\bar{q},\bar{p})$ which extremizes $S(q(u),p(u))$ is a critical point of the action.

$$\partial^2 S/\partial u^2\Big|_{u=0} = \int_T (h(t),j(t)\begin{bmatrix} -\partial^2 H/\partial\bar{q}\partial\bar{q} & -\partial_t-\partial^2 H/\partial\bar{q}\partial\bar{p} \\ \partial_t-\partial^2 H/\partial\bar{p}\partial\bar{q} & -\partial^2 H/\partial\bar{p}\partial\bar{p} \end{bmatrix}\begin{bmatrix} h(t) \\ j(t) \end{bmatrix})dt+$$

$$+ \text{ boundary terms}. \tag{A.8}$$

The operator

$$J(\bar{q}(t),\bar{p}(t)) \equiv \begin{bmatrix} -\partial^2 H/\partial\bar{q}\partial\bar{q} & -\partial_t - \partial^2 H/\partial\bar{q}\partial\bar{p} \\ \partial_t - \partial^2 H/\partial\bar{p}\partial\bar{q} & -\partial^2 H/\partial\bar{p}\partial\bar{p} \end{bmatrix} \tag{A.9}$$

is called the Jacobi operator, or the small disturbance operator, or the operator of geodetic deviation if $(\bar{q},\bar{p})$ is a geodesic . It is clear from the construction of the Jacobi operator that

$$J(\bar{q}(t),\bar{p}(t)) \begin{bmatrix} K(t,t_a) \\ L(t,t_a) \end{bmatrix} = 0, \text{ and } J(\bar{q}(t),\bar{p}(t)) \begin{bmatrix} J(t,t_a) \\ \tilde{K}(t,t_a) \end{bmatrix} = 0 . \tag{A.10}$$

There are at most 2n linearly independent Jacobi fields along a classical path $(\bar{q},\bar{p})$. If there are less than 2n, the classical path $(\bar{q},\bar{p})$ is said to be a degenerate critical point of the action functional. Equivalently (see[2c] and also [2e]) $(\bar{q},\bar{p})$ is said to be a degenerate critical point if there is at least one non zero Jacobi field along $(\bar{q},\bar{p})$ with vanishing boundary conditions. Degenerate critical points are discussed in paragraph 2. The initial and final points of $(\bar{q},\bar{p})$ are said to be conjugate to each other (see eq.A12).

Let $J(\bar{q}(t))$ be the Jacobi operator in configuration space, corresponding to $J(\bar{q}(t),\bar{p}(t))$ in phase space; then

$$J(\bar{q}(t))K(t,t_a) = 0 \quad \text{and} \quad J(\bar{q}(t))J(t,t_a) = 0 \quad , \tag{A.11}$$

$$L(t,t_a) = \partial_t K(t,t_a) \quad \text{and} \quad \tilde{K}(t,t_a) = \partial_t J(t,t_a) \ . \tag{A.12}$$

We shall refer to $J,K,\tilde{K},L$ as Jacobi matrices, although they are bivectors and L and $\tilde{K}$ do not satisfy the Jacobi equation in configuration space.

We list below the properties of the Jacobi matrices needed in this article and refer the reader to [2c,2e,4b] for other properties.

$$\text{i)}\quad J^{\alpha\beta}(t,t_a) = -J^{\beta\alpha}(t_a,t)$$
$$\tilde{K}_\alpha{}^\beta(t,t_a) = K^\beta{}_\alpha(t_a,t)$$
$$L_{\alpha\beta}(t,t_a) = -L_{\beta\alpha}(t_a,t) \tag{A.13}$$

ii) Van Vleck determinants

If they exist, the inverses M, N, $\tilde{N}$, P of the Jacobi matrices, defined as follows:

$$M_{\alpha\beta}(t_a,t)J^{\beta\gamma}(t,t_a) = \delta^\gamma_\alpha \tag{A.14a}$$
$$N^\alpha{}_\beta(t_a,t)K^\beta{}_\gamma(t,t_a) = \delta^\alpha_\gamma \tag{A.14b}$$
$$\tilde{N}_\alpha{}^\beta(t_a,t)\tilde{K}_\beta{}^\gamma(t,t_a) = \delta^\gamma_\alpha \tag{A.14c}$$
$$P^{\alpha\beta}(t_a,t)L_{\beta\gamma}(t,t_a) = \delta^\alpha_\gamma \quad , \tag{A.14d}$$

are elements of the Hessian of the action function, known as the Van Vleck matrices

$$M_{\alpha\beta}(t_a,t_b) = -\partial^2 S(b;a)/\partial b^\beta \partial a^\alpha \tag{A.15a}$$
$$N^\alpha{}_\beta(t_a,t_b) = \partial^2 S(b;p_a)/\partial b^\beta \partial p_{a\alpha} \tag{A.15b}$$
$$\tilde{N}_\alpha{}^\beta(t_a,t_b) = -\partial^2 S(p_b;a)/\partial p_{b\beta}\partial a^\alpha \tag{A.15c}$$
$$P^{\alpha\beta}(t_a,t_b) = \partial^2 S(p_b;p_a)/\partial p_{b\beta}\partial p_{a\alpha} \quad . \tag{A.15d}$$

Recall that the action function (A.2) is also function of t_a and t_b and that the Jacobi fields (A.5) are defined along a classical path characterized by boundary conditions, thus equations (A.15) make sense. The proof of (A.15) can be found in the appendix of [4b]. It consists in writing the action functions in term of the action functional (A.2) for a classical path $(\bar{q}(u),\bar{p}(u))$ and in expanding both sides of the equations in powers of u. Equation terms in u proves that the action functions defined by (A.2) are indeed the generating functions (A.2). Equating terms in u^2 gives (A.15) as well

as all components of the Hessians of all action functions in terms of the Jacobi matrices and their inverses.
The fact that the Van Vleck determinants are the inverses of the determinants of the Jacobi matrices gives a simple physical interpretation of the Van Vleck determinants, and a simple method for computing them.

iii) Green functions of the Jacobi operator.
If $(\bar{q},\bar{p})$ is not degenerate for a given set of boundary conditions, the Jacobi operator has a unique inverse in the space of paths (q,p) satisfying these boundary conditions. These inverses, i.e. the Green functions of the Jacobi operator with a set of boundary conditions, are easily expressible in terms of the Jacobi matrices. The Green functions of the Jacobi operator consist of 4 $n \times n$ matrices,

$$J(\bar{q}(t),\bar{p}(t)) \begin{bmatrix} {}^1G(t,s) & {}^2G(t,s) \\ {}^3G(t,s) & {}^4G(t,s) \end{bmatrix} = \begin{bmatrix} \delta_s(t) & 0 \\ 0 & \delta_s(t) \end{bmatrix} . \tag{A.16}$$

These 4 coupled equations can be decoupled. Let $J(\bar{q}(t))$ be the Jacobi operator in configuration space, then

$$J(\bar{q}(t))\ {}^1G(t,s) = \delta_s(t) , \tag{A.17a}$$

$${}^2G(t,s)=\partial_s\ {}^1G(t,s),\ {}^3G(t,s)=\partial_t\ {}^1G(t,s),\ {}^4G=\partial_t\partial_s\ {}^1G(t,s)-\delta(t-s). \tag{A.17b}$$

We shall give only ${}^1G(t,s)$ for the 4 cases a,b,c,d, and refer to (A.17b) for the other components.

Case a:

$$\begin{aligned} {}^1G(t,s)=&\theta(s-t)J(t,t_a)M(t_a,t_b)J(t_b,s)- \\ &-\theta(t-s)J(t,t_b)M(t_b,t_a)J(t_a,s) ; \end{aligned} \tag{A.18a}$$

case b:

$$^{1}G(t,s)=\theta(s-t)K(t,t_a)N(t_a,t_b)J(t_b,s)-$$
$$-\theta(t-s)J(t,t_b)\tilde{N}(t_b,t_a)\tilde{K}(t_a,s) \quad ; \tag{A.18b}$$

case c:

$$^{1}G(t,s)=\theta(s-t)J(t,t_a)\tilde{N}(t_a,t_b)\tilde{K}(t_b,s)-$$
$$-\theta(t-s)K(t,t_b)N(t_b,t_a)J(t_a,s) \quad ; \tag{A.18c}$$

case d:

$$^{1}G(t,s)=\theta(s-t)K(t,t_a)P(t_a,t_b)\tilde{K}(t_b,s)+$$
$$+ \ (t-s)K(t,t_b)\tilde{P}(t_b,t_a)\tilde{K}(t_a,s) \quad , \tag{A.18d}$$

where θ is the step function equal to 1 for positive arguments and 0 otherwise.

iv) Composition law of the WKB approximations.

The properties of the Jacobi fields provide a proof of the composition law of the WKB approximation. All WKB approximations are of the form (40)

$$K(\beta,t_b;\alpha,t_a) = C(\beta,\alpha)\left|\det_{\mu\nu}\partial^2 S(\beta,t_b;\alpha,t_a)/\partial\beta^\mu\partial\alpha^\nu\right|^{1/2}$$
$$\cdot\exp\left(\frac{i}{\hbar}S(\beta,t_b;\alpha,t_a)\right) \quad .$$

The composition law says that for $t_a < t < t_b$,

$$K_{WKB}(\beta,t_b;\alpha,t_a) = \text{s.p.a.}\int K_{WKB}(\beta,t_b;\gamma,t)K_{WKB}(\gamma,t;\alpha,t_a)d\gamma \quad . \tag{A.19}$$

This is true whether or not α,β,γ belong to the same representation. It is clear that for γ a critical point of the phase in (A.19)

$$S(\beta,t_b;\gamma,t) + S(\gamma,t;\alpha,t_a) = S(\beta,t_b;\alpha,t_a) \quad . \tag{A.20}$$

Indeed for γ a critical point the piecewise classical path characterized by (α,t_a), (γ,t), and (β,t_b) must be the classical path characterized by (α,t_a), (β,t_b). The additive property of the different action functions follows from (A.2).

The combination law of the Van Vleck determinants follows from the properties of the Jacobi fields and is proven in [4b]. The proof is valid whether or not γ is a conjugate point to α or β. If γ is not a conjugate point, a quicker proof+) consists in proving that

$$\det \partial^2 S(\beta;\alpha)/\partial\beta\partial\alpha =$$

$$= -\det\partial^2 S(\beta;\gamma)/\partial\beta\partial\gamma(\det\partial^2 S(\beta;\gamma)/\partial\gamma\partial\gamma+\det\partial^2 S(\gamma;\alpha)/\partial\gamma\partial\gamma)^{-1}$$

$$\cdot\det\partial^2 S(\gamma;\alpha)\partial\gamma\partial\alpha \ . \qquad (A.21)$$

This equation can be proved by computing the second derivatives of both sides of (A.20) when γ is a function of α and β.

2. Degenerate critical points of the action functionals

Consider the following families of classical paths.

Cases a,c: Family of classical paths $(\bar{q}(u), \bar{p}(u))$ with fixed initial point $\bar{q}(u,t_a) = a$.

Cases b,d: Family of classical paths with fixed initial momentum $\bar{p}(u,t_a) = p_a$.

It can happen that the projection on the configuration space has an envelope, or it can happen that the projection on momentum space has a limit point.

Case a: $\bar{q}(u,t_a)=a$, $\bar{q}(u,t_b)=b_u$; $\partial\bar{q}(u,t_a)/\partial u=h(t_a)=0$ for any u.

+) B.S. DeWitt, private communication.

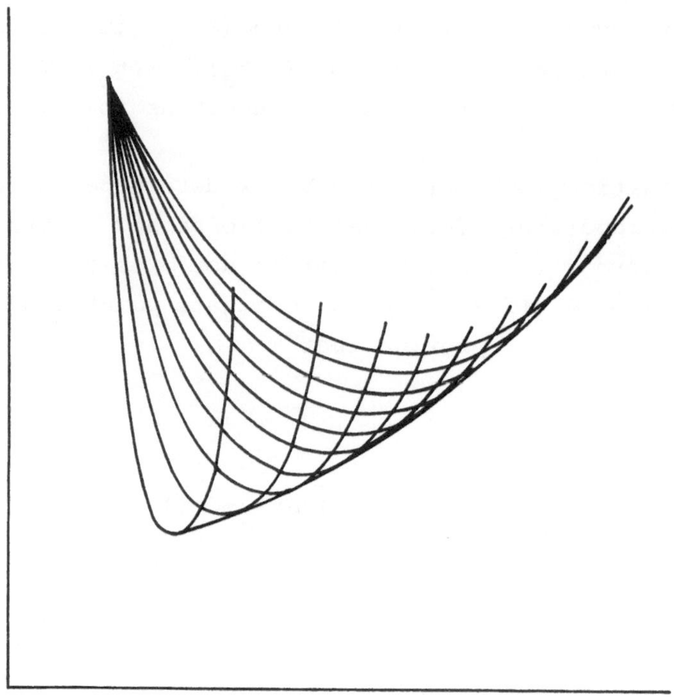

Fig.A.a: For a point in the "dark" side of the caustic there is no classical path; for a point on the "bright" side there are 2 classical paths which coalesce into a single one as the intersection of the 2 paths approaches the caustic. Note that the paths do not arrive at an intersection at the same time, the paths do not intersect in a space time diagram.

Case b: $\bar{p}(u,t_a) = p_a$, $\bar{q}(u,t_b) = b_u$, $j(t_a) = 0$. A flow on configuration space of charged particles in a repulsive Coulomb potential.

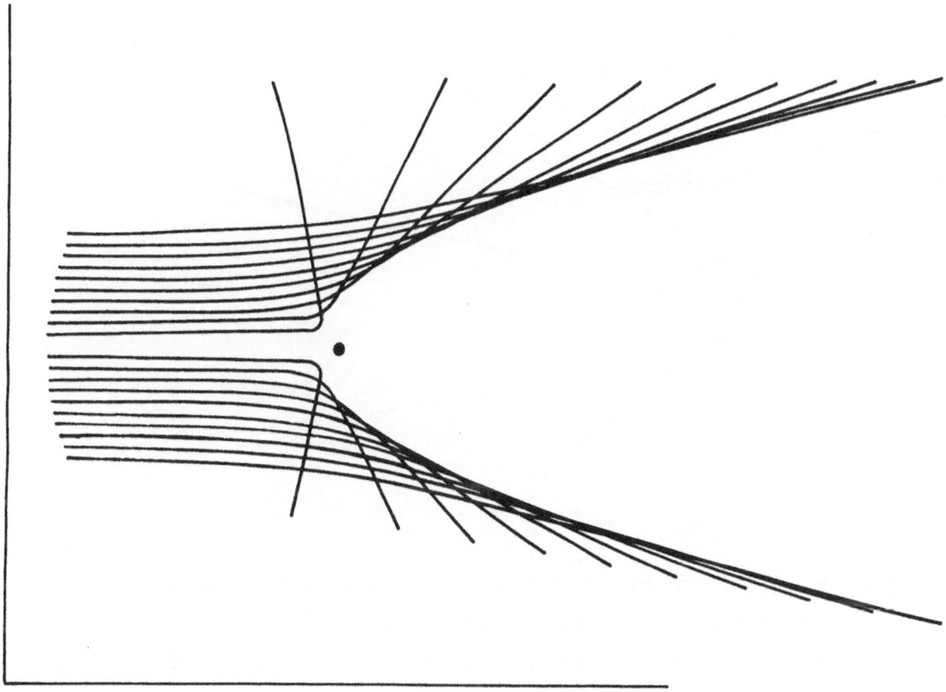

Fig.A.b: Graphics by Alice Young.

<u>Case c:</u> $\bar{q}(u,t_a) = a$, $\bar{p}(u.t_b) = p_{bu}$; $j(t_a) = 0$. Rainbow scattering from a point source.

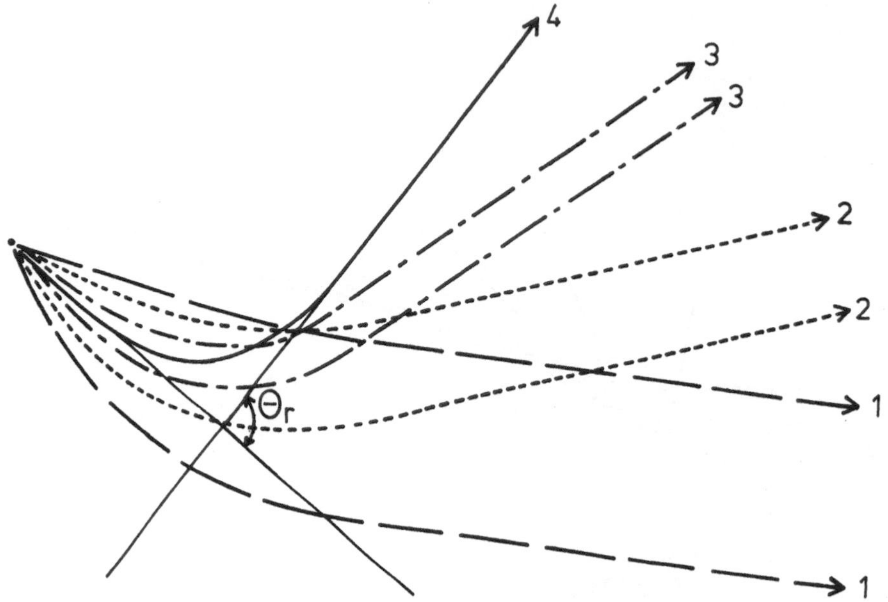

<u>Fig.A.c:</u> Projection on configuration space. Two paths with the same number exit the interaction domain with equal momenta. The pairs coalesce into a single path (path 4) for a certain value Θ_r of the scattering angle.

The projection on momentum space has a limit point.

<u>Case d:</u> $\bar{p}(u,t_a) = p_a$, $\bar{p}(u,t_b) = p_{bu}$, $j(t_a) = 0$. Rainbow scattering from a source at infinity.

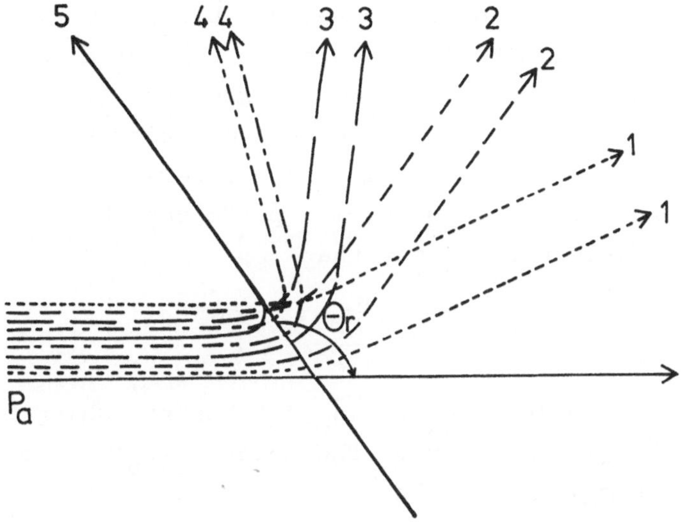

<u>Fig.A.d:</u> On configuration space, two paths with the same number exit the interaction domain with equal momenta. The pairs coalesce into a single path (path 5) for a certain value of the scattering angle Θ_r.

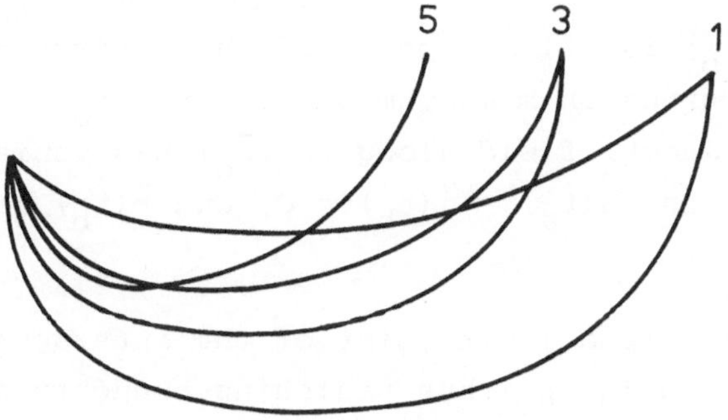

The projection on momentum space has a limit point, namely the momentum of path 5.

In the situations described in figures A.a, A.b, A.c, A.d, the Jacobi fields along any path have n vanishing components at t_a. If there is path $(\bar{q}_o, \bar{p}_o) : T \to R^{2n}$ in the family such that the Jacobi field $(\overset{o}{h}, \overset{o}{j})$ along $(\bar{a}_o, \bar{p}_o)$ determined by this family has also n vanishing components at t_b, then $(\bar{q}_o, \bar{p}_o)$ is a <u>degenerate critical point</u>.

<u>Case a</u>: If $\bar{q}_o(t_b)$ is on the envelope, $\overset{o}{h}(t_b) = 0$. Indeed an envelope is the limit of the intersections of a family of curves which, in the limit, are of order $k \geq 1$. (Recall that two curves have an intersection of order k if their first k derivatives are equal at the intersection.) Let $u \in [-1,1]$ and let $q(0) = \bar{q}$, $p(0) = \bar{p}$. To say that $\bar{q}_p(t_b)$ is on the envelope is to say that

$$\partial \bar{q}(u,t_b) \partial u \Big|_{u=0} \equiv \overset{o}{h}(t_b) = 0.$$

The Jacobi field $(\overset{o}{h}, \overset{o}{j})$ along $(\bar{q}_o, \bar{p}_o)$ has vanishing boundary conditions $\overset{o}{h}(t_a) = \overset{o}{h}(t_b) = 0$. The point $\bar{q}_o(t_b)$ is said to be <u>conjugate</u> to $\bar{q}_o(t_a)$. If there are k non zero Jacobi fields with vanishing boundary conditions, the conjugate point is said to be of <u>multiplicity</u> k. The envelope, also called the <u>caustic</u>, is the set of conjugate points.

<u>Case b</u>: If $\bar{q}_o(t_b)$ is on the caustic then the Jacobi field along $(\bar{q}_o, \bar{p}_o)$ has vanishing boundary conditions $\overset{o}{j}(t_a) = \overset{o}{h}(t_b) = 0$, and $\bar{q}_o(t_b)$ is conjugate to $\bar{p}_o(t_a)$.

<u>Case c</u>: If $\bar{p}_o(t_b)$ is a limit point of the intersections of the family of curves on momentum space, $\partial p(u,t_b)/\partial u\Big|_{u=0} \equiv \overset{o}{j}(t_b) = 0$. The Jacobi field along $(\bar{q}_o, \bar{p}_o)$ has vanishing boundary conditions $\overset{o}{h}(t_a) = \overset{o}{j}(t_b) = 0$, and $\bar{p}(t_b)$ is conjugate to $\bar{q}_o(t_a)$.

<u>Case d</u>: If $\bar{p}_o(t_b)$ is a limit point of the intersection, the Jacobi field along $(\bar{q}_o, \bar{p}_o)$ has vanishing boundary conditions $\overset{o}{j}(t_a) = \overset{o}{j}(t_b) = 0$, and $\bar{p}_o(t_b)$ is conjugate to $\bar{p}_o(t_a)$.

Summary.

There is a path $(\bar{q}_o,\bar{p}_o)$ such that the Jacobi field $(\overset{o}{h},\overset{o}{j})$ along $(\bar{q}_o,\bar{p}_o)$ has vanishing boundary conditons.

Case a: $\overset{o}{h}(t_a) = \overset{o}{h}(t_b) = 0, \det J^{\alpha\beta}(t_b,t_a) = 0,$

$$\det M_{\alpha\beta}(t_a,t_b) \text{ infinite ;} \qquad \text{(A.22a)}$$

case b: $\overset{o}{j}(t_a) = \overset{o}{h}(t_b) = 0 , \det \tilde{K}_{\alpha}{}^{\beta}(t_b,t_a) = 0,$

$$\det \tilde{N}^{\alpha}{}_{\beta}(t_a,t_b) \text{ infinite ;} \qquad \text{(A.22b)}$$

case c: $\overset{o}{h}(t_a) = \overset{o}{j}(t_b) = 0, \det K^{\alpha}{}_{\beta}(t_b,t_a) = 0,$

$$\det N_{\alpha}{}^{\beta}(t_a,t_b) \text{ infinite ;} \qquad \text{(A.22c)}$$

case d: $\overset{o}{j}(t_a) = \overset{o}{j}(t_b) = 0 , \det L_{\alpha\beta}(t_b,t_a) = 0 ,$

$$\det P^{\alpha\beta}(t_a,t_b) \text{ infinite ,} \qquad \text{(A.22d)}$$

where the Jacobi matrices and their inverses are given by (A.5) and (A.15).
In all these examples the degenerate critical point can be considered as a double root of the action functional on a space of paths defined on $T \equiv [t_a,t_b]$.
Multiple roots are not the only degenerate critical points of the action functional. It can happen that there is a one parameter family of critical points in the space of paths (or several), for instance, if the determinant of Jacobi matrix vanishes for all t, or for all t larger than a certain value. In such situations there are degenerate critical points of a different nature. Examples can be found in Chapter III and IV and in [4b,4d].

NOTATIONS

See tables 1 and 2, (II 40b), (A.19, A.20) .

$T = [t_a, t_b]$ a time interval

(q,p) a path in phase space, i.e. $(q,p): T \rightarrow R^{2n}$, but is not necessarily a solution of Hamilton Jacobi equations.

$(q,p) \in Z_{\alpha\beta}$ is a path in phase space which satisfies n initial conditions α and n final conditions β;
case a: Z_0 space of paths such that $q(t_a)=a$, $q(t_b)=b$;
case b: Z_+ space of paths such that $p(t_a)=p_a, q(t_b)=b$;
case c: Z_- space of paths such that $q(t_a)=a, p(t_b)=p_b$;
case d: $\bar{Z}$ space of paths such that $p(t_a)=p_a, p(t_b)=p_b$,

also referred to as position-to-position, momentum-to-position, position-to-momentum and momentum-to-momentum cases.

$(x,y) \in \Omega_{\alpha\beta}$ a path in phase space with vanishing boundary conditions.

S Action functional. A map from a space of paths into R. $S : Z_{\alpha\beta} \rightarrow R$.

S_m Abbreviated action functional.

$S'(q,p)$ Functional derivative of S at (q,p). Same notation for dervatives and functional derivatives. $S'(q,p)$: $T_{(q,p)} Z_{\alpha\beta} \rightarrow T_{S(q,p)} R$.

$(\bar{q},\bar{p})$ A path satisfying Hamilton-Jacobi equation. $(\bar{q},\bar{p}) \in Z_{\alpha\beta}$ is a critical point of S: $Z_{\alpha\beta} \rightarrow R$.

(h,j) a Jacobi field along $(\bar{q},\bar{p})$.

$(q_0,p_0) \in Z_{\alpha\beta}$ is a degenerate critical point of S: $Z_{\alpha\beta} \rightarrow R$.

$(\overset{o}{h},\overset{o}{j})$ a Jacobi field along $(\bar{q}_0,\bar{p}_0)$ with vanishing boundary conditions.

S Action function defined on $(\bar{q},\bar{p}) \in Z_{\alpha\beta}$. It is a solution of a Hamilton-Jacobi equation. It depends on the nature of $Z_{\alpha\beta}$. It should be given different labels in cases a,b,c,d (ss A.2), but it is expected that the context makes this added label unnecessary.

S_o initial manifold for a Hamilton-Jacobi equation.
S_m abbreviated action function for a system of mass m.

Latin indices are not component indices, e.g. $q(t_k) = q^k$. Greek indices run from 1 to 3 except in section IV.2 where they run from 0 to 4.

REFERENCES

1a. Robert Greenler:"Rainbows, halos, and glories" (Cambridge University Press, 1980).

1b. H.C. Bryant, and N. Jarmie,: "The Glory", Scientific American 231 (1974) 60.

1c. H.C. Bryant, and A.J. Cox: "Mie theory and the glory", J. Opt. Soc. Am. 56 (1966) 1529-1532.

1d. H.M. Nussenzveig: "Complex angular momentum theory of the rainbow and the glory", J. Opt. Soc. Am. 69 (1979) 1068-1079.

1e. W.Kenneth Ford, and John A. Wheeler: "Semiclassical description of scattering", Ann.Phys. 7 and references therein (1979) 259-286.

1f. M.V. Berry: "Uniform approximations for glory scattering and diffraction peaks" J.Phys.B. ser.2 vol.2 (1969) 381-392.

1g. F.A. Handler, and R.A. Matzner: "Gravitational wave scattering", Phys.Rev. D22 (1980) 2331.

2a. Cécile DeWitt-Morette: "Feynman's path integral definition without limiting procedure", Comm.Math.Phys. 28 (1972) 47-67.

2b. C. DeWitt-Morette: "Feynman path integrals: I.Linear and affine techniques; II. The Feynman-Green function", Comm.Math.Phys. 37 (1974) 63-81.

2c. C. DeWitt-Morette: "The semiclassical expansion", Ann. Phys. 97 (1976) 367-399.+)

2d. C. DeWitt-Morette, A. Maheshwari, and B. Nelson:

+) Note a misprint p.385, 1.5: the reference is not[4] but[5].

"Path integration in phase space", Gen. Rel. Grav. 8 (1977) 581-593.

2e. C. DeWitt-Morette, A. Maheshwari, and B. Nelson: "Path integration in non-relativistic quantum mechanics", Physics Rep.50 (1979) 255-372,and references therein.

3. C.R. Doering, and C. DeWitt-Morette: "The positivity of the Jacobi operator on configuration space and phase space", in The quantum theory of gravity, ed. S. Christensen (Adams Hilger, Ltd., 1984).

4a. C. DeWitt-Morette, Tian-Rong Zhang: "A Feynman-Kac formula in phase space with application to coherent state transitions", Phys.Rev. D28 (1983) 2517-2525.

4b. C. DeWitt-Morette, and Tian-Rong Zhang: "Path integrals and conservation laws", Phys.Rev. D28 (1983) 2503-2516.

4c. C. DeWitt-Morette, Bruce Nelson, and Tian-Rong Zhang, "The caustic problem in quantum mechanics with application to scattering theory", Phys.Rev. D28 (1983) 2526-2546.

4d. C. DeWitt-Morette, and Bruce Nelson: "Glories and other degenerate critical points of the action", Phys.Rev. D (1984).

4e. R. Matzner, C. DeWitt-Morette, B. Nelson and T.-R. Zhang: "Glory Scattering from Black Holes", Preprint,University of Texas, Center for Relativity (1984).

4f. T.-R. Zhang, and C. DeWitt-Morette: "WKB cross section for polarized glories", Preprint, University of Texas, Center for Relativity (1984).

5. S. Albeverio, and R. Høegh-Krohn: "Mathematical theory of Feynman path integrals", Lecture Notes in Mathematics, N. 523 (Springer Verlag, 1976).

6. N. Sanchez:"Elastic scattering of waves by a black hole", Phys.Rev. D18 (1978) 1798-1804, and references therein.

7. N. Bourbaki: "Eléments de mathématique", Chapter IX, Vol. VI, also referred to as Fascicule 35 or No.1343 of the Actualités Scientifiques et Industrielle (Hermann, Paris, 1969).

8a. L.S. Schulman:"Caustics and multivaluedness: two results of adding path amplitudes", in Functional integration and its application, edited by A.M. Arthurs (Oxford University Press, 1975).

8b. L.S. Schulman: Techniques and applications of path integration (J. Wiley and Sons, New York, 1981).

9. P. Pechukas: "Time-dependent semiclassical scattering theory: I. Potential scattering", Phys.Rev. 181 (1969) 166-185.

10. Sir Charles Darwin: "The gravity field of a particle", Proc. Roy. Soc. London A249,180-194.

11a. B. Nelson and B. Sheeks: "Path Integration for Time Dependent Metrics", J.Math.Phys. 22 (1982) 1944-1947.

11b. B. Nelson and B. Sheeks: "Path Integration for Velocity Dependent Potentials", Commun.Math.Phys. 84 (1982) 515-530.

12a. R.A. Isaacson: "Gravitational radiation in the limit of high frequency", Phys.Rev. 166 (1968) 1263-1280.

12b. Y. Choquet-Bruhat: "Construction de solutions radiatives approchées des équations d'Einstein", Commun.Math.Phys. 12 (1969) 16-35.

13. R.A. Matzner: "Scattering of Massless Scalar Waves by a Schwarzschild Singularity", J.Math.Phys. 9 (1969) 163-170.

14. C.W. Misner, K.S. Thorne, J.A. Wheeler:"Gravitation" (W.H. Freeman, San Francisco 1973).

15. B.S. DeWitt:"Dynamical Theory of Groups and Fields"(Gordon and Breach, New York 1965); also in"Relativity, Groups and Topology," eds. C.M. DeWitt and B.S. DeWitt (Gordon and Breach, New York 1964).

16a. R.P. Feynman: "Mathematical Formulation of the Quantum Theory of Electromagnetic Interaction", Phys.Rev. 80 (1950) 440-457, Appendix A.

16b. J. Schwinger: "On gauge invariance and vacuum polarization", Phys.Rev. 82 (1951) 664-679.

16c. The s-parametrization has been introduced and exploited in the following papers:
V. Fock, Phys. Zeit. Sow. Un. 12 (1937) 404;

E.C.C. Stueckelberg: "La Mécanique du point matériel en théorie de la relativité et en théorie des quanta", Helv.Phys. Acta 15 (1942) 23-37;
Y. Nambu: "The Use of the Proper Time in Quantum Electrodynamics", Prog. Theor. Phys. (Kyoto) 5 (1950) 82-94;
C. Garrod: "Hamiltonian path integral methods", Rev.Mod. Phys. 38 (1966) 483-494.

17a. C. DeWitt-Morette, K.D. Elworthy, B.L. Nelson, and G.S. Sammelman: "A stochastic scheme for constructing solutions of the Schrödinger equations", Ann. Inst. H. Poincaré XXXII, 4, (1980) 327-341.

17b. K.D. Elworthy and A. Truman: "The diffusion equation and classical mechanics: an elementary formula", pp. 136-146, in Stochastic processes in quantum theory and statistical physics eds. S. Albeverio, Ph. Combe, M. Sirugue-Collin, Springer-Verlag, Lecture Notes in Physics 172 (Springer-Verlag Berlin 1982). The elementary formula has recently been generalized by K. Watling.

18a. C. DeWitt-Morette: "Path Integration at the Crossroad of Stochastic and Differential Calculus", pp. 166-170, in Gauge Theory and Gravitation, eds. K. Kikkawa, N. Nakaniski, and H. Nariai, Lecture Notes in Physics 176 (Springer-Verlag, Berlin 1983).

18b. C. DeWitt-Morette: "Path integration quantization", to appear in the Proceedings of the III. Marcel Grossmann Meeting (Shanghai 1982).

19. See for instance Y. Choquet-Brahat, C. DeWitt-Morette: "Analysis, Manifolds and Physics", revised edition (North Holland, Amsterdam 1982).

20. See for instance L.D. Landau and E.M. Lifshitz: "The Classical Theory of Fields" (Pergamon Press, Oxford 1960).

21. See for instance L.D. Landau and E.M. Lifshitz: "Mechanics" (Pergamon Press, Oxford 1960).

22. T.-R. Zhang: Ph.D. dissertation.

23. G.N. Watson: "Theory of Bessel Functions" (Cambridge University Press 1844) pp. 178-179.

24. S. Weinberg, "Gravitation and Cosmology" (Wiley, New York 1972).

Acta Physica Austriaca, Suppl. XXVI, 171–184 (1984)

THE KNIFE EDGE PROBLEM[+]

by

Stephen G. LOW[++]
Dept. of Physics
The University of Texas at Austin
Austin, Texas 78712 USA

and

Zentrum für Interdisziplinäre Forschung
Universität Bielefeld
D-4800 Bielefeld 1, FR Germany

ABSTRACT

This paper is a discussion of the semi-classical limit of the propagator in 1 space, 1 time dimension in the case that there is no classical path to expand about.

1. INTRODUCTION

The strict WKB approximation of a non-relativistic propagator assumes that there is a classsical path, satisfying the appropriate boundary conditions, to expand about. There are many situations of physical interest where such an

---

[+]Seminar given at the XXIII. Internationale Universitätswochen für Kernphysik, Schladming, Austria, February 20- March 1, 1984

[++]Supported in part by NSF grant No. PHY81-07381 and a scholarship from the NSERC Canada.

extremal of the action functional does not exist and as a consequence, the strict WKB approximation breaks down. Perhaps the most familiar example of this is the "knife edge" problem which has been studied by Schulman [1]. In the shadow region of the knife edge where there is no classical path joining the points in question, the usual WKB approximation will not work.

A generalized semiclassical approximation to handle these cases would be very useful. As a first step in this direction, one can study the 1 space, 1 time dimensional problem and show that a simple generalization exists for the case where there is a path which minimizes the action functional which is a broken extremal with finitely many corners.

To make the above statements more precise, recall that the strict WKB approximation for a position-position propagator is given by:

$$K^{WKB}(b,a) = [\mathrm{Det}\,(\frac{1}{2\pi i\hbar}\ \frac{\partial^2}{\partial b\partial a}\,S(b,a))]^{1/2}\exp(\frac{i}{\hbar}S(b,a)). \qquad (1.1)$$

The classical action function is, as usual, defined by:

$$S(b,a) = S[\bar{q}] \qquad (1.2)$$

where $\bar{q}$ is the extremal which minimizes the action functional,

$$S[q] = \int_{t_a}^{t_b} L(q(t),\dot{q}(t))dt \quad , \qquad (1.3)$$

subject to the boundary conditions

$$q(t_a) = a\ ;\ q\,(t_b) = b\ . \qquad (1.4)$$

If constraints are present in the system, the path which

minimizes the action functional may not be an extremal. The simplest example of this is a free particle subject to the constraint that at the time t_c , $q(t_c \geq c$ $(t_b > t_c > t_a)$. This may be called the 1 space, 1 time dimensional knife edge in a manner completely analogous to the 1 space, 1 time dimensional single and double slits given by Feynman and Hibbs [2].

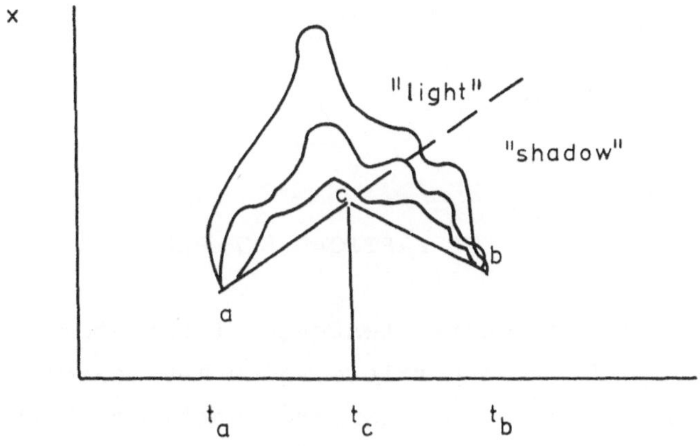

Fig.1.

As illustrated in figure 1, the path integral must be done over the subspace of the space of continuous paths from a to b which respect the constraint. If the point b lies in the shadow region, the path which minimizes the action functional is a broken extremal with a corner at c. In the semi-classical limit, the propagator for the shadow region will be shown to be:

$$\tilde{K}(b,a) = \frac{-i\ K^{WKB}(b,c)\ K^{WKB}(c,a)}{\left|\frac{\partial}{\partial c}(S(b,c)+S(c,a))\right|} \quad . \tag{1.5}$$

This result is interesting because it depends only on the fact that the path which minimizes the action functional is a broken extremal with a corner and not on the particular properties of the constraint. In fact, for an arbitrary action functional which has a minimizing path from c_o to c_{N+1} which is a broken extremal with N corners at $c_1 \ldots c_N$, it will be shown that in the semi-classical limit, the propagator is given by:

$$\tilde{K}(b,a) = \frac{(-i\hbar)^N K^{WKB}(c_{n+1},c_N) \; \ldots\ldots\ldots \; K^{WBK}(c_1,c_o)}{\prod_{i=0}^{N} \left|\frac{\partial}{\partial c_i}(S(c_{i+1},c_i)+S(c_i,c_{i-1}))\right|} \, . \tag{1.6}$$

2. FREE PARTICLE EXAMPLE

Many of the essential features of the problem are illustrated by the free particle which may be solved exactly in closed form in terms of standard functions. With no loss in generality, the knife edge constraint may be taken to be $1(0) \geq 0$. Then, using the semi-group property of the propagators, one has:

$$K(b,a) = \int_0^\infty K_F(b,u)K_F(u,a)\, du \quad , \tag{2.1}$$

where K_F is the free particle propagator:

$$K_F(b,a) := \frac{1}{\mu} \frac{1}{\sqrt{2\pi i(t_b-t_a)}} \exp\left(\frac{i}{\mu^2} \frac{(b-a)^2}{t_b-t_a}\right) \tag{2.2}$$

where $\mu^2 = \hbar/m$. This expression for K_F may be sustituted into equation (2.1) and simplified to yield:

$$K(b,a) = K_F(b,0)K_F(0,a)I \tag{2.3}$$

where

$$I = \int_0^\infty du \exp\left(\frac{i}{\mu^2}\left(u^2(1/t_b - 1/t_a) - 2u(b/t_b - a/t_a)\right)\right). \tag{2.4}$$

This integral is tabulated in Gradshteyn and Ryzhik [3]:

$$I = \mu\sqrt{\pi i\alpha}\left(\exp\frac{-i\alpha\Delta^2}{\mu^2}\right)\left(1 - \Phi\left(\frac{\Delta\sqrt{-i\alpha}}{\mu}\right)\right) \tag{2.5}$$

where

$$\Delta = b/t_b - a/t_a\;, \qquad \alpha = 1/t_b - 1/t_a\;; \tag{2.6}$$

Φ is the special function called the probability integral defined by:

$$\Phi(x) = \frac{2}{\sqrt{\pi}}\int_0^x e^{-t^2}\,dt\;.$$

The sign of Δ determines whether or not there is a classical path connecting the point a to the point b. If $\Delta > 0$, there is a classical path, whereas if $\Delta < 0$, the point b lies in the shadow region for which there is no classical path.

The semi-classical approximation is the limit $\mu \to 0$ with α and Δ fixed. As μ appears in the denominator of the argument of Φ, asymptotic expansions of the probability integral for large values of its argument are required. These expansions are well known [3]:

$$\Phi(x) \simeq 1 - \frac{1}{\sqrt{\pi}x}\exp(-x^2) + O(x^{-3})\;,\; x = |x|\,e^{i\phi}, 0<\phi<\pi \tag{2.7a}$$

$$\Phi(x) \simeq -1 + \frac{1}{\sqrt{\pi}x}\exp(-x^2) + O(x^{-3}),\; x = |x|\,e^{i\phi}, 0>\phi>-\pi\;. \tag{2.7b}$$

For the case in question, the argument x is defined to be:

$$x = \frac{\Delta\sqrt{-1}\alpha}{\mu} = \begin{cases} \left|\frac{\Delta\sqrt{\alpha}}{\mu}\right| e^{-i\pi/4} & , \Delta > 0 \\ \left|\frac{\Delta\sqrt{\alpha}}{\mu}\right| e^{i3\pi/4} & , \Delta < 0 \ . \end{cases} \tag{2.8}$$

It can be seen that which of the expansion in equations (2.7) is to be used depends on whether or not the point b lies in the shadow region. Note also that if $\Delta = 0$, corresponding to the point lying exactly on the boundary, the expansion is not defined. Considering first the case $\Delta < 0$ for which there is no classical path, one obtains:

$$\begin{aligned} I_{shadow} &= \mu\sqrt{\pi i\alpha}\, e^{x^2} \left(1- \left(1 - \frac{e^{-x^2}}{\sqrt{\pi}x} + O(x^{-3})\right)\right) \\ &\simeq \frac{\mu\sqrt{i\alpha}}{x} \simeq \frac{-i\mu}{|\Delta|} \ . \end{aligned} \tag{2.9}$$

Therefore:

$$\tilde{K}_{shadow}(b,a) = \frac{-i\mu}{|\Delta|} K_F(b,0)\, K_F(0,a) \ ,$$

where the $\sim$ indicates that this is valid only in the semi-classical limit $\hbar \to 0$. Noting that Δ/m is the discontinuity in the momentum at the time t_c, and that the WKB approximation for the free particle is exact, this may be written as:

$$\tilde{K}_{shadow} = \frac{-i\hbar\, K_F^{WKB}(b,0)\, K_F^{WKB}(0,a)}{\left|\frac{\partial}{\partial c}\,(S(b,c)+S(c,a))\right|_{c=0}} \ . \tag{2.10}$$

An exactly analogous calculation using the asymptotic expansion valid for $\Delta > 0$ shows that in the case where there exists a classical path, the semi-classical approximation is given by the strict WKB approximation as expected:

$$\tilde{K}_{light} = K^{WKB}(b,a) \quad . \tag{2.11}$$

The case of a particle in a simple harmonic oscillator potential with the knife edge constraint is soluble in closed form in a completely analogous manner and it can be verified that these formulae remain valid in that case.

3. THE GENERAL ACTION FUNCTION CASE

A derivation of the general case follows from a generalization of the proposition B3 in a paper by DeWitt-Morette [4]. This proposition shows how the definition of the path integral in terms of a gaussian pro-distribution with a covariance given by the Green's function of the Jacobi operator with vanishing boundary conditions is related to a path integral given in terms of a broken action function. This broken action function is defined to be the sum of the action functions for continuous piecewise classical paths from a to b. That is:

$$S_p(b....k....a) = \sum_{k=0}^{p} S_f(k + 1,k) \tag{3.1}$$

where

$$S_f(k + 1,k) = S(f(t_{k+1}),\ f(t_k)),\quad f(t_0) = a, f(t_{k+1}) = b \ , \tag{3.2}$$

and f is a continuous path which is a broken extremal of the action functional with corners at $t_1....t_p$. In this proposition, a classical path $\bar{q}$ connecting the point a to b is assumed to exist. Then each $S_f(k+1,k)$ can be regarded as a function of two variables which may be Taylor expanded about the corresponding action function for the classical path. That is:

$$S_f(k+1,k) = S_{\bar{q}}(k=1,k) + \left(\frac{\partial}{\partial u^k} S_f(k+1,k)u^k + \right.$$

$$\left. + \frac{\partial}{\partial u^{k+1}} S_f(k+1,k)u^{k+1}\right)\Big|_{u^k=u^{k+1}=0} + \ldots , \qquad (3.3)$$

where the u^k are given by:

$$u^k = f(t_k) - \bar{q}(t_k) . \qquad (3.4)$$

This expansion may be subsituted into equation (3.1) and the sums computed term by term. The linear terms sum to give the action function from a to b, and the linear terms vanish as the classical path is assumed to exist:

$$\sum_{k=0}^{p} S_{\bar{q}}(k+1,k) = S(b,a) \qquad (3.5)$$

and

$$\sum_{k=0}^{p} \left(\frac{\partial}{\partial u^k} S_f(k+1,k)u^k + \frac{\partial}{\partial u^{k+1}} S(k+1,k)u^{k+1}\right)\Big|_{u^k=u^{k+1}=0} = 0 \quad (3.6)$$

as

$$\left(\frac{\partial}{\partial u^k} S_f(k+1,k) + \frac{\partial}{\partial u^k} S_f(k,k-1)\right)\Big|_{u^k=0} = 0, \quad k = 1, \ldots, p , \qquad (3.7)$$

and

$$u^0 = u^{k+1} = 0 . \qquad (3.8)$$

Thus, using matrix notation, the broken action function may be written as:

$$S_p(b\ldots k\ldots a) = S_q(b,a) + {}^t uMu \tag{3.9}$$

where $u = (u^1\ldots u^p)$ and M is the $p\times p$ matrix with components M_{ij} defined by the second derivatives of the $S_f(k+1,k)$ appearing in the quadratic term of the Taylor expansion given in equation (3.4):

$$M_{kk} = \frac{\partial^2}{\partial u^k \partial u^{k+1}} \left(S(k+1,k) + S(k,k-1)\right)\Big|_{u^{k+1}=u^k}$$

$$M_{kk-1} = \frac{\partial^2}{\partial u^k \partial u^{k-1}} S(k,k-1) \Big|_{u^k=u^{k-1}=0} := M(k,k-1) \; . \tag{3.10}$$

The indices k run from 1 to p.

As we are interested here only in the semi-classical limit, the terms in the Taylor expansion higher than the quadratic power my be neglected (provided that the Jacobi fields are linearly independent) as these correspond to higher order corrections in $\hbar$.

In this approximation, as described in section IV of reference [4], the propagator is given by:

$$K(b,a) = \int_{R^p} \exp \frac{i}{\hbar} S_p(b\ldots k\ldots a)\mathcal{D}^p u \tag{3.11}$$

where

$$\mathcal{D}^p u = \prod_{k=1}^{p} du^k \left(\det \frac{M(k+1,k)}{2\pi i\hbar}\right)^{1/2} \; . \tag{3.12}$$

Substituting equation (3.7) into (3.11), the propagator

becomes:

$$K(b,a) = e^{i/\hbar\, S(b,a)} \int_{R^p} \exp\left(\frac{i}{2\hbar}\, {}^t uMu\right) \mathcal{D}^p u \ . \tag{3.13}$$

This integral may be evaluated using properties of the Jacobi fields detailed in the reference to give the Van Vleck-Morette determinant in the strict WKB approximation cited in equation (1.1).

This result may be generalized in the 1 space, 1 time dimensional case by Taylor expanding about a broken extremal q with a corner at t_c. Then, on inserting the expansion into the sum in equation (3.1), the linear term at the corner does not vanish and the zero order term may only be written as the sum of two action functions. That is:

$$\sum_{k=0}^{p} S_f(k+1,k) = S_q(b,c) + S_q(c,a) \tag{3.14}$$

and

$$\Delta = \frac{\partial}{\partial u^c} \left(S_f(c+1,c) + S_f(c,c-1)\right)\Big|_{u^c=0} \neq 0 \ . \tag{3.15}$$

If this broken extremal is a minimum, then the point c must lie on the boundary of the subspace of continuous paths from a to b which respect the constraint that produced the corner. Hence, the co-ordinate u^c in the path integral in equation (3.11) must be integrated only over the interval (c,∞) or $(c,-\infty)$. As we are free to choose the co-ordinate system, this interval may be taken to be $(0,\infty)$ with no loss of generality. In this case Δ is less than 0 if the path is a minimum.

Equation (3.9) must then be modified to read:

$$S_p(b\ldots k\ldots a) = S_q(b,c) + S_q(c,a) + \Delta u^c + \frac{1}{2}\, {}^t uMu \ . \tag{3.16}$$

This equation may be inserted into the expression for the propagator given in equation (3.11) to obtain:

$$K(b,a) = e^{i/\hbar(S(b,c)+S(c,a))}\int_{\tilde{R}^p} \exp\left(\frac{i}{\hbar}\left(\frac{1}{2}\, {}^t u\, M u + \Delta u^c\right)\right) \mathcal{D}^p u \tag{3.17}$$

where

$$\tilde{R}^p = R_1 \times R_2 \times \dots R_c^+ \times \dots R_p \quad . \tag{3.18}$$

It is necessary to decouple the integration over the u^c-co-ordinate from the rest in order to be able to compute this integral. Setting

$$u = (u^1, \dots u^c \dots u^p) = (v^1 \dots v^{c-1}, u^c, w^1 \dots w^{p-c}) \ ,$$

the quadratic form may be written as:

$${}^t u\, M u = {}^t v\, M_v\, v + {}^t w\, M_w w + u^c((B_w - {}^t B_w)w + (B_v - {}^t B_v)v) + M_u (u^c)^2 \tag{3.19}$$

where the p×p matrix M has been decomposed into sub-matrices:

$$M = \begin{array}{c} \\ c-1 \\ 1 \\ p-c \end{array} \left[\begin{array}{c|c|c} c-1 & 1 & p-c \\ \hline M_v & B_v & O \\ \hline {}^t B_v & M_u & -{}^t B_w \\ \hline O & B_w & M_w \end{array}\right] \quad . \tag{3.20}$$

Then, as iM_v and iM_w are Hermitian, they may be diagonalized with the unitary matrixes U_v and U_w. That is:

$$U_v i M_v U_v^{-1} = iD_v$$

$$U_w i M_w U_w^{-1} = iD_w \quad . \tag{3.21}$$

The corresponding eigenbases are:

$$\xi_v = \mathcal{U}_v\, v \qquad \xi_w = \mathcal{U}_w\, w$$

and (3.22)

$$\tilde{B}_v = B_v \mathcal{U}_v^{-1} \qquad \tilde{B}_w = B_w \mathcal{U}_w^{-1} \quad .$$

Thus

$$ {}^t u i \mathcal{M} u = {}^t\xi_v\, iD_v \xi_v + {}^t\xi_w\, iD_w \xi_w + iu^c((\tilde{B}_v - {}^t\tilde{B}_v)\xi_v + $$

$$ + (\tilde{B}_w - {}^t\tilde{B}_w)\xi_w) + i\mathcal{M}_u (u^c)^2 \quad . \tag{3.23}$$

Inserting this into equation (3.17), the expression for the propagator becomes:

$$K(b,a) = e^{\frac{i}{\hbar}(S(b,c)+S(c,a))} \int_{\tilde{R}^p} \exp(\tfrac{i}{\hbar}\,(\tfrac{1}{2}({}^t\xi_v iD_v \xi_v + {}^t\xi_w iD_w \xi_w +$$

$$ + iu^c((\tilde{B}_v - {}^t\tilde{B}_v)\xi_v + (\tilde{B}_w - {}^t\tilde{B}_w)\xi_w + \Delta u^c))\mathcal{D}^{c-1}\xi_v \mathcal{D}^{p-c}\xi_w du^c \; . \tag{3.24}$$

Then, completing the square and translating the co-ordinates ξ_v and ξ_w, this integral may be written as:

$$K(b,a) = \int_0^\infty du \exp(\tfrac{i}{\hbar}\,(\Delta u + \tfrac{1}{2}\,\Gamma u^2)) e^{\frac{i}{\hbar}S(c,a)} \int_{R^{c-1}} \exp(\tfrac{i}{\hbar}{}^t\eta_v D_v \eta_v \mathcal{D}^{c-1}\eta_v) \;\cdot$$

$$\cdot\; e^{\frac{i}{\hbar}S(b,c)} \int_{R^{p-c}} \exp(\tfrac{i}{\hbar}\,{}^t\eta_w\, D_w \eta_w\, \mathcal{D}^{p-c}\eta_w) \quad , \tag{3.25}$$

where

$$\eta_v^i = \xi_v^i + \tfrac{u}{2}(\tilde{B}_v - {}^t\tilde{B}_v)^i / D_v^{ii} \quad , \; i = 1,\dots,c-1 \; ,$$

$$\eta_w^j = \xi_w^j + \frac{u}{2}(\tilde{B}_w - {}^t\tilde{B}_w)^j / D_w^{jj} \quad , \quad j = 1,\dots,p-c \ ,$$

$$\Gamma = \frac{1}{4}\left\{\sum_{i=1}^{c-1}\left[\frac{(B_v - {}^tB_v)^i}{D_v^{ii}}\right]^2 + \sum_{i=1}^{p-c}\left[\frac{(B_w - {}^tB_w)^i}{D_w^{ii}}\right]^2 + M_u\right\} . \qquad (3.26)$$

The second and third integrals are of the form given in equation (3.13) which , upon evaluation,give the strict WKB expressions for the propagator from a to c and c to b , respectively. The first integral has been encountered previously in the discussion of the free particle case. If the broken extremal is a minimum, $\Delta < 0$ and this integral can be evaluated as in equations (2.7) to (2.9) to give:

$$\int_0^\infty du \exp\left(\frac{i}{\hbar}\left(\Delta u + \frac{1}{2}\Gamma u^2\right)\right) \simeq \frac{-i\hbar}{|\Delta|} \ . \qquad (3.27)$$

Putting this together, the expression for the propagator becomes:

$$K(b,a) = \frac{-i\hbar}{|\Delta|} K^{WKB}(b,c)\, K^{WKB}(c,a) \ . \qquad (3.28)$$

The remarkable fact is that in the WKB approximation, the term Γ which depends on the transformation U does not appear and so the semi-classical approximation to the propagator is given in explicit form. Using the semi-group property of the propagators in the semi-classical limit, this result extends immediately to the case where the minimizing path is a broken extremal with N corners cited in equation (1.6).

REFERENCES

1. L.S. Schulman, "Ray Optics for Diffraction: A Useful Paradox in a Path Integral Context" ,in: Le Dualisme Onde-Corpuscule, Eds. S. Diner, D. Fargue, G. Lochak, and F. Selleri (Reidel, Dordrecht , 1983).

2. L.S. Schulman, "Exact time-dependent Green 's function for the half-plane barrier", Phys. Rev. Letters 49 (1982) 599-601.
3. R.P. Feynman and A.R. Hibbs, "Quantum Mechanics and Path Integrals" (McGraw-Hill, New York, 1965), pp. 47-56.
4. I.S. Gradshteyn and I.M. Ryzhik, "Tables of Integrals, Series, and Products" (Academic Press, New York, 1980).
5. C. DeWitt-Morette, "The Semiclassical Expansion", Annals of Physics 97 (1975) 367-399.

Acta Physica Austriaca, Suppl. XXVI, 185–209 (1984)

TRAPPING FOR NEWTONIAN DIFFUSION PROCESSES+

by

Ph. BLANCHARD
Theoretische Physik
and BiBoS
Universität Bielefeld
D-4800 Bielefeld 1

I. INTRODUCTION

In the first lecture I will try to explain a general mechanism for the formation of impenetrable barriers for diffusion processes (see [1]). I will consider a special class of diffusion processes, which we call Newtonian diffusions. This name is justified by the fact that such a diffusion process satisfies a Newton law in the mean. We will see how it is possible to define a mean stochastic acceleration a for diffusion processes. The Newton law in the mean

$$\mu a = F = -\nabla V \quad ,$$

where μ is the mass of a test-particle, plays the role of a constraint and allows to construct under some assumptions

---

+Lecture given at the XXIII. Internationale Universitätswochen für Kernphysik, Schladming, Austria, February 20-March 1, 1984.

a family of possible probability distributions ρ of the stochastic diffusion process. We will also show that the nodal surfaces of $N_\rho = \{x \in R^d | \rho(t,x) = 0\}$ can also be impenetrable barriers, splitting the family of typical particles into several groups. Since no particle can pass from one group to another we have to do with a segregation (confinement, trapping) mechanism.
In the second lecture I will discuss possible applications of this confinement mechanism:

- Formation of jet-streams in the protosolar nebula
- Zonal wind structure of the planetary atmospheres
- Formation of the Van Allen's radiation belts around the Earth.

For further details and references we refer to [2,3,4,5,6,7, 18]. The material discussed here was developed in collaboration with S. Albeverio, Ph. Combe, R. Høegh-Krohn, R. Rodriguez, M. Sirugue and M. Sirugue-Collin.

II. NEWTONIAN DIFFUSION

II.1. Definition and properties of Newtonian diffusions

In classical mechanics the Newton equation $\mu \ddot{X}_t = F = -\nabla V$, given the initial conditions, allows to construct the path of the particle on which acts a force $F = -\nabla V$. In our framework the main assumption is that the motion of a test particle is modelled by a stochastic diffusion process taking its values in R^d. In such a situation the position $X_t(\omega) \in R^d$ at time t of the particle depends on a random element $\omega \in \Omega$ where $(\Omega, \mathcal{B}, P)$ is a probability space. Accordingly X_t does not satisfy a deterministic equation but a so-called stochastic differential equation

$$dX_t = \beta_t(X_t,t)dt + \sigma(X_t)dW_t \tag{1}$$

where β_t is the drift vector, $\sigma(\cdot)$ are diffusion coefficients and W_t the standard Wiener process in R^d. For $dt > 0$, dW_t is defined by $dW_t \equiv W(t+dt)-W(t)$. We recall that $dW_t \sim (dt)^{1/2}$, which implies among other things that the trajectories are continuous but nowhere differentiable. We will consider the case where the diffusion matrix is constant and of the form $\sigma \equiv \sigma.1_d$. Physically this assumption expresses the isotropy and the homogeneity of the medium. If σ is not constant we can introduce in R^d the Riemannian metric $g_{ij} \equiv (\sigma^t\sigma)^{-1}_{ij}$ and rewrite the stochastic differential (1) as a stochastic differential equation for a process $\hat{X}_t$ with values in the Riemannian manifold $M = (R^d,g)$. For more details about this general case see e.g. [3,4,8]. The first step is to define an "Ersatz" for the time derivative. Following E. Nelson [9], this can be done only in the mean. We introduce the mean forward derivative D_+ by

$$D_+f(x,t) = \lim_{\Delta t \to 0} E\left[\frac{f(X_{t+\Delta t}, t+\Delta t)-f(X_t,t)}{\Delta t} \,\middle|\, X_t = x\right] \qquad (2)$$

where $E[.|X_t = x]$ is the conditional expectation given $X_t = x$. This is the relevant information available at time t to define a substitute for the time derivative. One can as well introduce a total derivative D_-, called the mean backward derivative, using the following definition:

$$D_-f(x,t) = \lim_{\Delta t \to 0} E\left[\frac{f(X_t,t)-f(X_{t-\Delta t}, t-\Delta t)}{\Delta t} \,\middle|\, X_t = x\right] . \qquad (3)$$

Considering now the time-reversed process $\hat{X}_t = X_{-t}$ ($\hat{X}_{-t}$ has the same law as X_t) it is well known that $\hat{X}_t$ is also solution of a stochastic differential equation, but with a drift β_-, which is in general different from $-\beta_+$):

$$d\hat{X}_t = \beta_- dt + \sigma dW_-(t) \tag{4}$$

with $dW_-(t) = W(t)-W(t-dt)$ $(dt > 0)$. It follows from the definitions of D_+ and D_- that

$$D_\pm X_t = \beta_\pm \quad . \tag{5}$$

The physical interpretations of β_+ and β_- are the following: β_+ is the mean velocity of the particles leaving x at time t and β_- is the mean velocity of particles entering x at time t.

If $f \in C^2$ $(R^d \times R, R)$ one can show that

$$(D_\pm f)(x,t) = \frac{\partial f}{\partial t} + \beta_\pm \cdot \nabla f \pm \frac{\sigma^2}{2} \Delta f \tag{6}$$

where Δ is the Laplace operator in R^d.

We define now the current velocity v and the osmotic velocity u by

$$v \equiv \frac{1}{2}(\beta_+ + \beta_-) \tag{7}$$

$$u \equiv \frac{1}{2}(\beta_+ - \beta_-) \quad . \tag{8}$$

It must be remarked that under time reversal $(t \to -t)$, v changes sign as velocity in a fluid and u does not. The role of u is to damp the fluctuations.

We assume now the existence of a smooth density $\rho(t,x)$ of the law of X_t with respect to the Lebesgue measure dx,

$$E[f(X_t)] = \int_\Omega f(X_t(\omega))P(d\omega) = \int_{R^d} \rho(t,x) f(x) dx \quad . \tag{9}$$

Due to the fact that the diffusion process X_t, as a function of time, has the same local behaviour as the Brownian motion, $f(X_t)$ is not differentiable but its expectation is differentiable:

$$\frac{d}{dt} E[f(X_t)] = \int_{R^d} \frac{\partial \rho}{\partial t} (t,x) f(x) dx = E[D_{\pm} f(X_t)] \quad . \tag{10}$$

Integrating (10) by parts we obtain the following equations

$$\frac{\partial \rho}{\partial t} = -\nabla(\beta_+ \cdot \rho) + \frac{\sigma^2}{2} \Delta\rho = -\nabla(\beta_- \rho) - \frac{\sigma^2}{2} \Delta\rho \tag{11}$$

which are called by the probabilists the Kolmogorov equations and by the physicists the Fokker-Planck equations. From (11) we deduce for u and v

$$\frac{\partial \rho}{\partial t} = -\nabla(\rho v) \quad \text{(continuity equation)} \tag{12}$$

$$\frac{\sigma^2}{2} \Delta\rho = \nabla(\rho u) \quad \text{(osmotic equation)} \quad . \tag{13}$$

In particular if $v = 0$ the probability distribution ρ is stationary. Using the following integration by part formula for functions f and g with compact support in time

$$\int E[D_+ f \cdot g] dt = -\int E[f \cdot D_- g] dt \quad ,$$

the definitions of D_+, D_- and ρ, it can be shown that the osmotic equation can always be integrated (see e.g.[9]),

$$u = \frac{\sigma^2}{2} \nabla \log\rho = \frac{\sigma^2}{2} \frac{\nabla\rho}{\rho} \quad . \tag{14}$$

Using the mean forward derivative D_+ and the mean backward derivative D_- it is now possible to define the mean second derivative or mean acceleration by

$$a = \frac{1}{2}(D_+ D_- + D_- D_+) X_t \quad . \tag{15}$$

We are now in position to define a Newtonian diffusion process. A diffusion process X_t satisfying the stochastic differential (1) is called a Newtonian diffusion if X_t satisfies "Newton's law in the mean", in the sense that there exists a positive constant μ and a real valued function V on R^d such that

$$\mu a = -\nabla V \quad . \tag{16}$$

We then say that a is the acceleration of the Newtonian diffusion. We shall see below that there exist Newtonian diffusions. The equation (16) plays the role of a constraint and allows to determine the drift β_+ of the diffusion process X_t. It is easy to rewrite the Newton law (16) in terms of the velocities u and v ,

$$\frac{\partial v}{\partial t} + v.\nabla v - u.\nabla u - \frac{\sigma^2}{2}\,\Delta u = -\frac{1}{\mu}\,\nabla V \quad ,$$
$$u = \frac{\sigma^2}{2}\,\nabla \log\rho \quad . \tag{17}$$

It must be remarked that in (16) and (17) - ∇V is an external (deterministic) force acting on the diffusing particles. To discuss (17) the method consists to transform the complicated nonlinear problem for ρ to a simpler linear elliptic problem. To do this we assume that the current velocity v is also a gradient

$$v = \nabla S \quad , \tag{18}$$

$$\frac{\partial S}{\partial t} - \frac{\sigma^4}{4}\,\Delta\,\log\rho + \frac{1}{2}(\nabla S)^2 - \frac{\partial^4}{8}\,(\nabla\log\rho)^2 + \frac{1}{\mu}\,v = 0 \quad , \tag{19}$$

and the continuity equation becomes

$$\frac{\partial\rho}{\partial t} = -\nabla(\rho\nabla S) \quad . \tag{20}$$

Using the "Ansatz" $\psi(x,t) \equiv \rho^{1/2} e^{\frac{i}{\sigma^2} S}$ it follows from (19) and (20) that Ψ satisfies the following "Schrödinger type" equation

$$i\mu\sigma^2 \frac{\partial\psi}{\partial t} = -\frac{\mu\sigma^4}{4}\Delta\psi + V\psi = H\psi \quad . \tag{21}$$

Vice et versa given $\psi \in L^2(R^d, dx)$ solution of (21), one can reconstruct ρ, u and v via

$$\rho = |\psi|^2 \tag{22}$$

$$u = \frac{\mu^2}{2} \nabla \log|\psi|^2 \tag{23}$$

$$v = \sigma^2 \nabla(\mathrm{Im} \log\psi) \tag{24}$$

and hence the Newtonian diffusion X_t with distribution given by ρ and drift $\beta_+ = v+u$.

Example:

We will consider the simplest case where $v = 0$, which implies that ρ is stationary and $\beta_+ = -\beta_- \equiv \beta$. The stochastic acceleration a is then given by $a = -\beta.\nabla\beta - \frac{1}{2}\sigma^2\Delta\beta$. Setting $\rho = \psi^2$ and using $\Delta\beta = \mathrm{grad\ div}\ \beta - \mathrm{rot\ rot}\ \beta = \mathrm{grad\ div}\ \beta$, since β is a gradient, we see that the Newton law

$$\mu(-\beta.\nabla\beta - \frac{\sigma^2}{2}\Delta\beta) = -\nabla V$$

can be integrated since $-\beta.\nabla\beta - \frac{\sigma^2}{2}\Delta\beta = \nabla(-\frac{1}{2}\beta^2 - \frac{\sigma^2}{2}\nabla\beta)$. We obtain then

$$\mu(\frac{1}{2}\beta^2 - \frac{\sigma^2}{2}\nabla\beta) - V = E \tag{25}$$

where E is a real constant. Using $\beta = \sigma^2 \frac{\nabla\phi}{\phi}$ and $\nabla\beta = \sigma^2\nabla\frac{\nabla\phi}{\phi} = \sigma^2 \frac{\phi\Delta\phi - (\nabla\phi)^2}{\phi^2}$, (25) is equivalent to

$$- \mu\frac{\sigma^4}{2} \Delta\phi + V\phi = E\phi \quad . \tag{26}$$

Let E be an eigenvalue and ψ_E an eigenfunction of the following eigenvalue problem:

$$- \mu\frac{\sigma^4}{2} \Delta\psi_E + V\psi_E = E\psi_E \ ; \tag{27}$$

then $\psi_E(x,t) = e^{-i\frac{E}{\mu\sigma^2}.t} \psi_E(x,0)$ is a solution of (21). With this choice for $\psi_E(x,t)$, v and u are time independent and the process is stationary with probability distribution ρ_E given by

$$\rho_E = |\psi_E(x,0)|^2 \quad . \tag{28}$$

For Newtonian diffusions ρ is given $\forall x \in R^d, \forall t \in [0,\infty]$, by $\rho = |\psi|^2$, ψ solution of the Schrödinger type equation (21) with initial condition $\psi(x,0)$ such that $|\psi(x,0)|^2$ gives the initial distribution of the process.

II.2. Impenetrable barriers for Newtonian diffusions

Let ψ_E be an eigenfunction of (21) to eigenvalue E,

$$H\psi_E = E\psi_E \ .$$

The nodal surface N_ρ of the associated probability distribution $\rho = |\psi_E|^2$ is defined by

$$N_\rho = \{x \in R^d | \psi_E(x,0) = 0\} \quad . \tag{29}$$

N_ρ is a union of (d-1) dimensional surfaces and split R^d into disjoint domains Γ_ρ^i, i = 1...n,

$$R^d = N_\rho \bigcup_{i=1}^{n} \Gamma_\rho^i \quad . \tag{30}$$

On N_ρ is the probability distribution $\rho = 0$ for all time, and this can be interpreted as an indication for the existence of an impenetrable barrier. On the other hand the osmotic velocity $u = \frac{\sigma^2}{2} \nabla \log\rho = \frac{\sigma^2}{2} \frac{\nabla\rho}{\rho}$ becomes singular on N_ρ, since on N_ρ $\nabla\rho \neq 0$. The problem of the singularity of u on N_ρ is its own solution: the singularity of the drift β_+ on N_ρ produces a repulsion strong enough to keep the configuration from ever reaching the nodal surface N_ρ. If we remember that the heuristic interpretation of $\beta_+(x,t)$ is the mean velocity of particles which leave the point x at time t, then the typcial trajectories of X_t in the neighbourhood of the nodal surface N_ρ are as depicted in Fig.1.

If the process X_t was started in $x \in \Gamma_\rho^i$ for some i, it will never reach the nodal surface N_ρ and will stay in Γ_ρ^i for all time. The nodal surface N_ρ acts also as impenetrable barrier for the process and X_t is confined in Γ_ρ^i. In conclusion we can say that the family of typical particles is split into several groups by the nodal surface N_ρ and no particle from one group can pass to another.

Remark

If σ is not constant we must consider the Riemannian manifold $M = (R^d, g)$, where the metric g is given in terms of the diffusion coefficients σ by

$$g_{ij} \equiv (\sigma^t\sigma)^{-1}_{ij}$$

and a Newtonian diffusion process with values in M. All the results discussed in the case where $\sigma = \sigma 1_d$ are again valid. For example in the Schrödinger type equation (21) the Laplace operator Δ must be replaced by the Laplace-

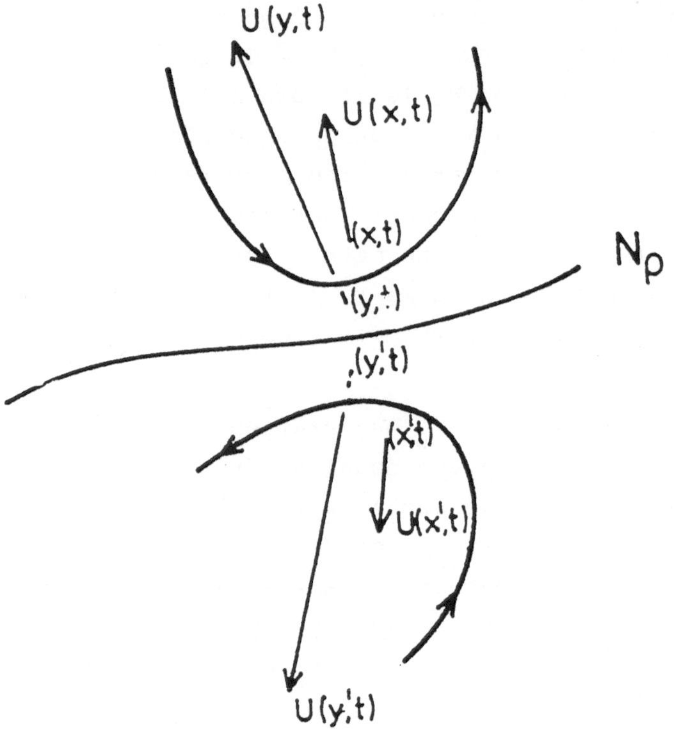

Fig. 1. Trajectories in the neighbourhood of the nodal surface N_ρ.

Beltrami operator Δ_σ .

The mathematical theory for the formation of impenetrable barriers for diffusion processes is based on the close connection between the theory of diffusion processes, the potential theory and the theory of Dirichlet forms. For a discussion of such connections see e.g. [1,10,11]. Under Dirichlet form we understand a form given for f in the domain D(E) of the form by

$$E(f,f) = \int_{R^d} (\nabla f)^2 \rho dx \quad . \tag{31}$$

Associated to the Dirichlet form we can introduce a notion of capacity (see e.g.[10]). Let $U \in R^d$ open, then the capacity of U is defined in the following way:

$$\mathrm{cap}(U) = \begin{cases} 0 \text{ if } \{u \in D(E), u \geq 1 \text{ a.s. } U\} = \phi \\ \inf\limits_{u \restriction U \geq 1} E_1(u,u) \end{cases} \tag{32}$$

with

$$E_1(u,u) = E(u,u) + ||u||^2_{L^2(R^d,\rho dx)}$$

The equilibrium potential e_U of U is exactly the function for which the minimum is reached ,

$$\mathrm{cap}(U) = E_1(e_U,e_U) \quad . \tag{33}$$

To illustrate this abstract notion we consider the example of the Newtonian capacity of the sphere of radius

$S_\varepsilon = \{X \in R^3 \mid ||X|| < \varepsilon\}$ in R^3. The equilibrium potential e_{S_ε} is in this case given by

$$e_{S_\varepsilon}(x) = \begin{cases} \frac{\varepsilon}{r}, & r \geq \varepsilon \\ 1, & x \in S_\varepsilon \end{cases} . \tag{34}$$

This is a function equal to 1 in S_ε and harmonic in $R^3 \setminus S_\varepsilon$. The simplest way to understand the connection between the theory of diffusion processes and the classical potential theory is to consider the case of the Wiener process (the Brownian motion of the mathematicians). The probability distribution ρ_W of the Brownian motion is given by (in three dimensions)

$$\rho_W = \frac{1}{(2\pi t)^{3/2}} e^{-\frac{|x|^2}{2t}} . \tag{35}$$

The number

$$P_W[W_t \in A] = \int_A \rho_W(x,t)\,dx \tag{36}$$

is the probability that our Brownian particle shall be in the region A at time t. It is not difficult to verify that

$$\int_0^{+\infty} dt\, P[W_t \in A] = \int_0^t dt \int_A \rho_W(x,t)\,dx = \frac{1}{2\pi} \int_A \frac{dx}{|x|} . \tag{37}$$

Here the left side is the mean time which the Brownian particle spends in A and the right side is the Newtonian potential at the origin of a uniform mass distribution in A. This kind of a connection between the theory of diffusion processes and potential theory has been used to guess and prove results. In particular the equilibrium potential

e_V has a probabilistic interpretation

$$e_U(x) = E_x[e^{-\sigma_U}] \quad (38)$$

where σ_U is the first hitting time of U and E_x denotes the expectation conditioned by the fact that the process was started in x. In particular if $e_U(x) = 0, \sigma_U = +\infty$. For more details about mathematical conditons for an barrier to be impenetrable see [1,3,5,19].

III. APPLICATIONS

III.1. Newtonian diffussions as a model of the formation of jet-streams in the protosolar nebula

It is an old hypothesis that the solar system was formed from a protosolar nebular consisting essentially of a gas of small particles (dust) of stellar or interstellar origin. In one form or another this hypothesis was discussed originally by Descartes (1644), Kant (1755) and Laplace (1796) and has been steadily accompanying all later development in the discussion of the origin of the solar system. There have been many earlier attempts to explain the origin of the regularity in the distances R_n of the planets from the sun. This regularity was described classically by the Titius-Bode law (1766), given R_n in the form

$$R_n = a + bc^n \quad (39)$$

for suitable constants a,b,c. We refer to [12] for the history of attempts to explain Titius-Bode's law and to [13] for new interpretations. One idea which has been intensively discussed recently is a sort of modern version of the Kant-Laplace ring formation: namely that, before the aggregation into planets, concentric roughly planar rings of dust were formed. The rings consist of gas, ice particles and dust, circulating inside the rings but with no communication with neighbouring rings. In Alfven's theory these rings, the so-called "jet-streams", are essentially of electromagnetic origin. The formation of the planets should then have happened in a later state by aggregation from the jet-streams. We shall now explain the formation of these jet-streams from Newtonian diffusions, resulting in distances in accordance with the actual distances of the planets (and hence, to a good extent, with Titius-Bode law). The same kinds of ideas can be applied to the formation of jet-streams around the Jovian planets. Our stochastic model provides a general mechanism capable of explaining the formation of the jet-streams around a main body: mutual chaotic collisions between dust grains moving in the gravitational field of the central body tend to focus into a number of jet-streams with toroidal shapes centered on the central body.

Let us now shortly describe the model. Our model presupposes the formation from some nebula of an attracting central body and the presence of material of stellar or interstellar origin around it. We imagine a central body of mass M at the origin acting by some spherical symmetric potential $V(|x|)$. All one needs is a potential giving rise to some quantum mechanical bound states. M is immersed in some disordered gas of small particles acted upon by V and interacting by collisions, e.g. like in the protosolar nebula of the most common cosmological models [14]. During the era when the matter of the nebula was dispersed, we can expect that the interaction between particles was of decisive importance. Let us consider the motion of one such particle. It is under the steady influence of V and moves accordingly to classical mechanics until it hits some other particle, bouncing off to move again under the influence of V until next collision and so on.

The basic idea consists in thinking of a typical particle in the nebula as performing, under the steady influence of the attraction of the central body and innumerous chaotic collisions with other particles a stochastic diffusion process. We assume a typical particle to move along the trajectories of a diffusion process, i.e. its position X_t satisfies the stochastic differential equation

$$dX_t = \beta_+(X_t,t)dt + \sigma dW_t \quad . \tag{40}$$

We assume that the diffusion coefficient is constant for isotropy and homogeneity reasons (a spatial dependence can be anyhow accomodated as noted in Sect.II introducing a suitable Riemannian metric). There actually is hope that (40) can indeed be obtained from classical mechanics in a suitable limit (see e.g. for related problems [15] and references therein). It is reasonable to assume that the diffusion is Newtonian (i.e. the Newton's law holds in the average), with a potential V given approximately by the gravitational attraction, since it arises from classical motion, and that there exists an invariant distribution, as the potential is attractive and the time scale involved is large. Of course the invariant distribution is thought to hold as long as the diffusion approximation is valid.

From the results of Section II we then know that the invariant distribution $\rho=|\psi|^2$ is given by the solution of a Schrödinger type equation $H\psi = E\psi$, with $H = \frac{\mu\sigma^4}{2}\Delta + V$, and that the nodes of ψ act as barriers for the Newtonian diffusion process, hence yielding an explanation for the non-communicating rings in the nebula. The potential being central, the eigenfunctions $\psi_{n,\ell,m}(x)$ in $L^2(R^3)$ with $\ell = 0, \ldots n-1$, $m = -\ell, \ldots +\ell$, are of the form

$$\psi_{n,\ell,m}(x) = R_{n,\ell}(|x|)Y_\ell^m(\Theta,\phi) \quad , \tag{41}$$

with $R_{n,\ell}$ solution of the radial equation and $Y_\ell^m(\Theta,\phi)$ the spherical harmonics of indices 1,m, $0 \leq \pi$, $0 \leq \phi < 2\pi$. The $|x|$ dependence of the zeros of $\psi_{n,\ell,m}$ is determined by the zeros of the radial function $R_{n,1}$. Setting $\rho_{n,\ell,m} = |\psi_{n,\ell,m}|^2$ we can calculate the associated current velocity $v_{n,\ell,m}$. The angular momentum in the z-direction is given by

$$L_z = \int_{R^3} dx\, e_z.X \wedge v_{n,\ell,m}$$

$$= c.m \quad , \text{ with c constant.} \tag{42}$$

This is the classical angular momentum of the nebula. Using the conservation of the total classical momentum and choosing Oz along this conserved quantity we have that the invariant measures to be considered are of the form

$$\rho_{n,\ell,\ell}(x) = |\psi_{n,\ell,\ell}(x)|^2 \quad . \tag{43}$$

Recalling now that $Y_\ell^\ell(\Theta,\phi)$ is proportional to $e^{i\ell\phi}(\sin\Theta)^\ell$ we see that the invariant measure $\rho_{n,\ell,\ell}(x)$ is confined for 1 large to a small angular region around the equatorial plane. This correponds to the fact that the planetary system is essentially two-dimensional. The confinement regions ("jet-streams") are regions confined between concentric spheres centered at the center of the main body (Sun, Jupiter,...) and two cones. The confinement of dust particles in these regions provides an (idealized) probabilistic model of those phenomena which cause the particles in a jet-stream to keep together. The node structure is particularly easy to discuss if V is the Newton gravitational potential of the central body. In this case the zeros of the $R_{n,\ell}$ are determined by the zeros of the associated Laguerre polynomial $L_{n+1}^{2\ell+1}$, which has $n-\ell-1$ zeros as a function of x. For more details see [4]. In this probabilistic framework we thus get a picture of planetary system and satellite systems as atoms with the main body as nucleus and the planets resp. satellites as electrons.

III.2 Large scale flow patterns of the mean global circulation of the planetary atmospheres

In the solar system one of the most prominent features of the planetary atmospheres is the zonal structure of the large scale flow patterns of the mean circulation. This zonal structure is stable and roughly syymmetric across the equator except for seasonal variations. An evidence of the subtropical zonal cell (Hadley cell) is displayed by the global pictures of the development of the cloud released by the eruption of the volcane El Chichon (March-April 1982) (see e.g.[16], [17] and references therein).

In our stochastic framework the main idea is to interpret the zone boundaries as barriers from a Newtonian diffusion. The typical particles are droplets, icy particles or grains of sand, which undergo a Markov diffusion process under the influence of collisions and feel the gravitation field of the planet. The purpose of the model is to give a unified description of the planetary atmospheres and especially of their zonal structure. On dimensional ground let us deserve that the diffusion constant σ can be taken as

$$\sigma^2 = (GML)^{1/2} \tag{44}$$

where G is the gravitational constant, M the mass of the planet, and L some characteristic length which can be linked to the width of the atmosphere. The classical potential V is only a function of the distance r to the center of the planet. A plausible shape for the potential V could be

$$V(r) = \begin{cases} +\infty, r \le R_0 & \text{(radius of the planet)} \\ \frac{GM}{r}, r \ge R_1 & \text{(upper limit of the clouds)} \end{cases} \tag{45}$$

and in between some smooth function of r not too different from the gravitational potential.

Then according to Section II the invariant distributions are of the form $\rho_{n,\ell,m} = |\psi_{n,\ell,m}|^2$ with $\psi_{n,\ell,m}$ given in spherical coordinates (r,Θ,ϕ) by

$$\Psi_{n,\ell,m} = R_{n,\ell}(r) P_\ell^m(\cos\Theta) e^{im\phi} \tag{46}$$

where P_ℓ^m is the associated Legendre function, which have $2(\ell-|m|)$ zeros corresponding to latitudes $\pm\Theta_1$, $\pm\Theta_2,\ldots,$ and consequently the nodal surface intercept a sphere of given radius on parallels leaving zonal structures in between. The current velocity v can be extracted from Ψ and is directed along the parallels, orthogonal to the radius so that the model describes winds in a zonal region either west or east winds (see Fig.2). The P_ℓ^m are real functions which furthermore satisfy

$$P_\ell^m(.) = P_\ell^{-m}(.) \quad . \tag{47}$$

Consequently on both sides of a nodal surface defined by a zero of P_ℓ^m we have the possibility to patch two solutions of the form (46) corresponding to the value $\pm m$. Consequently the current velocity might have opposite values on both sides of the zonal surface. For more details about the application of this model to the atmosphere of Venus, the Earth, Mars, Jupiter and Saturn see [6].

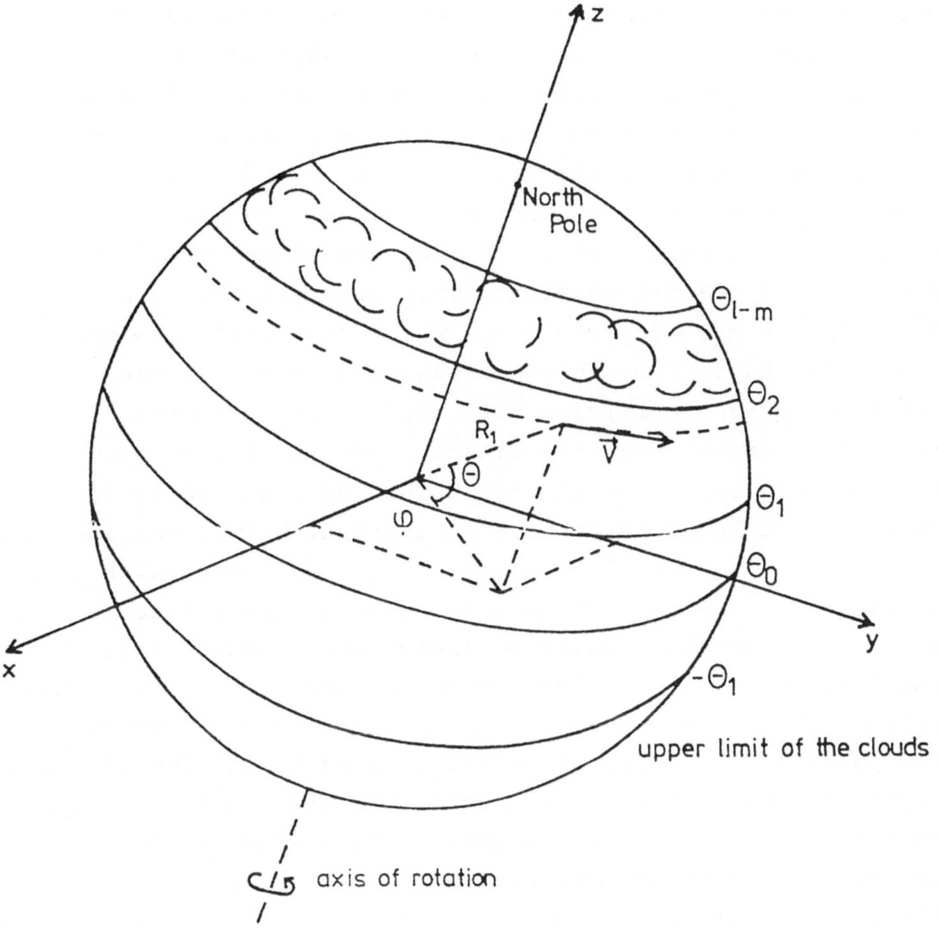

Fig. 2. Model describing winds in a zonal region, either west or east winds.

III.3. Newtonian diffusion as a model of the Van Allen radiation belts

It has been known or suspected for centuries that a magnetic field is associated with the Earth. One of the most entertaining results of rockets and satellite investigations has been the discovery in 1958 by Van Allen of zones of radiation which surround our planet. It was soon established that the source of the observed radiation must be charged particles (i.e. electrons and protons) trapped in the Earth's magnetic field. If the magnetic field B varies slowly with position the magnetic moment of a particle is nearly constant. This magnetic moment μ is given by $\mu = mv_{\perp}^2/2B$, where $v_{\perp}$ is the component of v perpendicular to B. From this expression we see that the gyrating charged particles will tend to be reflected from regions of higher magnetic fields. Suppose that $v_{\parallel}$ takes the particle into a region of increasing magnetic field. Then $v_{\perp}$ must increase to keep μ constant, but it can increase only until it is equal to the total velocity, at which time $v_{\parallel}$ has fallen to zero. Thus the particles are reflected back into region of lower magnetic field. This kind of region where the particles are reflected may be called a magnetic mirror. The Van Allen belts are actually toroidal because the particles drift longitudinally owing to the inhomogeneous magnetic field. In a dipole magnetic field electrons drift eastward and protons drift westward. This suggests that the charged particles are diffused by the irregularities of the magnetic field.

The results of Section II.1 can be extended to take into account the motion of a particle in an external magnetic field B=rot A, where A is the vector potential. The Lorentz force on a particle of charge q is given classically by

$$F = qv \wedge B \tag{48}$$

where v is the velocity. We consider (48) as the force acting on a charged particle undergoing a Newtonian diffusion (1) with v the current velocity. We do this because the force should be invariant under time inversion. For the stochastic acceleration a we substitute the Lorentz force divided by the mass μ. Assuming now that the generalized momentum $\mu v + qA$ is a gradient then $\Psi = \rho^{1/2} e^{iS}$ satisfies the Schrödinger equation

$$i\mu\sigma^2 \frac{\partial\psi}{\partial t} = \frac{1}{2\mu} (-i\mu\sigma^2\nabla - qA)^2 \quad . \tag{49}$$

For more details see e.g. [7,18]. We consider an electromagnetic potential A with axial symmetry producing a magnetic field in the meridian plane. The most general magnetic field satisfying this requirement is of the form

$$B_x = -x \frac{\partial A}{\partial \rho}$$

$$B_y = -y \frac{\partial A}{\partial z}$$

$$B_z = 2A + \rho \frac{\partial A}{\partial \rho} \tag{50}$$

with A some function of z and $\rho = (x^2+y^2)^{1/2}$. In cylindrical coordinates (z,ρ,ϕ) (49) reduces to a two-dimensional Schrödinger-like equation where the effective potential is given by

$$U^{eff}_{\pm|\ell|} = \frac{1}{2\mu\rho^2} [\mu\sigma^2|\ell| \pm q\rho^2 A(\rho,z)]^2 \quad . \tag{51}$$

Then if we assume that A goes to zero at infinity with $r = (x^2+y^2+z^2)^{1/2}$ going to infinity in a given direction, $U^{eff}_{\pm|\ell|}(x)$ have the shapes in fig.(3), depending on a relative sign of q and ℓ. From this it is quite obvious that for $q\ell \geq 0$ there is no bound state, whereas if $q\ell < 0$ there is

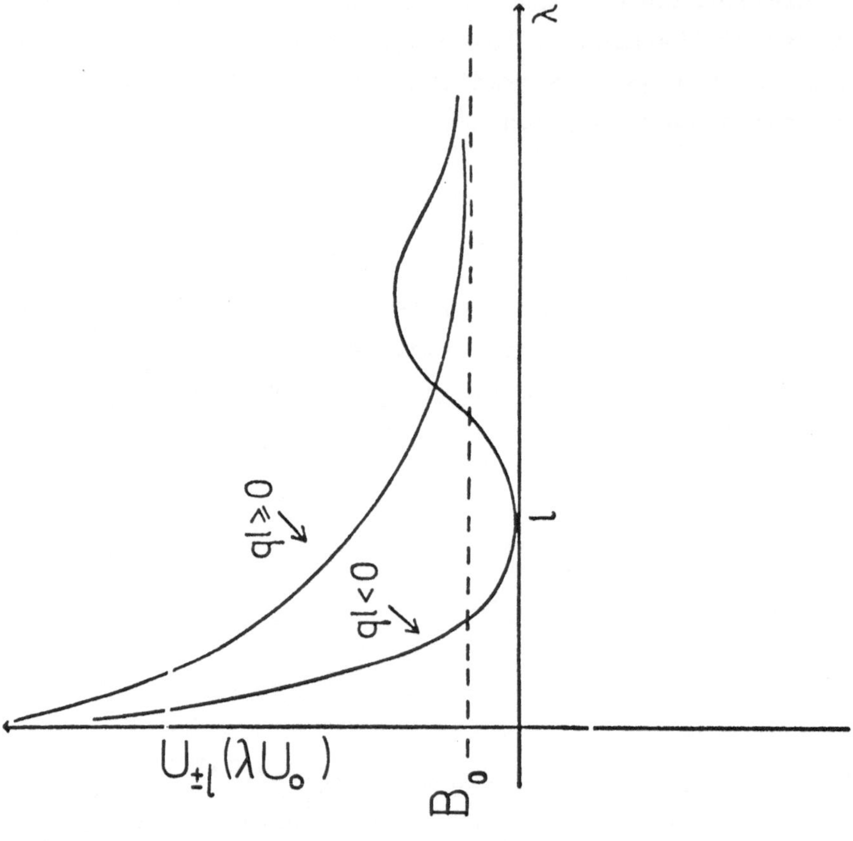

Fig.3. The shapes of the effective potentials $U^{eff}_{\pm|\ell|}$.

possibility for such bound state. To show that under some circumstances U^{eff} is confining we make assumptions on the fall-off of the magnetic field at large distance. The effectice potential U^{eff} behaves like $U^{eff} \simeq q^2\rho^2 A(\rho,z) \simeq B_0\rho^{-1}$. For large $r = (\rho^2+z^2)^{1/2}$ then the pure point spectrum of $(-i\mu\sigma^2\nabla - qA)^2$ is not empty.
The model accounts for the following observational facts:
1) There exists discrete zone of confinements related to the nodal surfaces of a probability density.
2) In each zone, particles show a general drift either westward or eastward according to the charge ($q\ell$ must be positive). Proton drift to the west, electrons to the east.
3) Confinement zones are roughly shaped according to the lines of the field.
4) Inner belts contain more energetic particles than outer belts.
For a general description of the model we refer to [18].

ACKNOWLEDGEMENTS

I am very grateful to S. Albeverio, Ph. Combe, R. Høegh-Krohn, M. Sirugue, M. Sirugue-Collin, W. Schneider, L. Streit for the joy of collaboration and to F. Gesztesy, H. Grosse, R. Jost, O. Steinmann for very helpful discussions.
Various stays at the Université d'Aix-Marseille II, CNRS (Marseille), ZiF (Universität Bielefeld) and at the Mathematics Department of the Ruhr-Universität, Bochum and of Oslo University are gratefully acknowledged, as well as the partial support of DFG (W.-Germany) and NAVF (Norway).

REFERENCES

1. S. Albeverio, M. Fukushima, W. Karwowski, L. Streit, Capacity and Quantum Mechanical Tunneling, Commun. Math. Phys. 81 (1981) 501-513.
2. S. Albeverio, Ph. Blanchard, R. Høegh-Krohn, Processus de diffusion, confinement et formation de jet-streams dans la nébuleuse protosolaire, Encontro de Fisica e Matematica Coimbra (1982), Preprint CERN TH 3536.
3. S. Albeverio, Ph. Blanchard, R. Høegh-Krohn, Newtonian Diffusion and Planets with a remark on non-standard Dirichlet forms and polymers, Symposium on Stochastic Analysis (Swansea), 1983, London Mathematical Society, Lecture Notes in Mathematics (Springer Verlag, 1984).
4. S. Albeverio, Ph. Blanchard, R. Høegh-Krohn, A Stochastic Model for the orbits of planets and satellites. An interpretation of Titius-Bode law, Exp. Math. 4 (1983) 365-373.
5. S. Albeverio, Ph. Blanchard, R. Høegh-Krohn, Diffusion sur une variété riémanienne. Barriéres infranchissables et applications. Colloque en l'honneur de Laurent Schwartz. Ecole Polytechnique Juin 1983, Astérique Société Mathematique de France (1984).
6. S. Albeverio, Ph. Blanchard, Ph. Combe, R. Høegh-Krohn, R. Rodriguez, M. Sirugue, M. Sirugue-Collin, Zonal wind structure of the planetary atmospheres. A unified stochastic approach, ZiF preprint 40, Feb. 1984.
7. S. Albeverio, Ph. Blanchard, M. Sirugue, M. Sirugue-Collin, work in preparation.
8. P.A. Meyer, Geometrie differentiélle stochastique (bis), Lecture Notes in Mathematics (Springer, 1982).
9. E. Nelson, Dynamical Theories of Brownian Motion, Mathematical Notes (Princeton University Press, 1967).
10. M. Fukushima, Dirichlet Forms and Markov Processes (North Holland, 1980).
11. L. Streit, Schladming Lecture Notes, 1984.

12. M. Nieto, The Titius Bode law of planetary distances, its history and theory (Pergamon Press, Oxford, 1972).
13. G. Louise, Quantum formalism in gravitation quantitative application to the Titius-Bode law. The Moon and the Planets 27 (1982) 59-62.
14. H. Reeves. L'origine du systême solaire, La Recherche 6 (1975) 807-818.
15. D. Dürr, All that Brownian Motion, Lecture Notes in Physics (1983) 173.
16. R.R. Ramfino, S. Self, The atmospheric effects of El-Chichon, Sc. American 250 (1984) 34-43.
17. A.P. Ingersoll, The atmosphere, Sc. American I (1983) 114-130.
18. S. Albeverio, Ph. Blanchard, Ph. Combe, R. Høegh-Krohn, R. Rodriguez, M. Sirugue, M. Sirugue-Collin, Magnetic bottles in a dirty environment. A stochastic model for radiation belts, ZiF preprint No. 54, April 1984.
19. W.A. Zheng, P.A. Meyer, Construction de processus de Nelson stationnaires Preprint, IRMA, Strasbourg, Feb. 1984.
20. M. Nagasawa, Segregation of a population in an environment, J. Math. Biology 9 (1980) 213-235.

Acta Physica Austriaca, Suppl. XXVI, 211–231 (1984)

MARKOV COSURFACES AND GAUGE FIELDS[+*]

by

S. ALBEVERIO [+], R. HØEGH-KROHN[++], H. HOLDEN[++]
Mathematisches Institut
Ruhr-Universität
D-4630 Bochum 1, BRD

ABSTRACT

We give a general construction of homogeneous Markov fields associated with d-1 hypersurfaces on a d-dimensional Riemannian manifold, with values in a Lie group G. In the case d = 2 these "Markov cosurfaces" coincide with pure continuum gauge fields with values in G. The constructed Markov cosurfaces have properties of the type of those postulated for Wilson loops Schwinger functions in the case of gauge fields and lead to relativistic quantum fields. The Markov cosurfaces provide an extension of Markov processes for the case where time points are replaced by d-1-dimensional hypersurfaces, and a corresponding extension of stochastic differential equations if given. The Markov cosurfaces are also shown to be obtainable from lattice models.

[+]Lecture given at the XXIII. Internationale Universitätswochen für Kernphysik, Schladming, Austria, February 20–March 1, 1984.

[++]Matematisk Institutt, Universitetet i Oslo.

[*]This work has been partly supported by the Norwegian Research Council for Science and the Humanities (Program Oslo Seminar).

1. INTRODUCTION

This Meeting is on stochastic processes and their applications. Let us first recall shortly the kind of situations one studies in the theory of stochastic processes. A general stochastic process is a family of random variables $X(t,\omega)$ with values in a space Z ("state space"), t running over some index set T ("time") and ω over some probability space (Ω, A, P). One is of course interested in studying situations where on T, Z, P one has additional structures, e.g.

1) $T \subset R$, Z ; $\left\{\begin{array}{l} Z = R,\ R^n,\ Z^d,\ \ldots \\ \text{Banach space, } \ldots \\ \text{distributional spaces} \end{array}\right.$ — the first two: ("ordinary process"); distributional spaces: ("generalized process")

   Additional structure for the process itself: Markov,... homogeneous, ...

2) $T \subset R^d$, Z^d, M(manifold) (T is then thought as "space" or "space-time"); Z as in 1). One speaks then of random fields ("ordinary resp. generalized random fields") on T, rather than processes. Again additional features are interesting: Markov ... homogeneous ...

3) T as in 1), Z = group
   The study of these processes resp. fields is part of the theory of representations of infinite-dimensional Lie groups, see e.g. [1], [2].

4) T as in 1), 2), Z = operator algebra: The study of this processes belongs to the study of "non-commutative probability theory", see e.g. R. Streater's lectures at this Meeting and e.g. [3].

5) T set of lower dimensional subsets of a manifold, Z = group. This will be the topic of these lectures.

Before we come to a discussion of the situation 5) from

Section 2 on, let us make a few historical remarks and describe the motivation and content of this lecture.

In recent years there has been a great interest in constructing homogeneous Markov random fields on R^d and associated quantum fields, see e.g. [2] and references therein.

No non-Gaussian model for $d \geq 4$ is known at present, for some partial results see S. Albeverio's second lecture at this meeting. For d = 2,3 models of quantized gauge fields have also been constructed, see e.g. [4], [5] and references therein.

In the simple case of the so-called pure Yang-Mills fields the construction of the quantized field goes through the construction of a suitable measure associated with fields with values in the Lie algebra of a compact Lie group, invariant under Euclidean transformations and satisfying an Osterwalder-Schrader positivity condition (a weaker form of "markovicity"). Basic quantities in this construction are expectations of products of "Wilson loops", essentially obtained by integrating fields (understood as Lie algebra-valued 1-forms on R^d) along 1-dimensional closed chains in R^d.

Most work on quantum gauge fields has however been performed in the case of lattice models in which R^d is replaced by a lattice εZ^d, $\varepsilon > 0$, because of the well known divergence difficulties associated with fields on R^d. In the present work we shall introduce a class of models on R^d. In the case d = 2 the models contain as particular cases the continuum pure gauge fields, with gauge group a (locally) compact Lie group G. Let

$$A = \sum_{i=1}^{2} A_i(x)dx_i \ ,$$

$x \in R^2$, be a one-form from R^2 with values in the Lie algebra g of G; then for any closed 1-chain S we can consider $\int_S A$.

Such quantities are distributed according to an Euclidean invariant measure μ, with formal density $\exp(-\frac{1}{2} g\|dA\|^2)$ with respect to flat measure, which can be constructed in a suitable continuum limit $\varepsilon \downarrow 0$ from corresponding quantities on the lattice εZ^2. The objects we construct are d-dimensional extensions of the above (A,μ), the role of 1-chains S being then played by d-1-chains.

We prove (for G abelian if $d > 2$, for general compact G if $d = 2$) that the expectations with respect to μ of objects corresponding to the Wilson loops of the 2-dimensional theory do have all properties (symmetry, Euclidean invariance, Osterwalder-Schrader positivity, clustering, growth conditions) which have been postulated for gauge theories in [4], [22]. In fact we have the Markov property (stronger than Osterwalder-Schrader positivity) with respect to half planes (much in the same spirit of Nelson's postulate [11], as had been previously verified in two-dimensional models [6], [7], [9]). The measure μ is entirely given by a convolution semigroup of Markov kernels on the locally compact abelian group G, and any such semigroup generates a Markov field with the above properties. By adapting results of [4], [22] we can associate to μ relativistic quantized fields, associated instead of with points, with d-1-dimensional hypersurfaces. For $d = 2$, G can be non-abelian.

Let us also stress the probabilistic meaning of the constructions of this paper. The classical Kolmogorov construction of Markov processes starts from a start measure and a Markov semigroup, the basic quantities being the joint distributions for the process $X(t)$ to be at times $t_1 \dots t_n$ in $B_1,\dots,B_n$, the B_i being Borel sets in the state space. The construction in this paper gives a similar result for processes (called Markov cosurfaces) with the parameter t (resp. intervals $[0,t]$) replaced by d-1 hypersurfaces S in a d-dimensional Riemannian manifold M and with state space replaced by a (compact) group G (abelian for $d > 2$). The so constructed Markov cosurfaces are again determined by a Markov convolution semigroup on G and have as start measure the Haar measure on G. In the special case $G = T^{\nu}$, the ν-dimensional torus, the Markov cosurfaces A are determined by R-valued Markov covector fields a. These give solutions to a stochastic equation of the form $da = \xi d\lambda$, with $d\lambda$ the Riemann-Lebesgue measure on M and ξ a generalized random field with independent values at every point. In this sense a and hence $C = \exp(2\pi ia)$ are a G-valued more-dimensional time-parameter version of stochastic integrals. Such stochastic integrals are also constructed in the non-abelian 2-dimensional case.

This lecture is structured as follows:

2 Definition and basic construction of Markov cosurfaces
3 Invariance properties of Markov cosurfaces
4 The case of Lie groups, stochastic differential equations for forms
5 Lattice cosurfaces with Gibbs distribution (lattice gauge fields for d = 2)
6 Markov cosurfaces on R^d as limits of lattice cosurfaces with Gibbs distribution
7 Markov semigroups and relativistic Markov fields associated with Markov cosurfaces

2. DEFINITION AND BASIC CONSTRUCTION OF MARKOV COSURFACES

Let M be an oriented Riemannian manifold of dimension d (e.g. $M = R^d$). Let G be a group, assumed to be abelian if $d > 2$ (G can be non-abelian if d = 1,2). In the following we shall illustrate all concepts and give all results for the abelian case, although generally speaking all we shall say holds also in the non-abelian case provided in this case d = 1,2.

Let H_M be the family of all oriented piecewise smooth connected (d-1)-dimensional hypersurfaces in M with no self-intersections and which are closed sets. For S_1, $S_2 \in H_M$

we shall define a product S_1S_2 if the endpoint of S_1 equals the start point of S_2 if $d = 2$ resp. $S_1 \cap S_2$ is (d-2)-dimensional with opposite orientation if $d > 2$. For such S_i, $i=1,2$, the product S_1S_2 is then by definition $S_1 \cup S_2$ as set and its orientation is the one inherited from the orientation of S_1 and S_2.
Let $S_1,\dots,S_n \in H_M$ be such that $S_1S_2, (S_1S_2)S_3, ((S_1S_2)S_3)S_4\dots$ defined. Then we define recursively $S_1 \dots S_n \equiv (S_1 \dots S_{n-1})S_n$. Let $\Sigma_M \equiv \{ S_1 \dots S_n | \; n \in N, \; S_i \in H_M\}$, i.e. Σ_M consists of all products of the form $S_1 \dots S_n$ with $S_i \in H_M$. We define S^{-1} (denoted also by $-S$ in the abelian case) to be equal to S as sets, but with opposite orientation.

E.g.

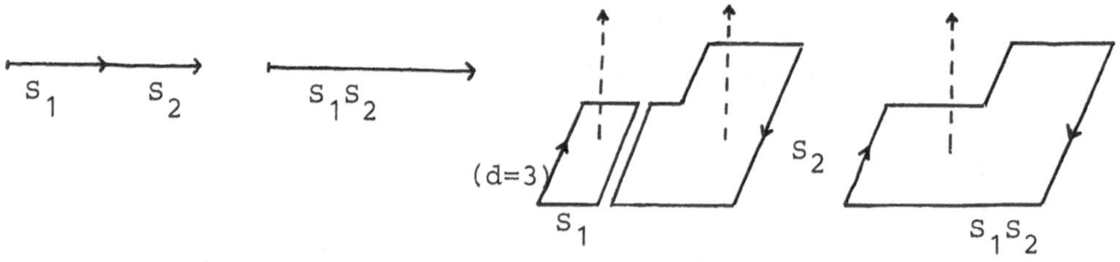

Rem.: For $d = 1$, Σ_M can be identified with the set of all points of M.

A cosurface C on M with values in the (abelian) group G is a map C from Σ_M into G, with:

1) $C(-S) = -C(S)$

2) $C(S_1 . S_2) = C(S_1) + C(S_2)$.

Let $\Gamma_{M,G}$ the set of all cosurfaces on M. We assume G has a measurable structure $(G, \mathcal{B}_G)$, then $\Gamma_{M,G}$ gets the measurable structure $\{\sigma(C(S)^{-1}(\mathcal{B}_G))$; $S \in \Sigma_M, \; C \in \Gamma_{M,G}\}$, i.e. $\Gamma_{M,G}$ is endowed with the smallest σ-algebra making all maps $C(S)$ measurable.

A stochastic cosurface on M is by definition a measurable map C from some probability space $(\Omega, \mathcal{B}, P)$ into $\Gamma_{M,G}$. For $M = R$ a stochastic cosurface is just a G-valued stochastic process indexed by R; thus a stochastic cosurface is an extension of the concept of stochastic process when the time points $t \in T$ are replaced by d-1-hypersurfaces S of M.

A complex K is an ordered n-tuple $K = \{S_1,\dots,S_n\}$, with $S_i \in \Sigma_M$. For d = 1, K is just an n-tuple of points ("times"). We shall use the notation $C(K) \equiv \{C(S_1),\dots,C(S_n)\}$. We can look upon C(K) as an element in G^K.
Two stochastic cosurfaces $C,\tilde{C}$ with underlying probability measures $P,\tilde{P}$ are called equivalent iff the induced image probability measures $P_C, \tilde{P}_{\tilde{C}}$ are equal as measures on G^K. We shall henceforth identify equivalent stochastic cosurfaces.

For the precise definition of Markov stochastic cosurfaces and their construction we need the concept of regular complex: a complex $K = \{S_1,\dots,S_n\}$ is called regular if $S_i \cap S_j \subset \partial S_i \cap \partial S_j$ when $i \neq j$.
Let Λ be a subset of M; we shall denote by $\Sigma(\Lambda)$ the σ-algebra generated by all stochastic variables C(S) with $S \in \Sigma_M$, $S \subset \Lambda$. We say that the stochastic cosurface C has the Markov property whenever for any regular complex $K = \{S_1,\dots,S_n\}$ with the properties that a subset $S_j,\dots,S_l$ of $S_1,\dots,S_n$ splits its complement $M - \bigcup_{i=j}^{l} S_j$ into 2 connected components $M^{\pm}$ s.t. $S_1,\dots,S_{j-1} \subset \overline{M^{+}}$ and $S_{l+1},\dots,S_n \subset \overline{M^{-}}$, where $^{-}$ means closure, and

$$E(f^+f^-|\Sigma(S)) = E(f^+|\Sigma(S))E(f^-|\Sigma(S)), \text{ with } S = \bigcup_{i=j}^{l} S_i,$$

for any f which are $\Sigma(M^{\pm} \cup S)$ measurable. E.g. for $M = R^2$, $n = 3$, $j = l = 2$, $S = S_2 = \{(x^1,x^2) \in R^2 | x^2 = 0\}$ this is equal to conditional independence "of functions of curve variables" in the upper halfplane and in the lower halfplane, given the curve variables on the line S.

For d = 1 this property corresponds to the Markov property of the G-valued processes. We shall call Markov cosurface any stochastic cosurface which has the above

Markov property.

Question: Do there exist Markov cosurfaces?

We shall now give a way to construct examples directly "in the continuum". Later on we shall give another procedure to construct Markov cosurfaces from "lattice Gibbs fields", and show that it yields the same result.

For the construction of a Markov cosurface we shall need the concept of <u>regular saturated complex</u>. We call in this way any regular complex $K = \{S_1,\dots,S_n\}$ s.t. there exists a partition $D_K = \{A_1,\dots,A_m\}$ of M into connected and simply connected subsets A_i of M s.t. M is the union of the A_i and if $A_i \cap A_j \neq \emptyset$ for some $i \neq j$ then either $A_i \cap A_j$ is d-2-dimensional or else $A_i \cap A_j$ is a piecewise smooth d-1-hypersurface which can be written as the union of some of the S_i and s.t.

$$\bigcup_{i\neq j} A_i \cap A_j = \bigcup_{i=1}^{n} S_i$$

. E.g. for $M = R^2$,

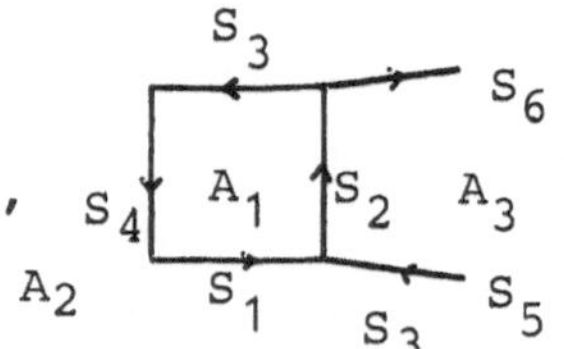

is saturated with $D_K = \{A_1,A_2,A_3\}$, whereas e.g.

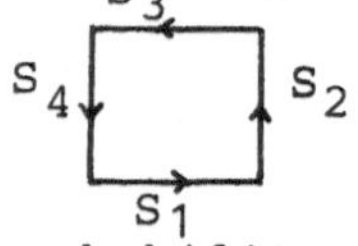

is not saturated.

We shall now define a coherent system of probability measures on regular saturated complexes, assuming that G is compact (or, with modifications, locally compact). The probability measures will be defined with the help of Haar measure dx on G and of a Markov convolution semigroup Q_t, $t \geq 0$, on G, with density $Q_t(x)$ s.t. $Q_t(x) \geq 0$,

$$\int_G Q_t(x-y)\,dy = 1, \qquad Q_t Q_s = Q_{t+s}$$

in the sense $\int Q_s(x-y)Q_t(y-z)\,dy = Q_{t+s}(x-z)$, and $Q_t(x-y) \to$ $\to \delta(x-y)$ weakly as $t \downarrow 0$. We define for any regular satured complex $\tilde{K} = \{S_1,\dots,S_n\}$ a probability measure $\mu^{Q}_{\tilde{K}}$ on $G^{\tilde{K}}$ by

$$d\mu^{Q}_{\tilde{K}}(C(\tilde{K})) \equiv \prod_{A\in D_{\tilde{K}}} Q_{|A|}\Big(\sum_{S\in\partial A\cap\tilde{K}} C(S)\Big) \prod_{S\in\tilde{K}} dC(S),$$

where $|A|$ is the Riemannian volume measure of A. We set $Q_\infty = 1$.

We can extend the definition of these probability measures to arbitrary complexes $K = \{S_1,\ldots,S_n\}$ contained in a regular saturated complex $\tilde{K} \equiv \{\tilde{S}_1,\ldots,\tilde{S}_n\}$ in a natural way, by setting

$$d\mu_K(C(K)) \equiv \int d\mu_{\tilde{K}}(C(\tilde{K}))\prod dC(\tilde{S}_i) ,$$

the integration being over the regular saturated complex $\tilde{K} \equiv \{\tilde{S}_1,\ldots,\tilde{S}_n\}$ and s.t. $C(S) = \sum C(\tilde{S}_i)$ whenever $S = \cup\tilde{S}_i$, $S \in K$; μ_K is independent of the choice of $\tilde{K} \supset K$. One proves that the system (μ_K, K complex) is a projective system. This uses the assumption G abelian for $d > 2$ and the semi-group property of Q_t. Typically for $M = R^2$,

$K = \{S_1,\ldots,S_4\}, \tilde{K}_1 = \{S_1\ldots S_6\}$,

$\tilde{K}_2 = \{S_1,\ldots,S_{13}\}$, we have on one hand

$$d\mu^Q_K(C(K)) = \int d\mu_{K_1}(C(\tilde{K}_1)) \prod_{i=1}^{6} dC(\tilde{S}_i)$$

$$= \int Q_{|A_1|}\Big(\sum_{i=1}^{4} (S_i)\Big)Q_{|A_5|}(-(S_2)+(S_5)+(S_6)) \prod_{i=1}^{6} d\,(S_i) =$$

$$= Q_{|A_1|}\Big(\sum_{i=1}^{4}(S_i)\Big) \prod_{i=1}^{4}(S_i)) \quad {}_{i=1}^{4}\; d\,(S_i) ,$$

(+)

on the other hand

$$\int d\mu_{K_2}(C(K_2)) \prod_{i=1}^{13} dC(S_i) =$$

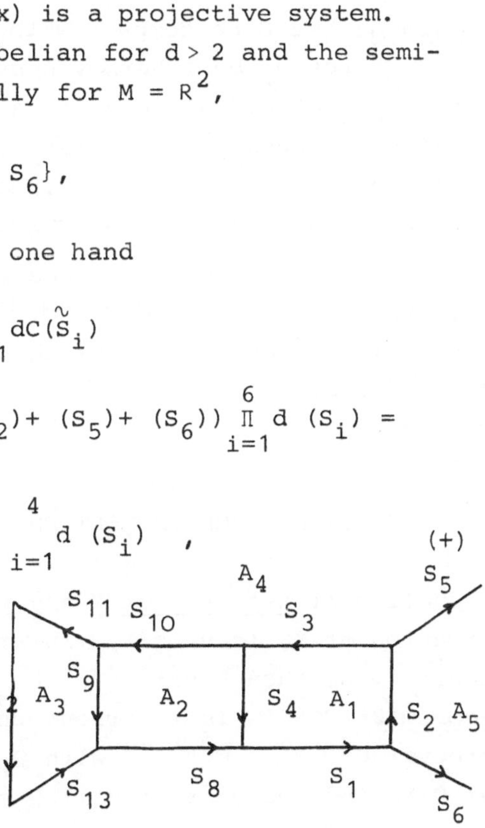

$$= \int Q_{|A_1|}\left(\sum_{i=1}^{4}(S_i)\right)Q_{|A_2|}(-(S_4)+(S_{10})+(S_9)+(S_8)) .$$

$$\cdot\ Q_{|A_3|}(-(S_9)+(S_{11})+(S_{12})+(S_{13}))\prod_{i=1}^{13} d(\tilde{S}_i) =$$

$$= \int Q_{|A_1|}\left(\sum_{i=1}^{4}(S_i)\right)Q_{|A_1 \cup A_3|}(-(S_4)+(S_{10})+(S_{13})+(S_{12}) +$$

$$+ (S_{11})+(S_8))\prod_{i\neq 9} dC(\tilde{S}_i) ,$$

where we have integrated with respect to $C(S_9)$ and used the semigroup property of Q_t. Integrating now with respect to $C(S_i)$, $i = 8,10,11,12,13$, using that $dC(S)$ is the normalized Haar measure and $Q_{|A4|} = 1$, as well as $C(S_8) = -C(S_2)$ and that $dC(S)$ is the normalized Haar measure and $Q_{|A4|}=1$, the semigroup property of Q, we get the equality of this with (+). We have the

<u>Theorem 1</u>: Let G be compact with countable base, abelian for $d > 2$. Let Q_t be a Markov convolution semigroup on G. Then $\{\mu^Q_K\}$ defined as above is a projective family of probability measures. There exists a probability space (Ω,A,P) and a stochastic Markov cosurface C on it s.t. $C(K)(P)=\mu^Q_K$. $\{(\Omega,A,P)C\}$ is unique up to equivalence.

The stochastic Markov cosurface C depends in general on Q, the volume measure $\lambda = |.|$ on M and the orientation σ on M. If we want to stress this dependence we shall write $C^{Q,\lambda,\sigma}$ for C.

3. INVARIANCE PROPERTIES OF MARKOV COSURFACES

We shall call <u>isometry of M</u> any piecewise smooth transformation of M, leaving λ,σ invariant. Let $C^{Q,\lambda,\sigma}$ as in Theor. 1. We then have:

<u>Theorem 2</u>: $C^{Q,\lambda,\sigma}$ is invariant under any isometry of M. Moreover $C^{Q,\lambda,\sigma} = C^{\tilde{Q},\lambda,-\sigma}$ with $\tilde{Q}_t(x) \equiv Q_t(-x)$. In particular C is indenpendent of σ if $\tilde{Q} = Q$.

The proof of this theorem uses the invariance of $\lambda(A) = |A|$ under isometries of M and the fact that μ_K is defined in term of $Q_{|A|}$.

We have a converse of this theorem for the case one starts from an invariant Markov stochastic cosurface which is jointly measurable in ω and S, with a suitable topology in Σ_M.

Theorem 3: If C is a jointly measurable Markov stochastic cosurface invariant under the isometries of M, then there exists a Markov semigroup Q_t on G s.t., if one assume this to have a density with respect to the Haar measure, $C = C^{Q,\lambda,\sigma}$.

The proof uses the assumed Markov property of the cosurface C to construct the Markov semigroup Q_t by which then $C^{Q,\lambda,\sigma}$ is constructed following theor. 1.

4. THE CASE OF LIE GROUPS, STOCHASTIC DIFFERENTIAL EQUATION FOR FORMS

Let now G be a connected compact Lie group (abelian for $d > 2$). We shall here give the results for G abelian i.e. $G = T^\nu$ (the ν-dimensional torus).

Any jointly measurable stochastic cosurface on M with values in G has the form

$$C(S) = (e^{2\pi i \int_S a^1}, \ldots, e^{2\pi i \int_S a^\nu}), \quad S \in \Sigma_M,$$

a^i a real-valued d-1 form on M (the integrals being understood in De Rham sense, as linear functionals). For short we shall write $C(S) = e^{2\pi i \int_S a}$, with $a = (a^1, \ldots, a^\nu)$, a R^ν-valued d-1 form on R ("d-1-vector form"). We have the following

Theorem 4: If C is a T^ν-valued, jointly measurable Markov cosurface on M invariant under isometries of M, then $C_\omega(S) = e^{2\pi i \int_S a_\omega}$, with a_ω a solution of the stochastic differential equation $da_\omega = \xi_\omega \, d\lambda$, with ξ_ω an infinitely divisible generalized stochastic random field on M with values in R^ν (i.e. ξ_ω given by a "Levy-Khinchin formula"). And conversely, given ξ_ω as above, then the above s.d.e. has a solution a_ω which is a Markov covector s.t. $C_\omega(S) \equiv$ $\equiv \exp(2\pi i \int_S a_\omega)$ is a T^ν-valued invariant Markov cosurface.

Rem.: For M = R this corresponds to the construction of time homogeneous processes in T^ν.

Proof: (sketch for ξ_ω = white noise) Let $Q_t = e^{\frac{t}{2}\Delta}$, Δ the Laplace-Beltrami operator on T. One has $E(\exp(i\langle\xi_\omega,\phi\rangle)) =$ $= \exp(-\frac{1}{2}(\phi,\phi))$, with $\langle T,\phi\rangle \equiv \int T(x)\phi(x)dx$, in distributional sense. From this one deduces $E(\exp i\langle\xi_\omega,\beta\,\chi_A\rangle) =$ $= \exp(-\frac{1}{2}|A|\,|\beta|^2)$, for any Borel $A \subset M$ and any $\beta \in R^\nu$. The process associated with Q_t is the Wiener process W_t on T^ν. By the construction of $\{(\Omega,A,P);\, C^Q\}$, Q determines the construction of the distribution of $\exp(2\pi i\int_S a_\omega)$.

$$\int_S a^i_\omega,\quad S = \partial A, \text{ is distributed as } \int_0^{|A|} \xi_\tau\, d\tau \text{ with } \xi_\tau$$

white noise on R.

Thus

$$E(\exp(2\pi i \int_S a,\beta)) = E(\exp(2\pi i \int_0^{|A|} \xi_\tau\, d\tau,\beta)) = \exp[-\tfrac{1}{2}|A|\beta^2],$$

which gives the identification $\int_S a$ with $\int_A \xi_\omega$, $\partial A = S$, and proves the theorem.

Remarks 1) There exist extensions of this result to non-invariant Markov cosurface, see [12].

2) There exist results on "smoothness" of the dependence of $C^Q(S)$ on S: roughly speaking, $S \to C^Q(A)$ has the same smoothness as the process X_t associated to the semigroup Q_t (e.g. for Q_t the heat semigroup, $S \to C^Q(S)$ will be Hölder-continuous of order <1/2).

5. LATTICE COSURFACES WITH GIBBS DISTRIBUTION (GAUGE FIELDS FOR d = 2)

Let us assume for simplicity $M = R^d$. As before G has to be abelian for d > 2, it can be non-abelian for d = 1,2.

Let $L_\varepsilon \equiv \varepsilon Z^d$, the lattice with spacing ε. We call cell γ any oriented elementary cell of L_ε and we call face any d-1-oriented hyperface belonging to the boundary $\partial\gamma$ of γ.

Let U be an invariant real-valued function on G ($U(gh) = U(hg)$ for all h, $g \in G$) and let B be a real constant. Let Λ be a finite union of cells. We define as Gibbs interaction on L_ε the quantity

$$W_\Lambda(\ ) \equiv -\beta \sum_{\gamma\subset\Lambda} U(\ (\partial\gamma))\ ,$$

where C is the quantity which corresponds to a cosurface on L_ε, i.e. C(F) is defined on faces of cells and satisfies properties 1), 2) in Sect. 2 of cosurfaces. We assume that $\int \exp(\beta U(g))dg \equiv C(\beta)$ exists. With the help of the Gibbs-interaction W we construct the Gibbs probability measure μ_Λ on the cosurface restricted to faces in Λ i.e.

$$d_{\mu_\Lambda}(C(F),\ F \subset \Lambda) \equiv Z_\Lambda^{-1} \exp(W_\Lambda(C)) \prod_{F\subset\Lambda} dC(F),$$

where Z_Λ is a normalizing factor. μ_Λ is a probability measure on G^{N_Λ}, with N_Λ the number of faces in Λ. We call Gibbs state $\mu_\varepsilon \equiv \mu_{L_\varepsilon}$ to the interaction W any probability measure on the product space $(\Omega_{L_\varepsilon}, \mathcal{B}_{L_\varepsilon})$ s.t.

$$E_{\mu_\varepsilon}(f|\Sigma(\Lambda^c)) = E_{\mu_\Lambda}(f|\Sigma(\Lambda^c))$$

for any $\Sigma(\Lambda_o)$-measurable function f, where $\Lambda^o \subset \Lambda$, $\Lambda^c \equiv M-\Lambda$. It is easily seen that μ_ε has the Markov property in a natural sense, corresponding to the one discussed above

(Sect. 2) for Markov cosurfaces. We call

$\{(\Omega_{L_\varepsilon}, \mathcal{B}_{L_\varepsilon}, \mu_\varepsilon); C(F), F \text{ face of } L_\varepsilon\}$

a "lattice cosurface with Gibbs distribution".

Rem.: For d = 2, $U(y) = \text{Re}\chi(y)$, χ charac er of a unitary representation of a compact Lie group G, we have that above lattice cosurface G describes a gauge field theory on the lattice L_ε. For $G = U(1) = T^1$, γ a plaquette, $\sum_{F \in \partial\gamma} C(F) = \exp(2\pi i \int_{\partial\gamma} a)$ is a Wilson loop, with

$$a = \sum_{i=1}^{2} a_i(x)dx_i$$

giving the connection.

6. MARKOV COSURFACES ON R^d AS LIMITS OF LATTICE COSURFACES WITH GIBBS DISTRIBUTION

Let $\varepsilon > 0$ be fixed and let $(\Omega_{L_\varepsilon}, \mathcal{B}_{L_\varepsilon}, \mu_\varepsilon; C)$, $L_\varepsilon = \varepsilon Z^d$, be the lattice cosurface with Gibbs distribution discussed in Sect.5 (lattice gauge fields for d=2). What happens for $\varepsilon \downarrow 0$? For simplicity we shall discuss here only the case G = U(1). Let $q_\beta(y) \equiv \exp(-\beta U(y))$, $y \in G$, U(y) being the real-valued function on G which gives the Gibbs interaction.
Let us set $\varepsilon_n \equiv 2^{-n}\varepsilon$. Let f be real-valued function on G^{2d} and let $F_1^\varepsilon, \ldots, F_{2d}^\varepsilon$ be the faces of an elementary cell γ of L_ε. We can look upon on $f(C(F_1^\varepsilon), \ldots, C(F_{2d}^\varepsilon))$ as a function of the $C(F_i^{j,\varepsilon_n})$, F_i^{j,ε_n} being faces of one of the elementary cells γ_{ε_n} into which γ is divided when we replace ε by ε_n.

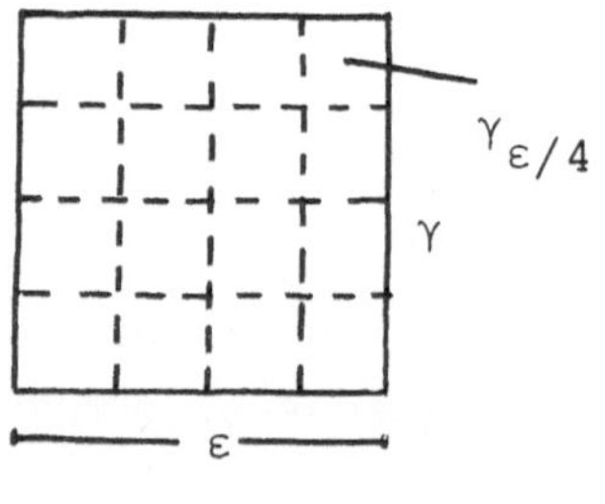

We then have for the expectation, by integrating with respect to the variables C(F), with F running over the interval faces of γ_{ε_n}: $Z_{\varepsilon_n} \equiv C(\beta)2^{dn}$,

$$E_{\mu_{\varepsilon_n}}(f(C(F_1^\varepsilon),\ldots,C(F_{2d}^\varepsilon)) = Z_{\varepsilon_n}^{-1} \int f\, q_s^{*\,2^{nd}} \prod_{F\in\partial\gamma_\varepsilon} dC(F)\,.\, q_\beta{*}\,k$$

means k-fold convolution of q with itself. By studying the r.h.s. one sees that it has a limit as $n \to \infty$, provided U is such that e.g. the Fourier transform $\hat{q}_\beta$ of q_β exists and one can choose $\beta(n)$s.t.

$$Z_{\varepsilon_n}^{-1} \quad \hat{q}_{\beta(n)}^{\;2^{nd}/\varepsilon^d}$$

converges pointwise as $n \to \infty$ to some function $\hat{h}$. Then letting Q_t be the Markov semigroup of densities on G defined by continuity from

$$\hat{Q}_{k/l} \equiv \hat{h}^{k/l},\ l \neq 0\ ,$$

$k,l \in N$, we get that the limit of the above right hand side is

$$\int f Q_{|\gamma_\varepsilon|}(C(\partial\gamma_\varepsilon)) \prod_{F\in\partial\gamma_\varepsilon} dC(F)\ ,$$

which is the expectation with respect to the probability space belonging to the Markov cosurface constructed from the Markov semigroup Q_t, according to Theor. 1 in Sect. 2. Corresponding results hold in the case where f is a function of C(F), with F running over different cells of L_ε. We arrive then at the following

Theorem 5: In the case $M = R^d$, the Markov cosurface $((\Omega,\mathcal{B},P);C^Q)$ of Theor. 1 in Sect. 2 can be constructed as limit of the lattice cosurfaces with Gibbs distributions of Sect. 5, provided Q can be obtained as limit for $n \to \infty$ in the above way.

Example : 1. $G = U(1) = \{e^{i\phi},\ \phi\in[0,2\pi]\}$: $U(e^{i\phi}) = \mathrm{Re}\ \chi(e^{i\phi}) =$

$$= \mathrm{Re}\ e^{i\phi} = \cos\phi\ ,\ \beta(n) = 2^{dn}.$$

In this case we get $Q_t = e^{\frac{t}{2}\Delta}$, with Δ the Laplace-Beltrami operator on G.
This is discussed in [12] (the discussion parallels the one for d = 2 given in [18], [21]).

2. $d = 2$, $G = SU(2)$, $U(g) = \sin t/(\sin t/2)$, $t \in [0,2\pi]$, with $e^{\pm it/2}$ the eigenvalues of $g \in SU(2)$, $\beta(n) = 2^{2n}$, in which case $Q_t = e^{\frac{t}{2}\Delta}$, with Δ the Laplace-Beltrami operator on G (cfr. [18], [21]).
3. $d = 2$, $G = Z^2 = \{\pm 1\}$, $U(g) = g$, $\beta(n) = \ln \sigma + n \ln 2$, in which case $Q_t(g) = ge^{-t/\sigma^2} + 1$.
4. For all $d \geq 2$, G abelian for $d > 2$, arbitrary for $d = 2$: Let Q_t be a convolution semigroup on G and define $q_\beta(C(\partial\gamma)) \equiv Q_{\beta|\gamma|}(C(\partial\gamma))$.
   In this case μ_{ε_n} converges weakly to the measure P constructed from Q according to Theor.,1 Sect. 2. This gives then, even for $d > 2$, examples of Markov cosurfaces which are limits of lattice cosurfaces with Gibbs distribution.

For d = 2 the above examples show in particular that Markov cosurfaces can be obtained as continuum limits of lattice gauge field models.
For the continuum limit of lattice gauge fields models for d = 3 see e.g. [5], [29] - [32]. For a non-standard analysis approach to the construction of the continuum limit see [8].

7. MARKOV SEMIGROUPS AND RELATIVISTIC QUANTUM FIELDS ASSOCIATED WITH MARKOV COSURFACES

We shall now see that in the case $M = R^d$, one can associate to the invariant Markov cosurfaces of Sect. 2.4 (also obtained in Sect. 6 as limit of lattice cosurfaces with Gibbs distribution) a Markov semigroup acting on a Hilbert space $\mathcal{H}$ of real (or complex)-valued functions. In fact, let $f^\pm$ be real-valued bounded continuous functions defined on $G^{K^{\pm}}$, with $K^{\pm} = \{S_1^{\pm},\ldots,S_{n^{\pm}}^{K^{\pm}}\}$ complexes and $S_i^{\pm} \subset M^{\pm}$, with $M^{+} \equiv \{(x^1,\ldots,x^d) \in R^d | x^1 > 0\}$, $M^{-} \equiv \{(x^1,\ldots,x^d) \in R^d | x^1 < 0\}$. From the Markov property of the Markov cosurface we have

$$E(f^{+}f^{-}|\Sigma(S_o)) = E(f^{+}|\Sigma(S_o))E(f^{-}|\Sigma(S_o)) ,$$

with $S_o \equiv \{(x^1,\ldots,x^d) \in R^d | x^1 = 0\}$. From this we get for $f^{-} \equiv f^{+} \circ T$, T the refelction with respect to S_o: $E(f^{+}f^{-}|\Sigma(S_o)) \geq 0$.
Let now H be the Hilbert space of (equivalence classes of) functions f^{+} with scalar product $(f^{+},g^{+})_H \equiv E(f^{+}g^{+} \circ T|\Sigma(S_o))$ modulo $(f^{+},f^{+})_H = 0$.

Define for $f \in H$, $T_t f(C(K)) \equiv f(C(K_t))$, where K_t is the complex K translated by the amount t in the direction x^1. Then T_t is a Markov semigroup on H. If $H \geq 0$ is its generator, then $T_t = e^{-tH}$, $t \geq 0$, and its analytic continuation e^{itH} is a unitary group on H. H has then the interpretation of an Hamiltonian. The invariance of the Markov cosurface $\{(\Omega,A,P),C\}$ under isometries of R^d means here translation and rotation invariance. One has also reflection invariance if the original semigroup Q was symmetric in the sense $Q_t(x) = Q_t(-x)$. The Osterwalder-Schrader positivity condition is here obviously satisfied, by the Markov property. Hence we are in the situation studied by Fröhlich, Osterwalder and Seiler ([4] and [22]) in their axiomatic framework (the growth condition holds as is easily seen; a strong continuity holds also, at least when Q_t is a diffusion semigroup on G). By those general results we then get a relativistic theory with Hamiltonian H, in which the relativistic quantum fields are associated with d-1-hypersurfaces, rather than points. This theory satisfied all postulates for a relativistic theory.

Remark: For d = 2 and G a compact group this theory coincides with continuum pure gauge field theory with gauge group G. For d > 2 it gives relativistic models in d space-time dimensions. The physics of these fields is not yet well studied (mass spectrum, e.g.?). We think it would be worthwhile to get more information about the properties of these fields, since they are in a natural sense extensions to higher dimensional parameter space of Brownian motion on a group (the "one-dimensional non-linear σ-model"). Mathematically the constructions sketched in this lecture yield the starting point for a stochastic calculus for d-1-stochastic forms and for an extension of multiplicative stochastic integrals to multiplicative stochastic integrals of currents. Finally it is hoped that they can add some useful stochastic analysis to the realm of stochastic geometry.

ACKNOWLEDGEMENTS

The first author is very grateful to the Organizing Committee of the XXIII. Internationale Universitätswochen für Kernphysik, Schladming, for a very kind invitation. The hospitality of Professor Mohammed Mebkhout and Ludwig Streit at the Centre de Physique Théorique, CNRS, Université d'Aix-Marseille II, and of the Centre for Interdisciplinary Research (ZiF), Universität Bielefeld, as well as of the Mathematics Institutes of the Universities at Bochum, Oslo, Roma and Trento is also gratefully acknowledged.

REFERENCES

1. S. Albeverio, R. Høegh-Krohn, Diffusions, quantum fields and groups of mappings, pp. 133-145, M. Fukushima, Edt., Functional Analysis in Markov Processes, Proc. Katata and Kyoto 1981, Lect. Notes Maths. 923, Springer (1982).
2. S. Albeverio, R. Høegh-Krohn, Diffusions, quantum fields and fields with values in Lie groups, to appear in

Stochastic Analysis and Applications, Ed. M. Pinsky, M. Dekker (1984).

3. S. Albeverio, R. Høegh-Krohn, Some Markov processes and Markov fields in quantum theory, group theory, hydrodynamics and C^*-algebras, pp. 497-540 in Stochastic Integrals, Proc. LMS Durham Symp., 1980, Edt. D. Williams, Lect. Notes Maths. 851, Springer, Berlin (1981).
4. E. Seiler, Gauge theories as a problem of constructive quantum field theory and statistical mechanics, Lect. Notes in Phys. 159, Springer-Verlag, Berlin (1982).
5. L. Gross, Convergence of the U(1)-lattice gauge theory to its continuum limit, Comm. Math. Phys. 92 (1983) 137-162.
6. S. Albeverio, R. Høegh-Krohn, Uniqueness and the global Markov property for Euclidean fields. The case of trigonometric interactions, Comm. Math. Phys. 68 (1979) 95-128.
7. R. Gielerak, Verification of the global Markov property in some class of strongly coupled exponential interactions, J. Math. Phys. 24 (1983) 347-355.
8. S. Albeverio, J.E. Fenstad, R. Høegh-Krohn, T. Lindstrøm, Nonstandard analysis methods in probability and mathematical physics, Acad. Press (1984).
9. B. Zegarlinski, Uniqueness and the global Markov property for Euclidean fields: the case of general interactions, Wroclav Preprint (1983).
10. J. Glimm, A. Jaffe, Quantum Physics, Springer-Verlag, New York (1981).
11. E. Nelson, Probability theory and Euclidean field theory, in "Constructive quantum field theory", Ed. G. Velo, A. Wightman, Lect. Notes in Phys., Springer Verlag (1973).
12. S. Albeverio, R. Høegh-Krohn, H. Holden, in preparation.
13. G.A. Hunt, Semi-groups of measures on Lie groups, Trans. Am. Math. Soc. 81 (1956) 264-319.
14. P. Feinsilver, Processes with independent increments on a Lie group, Tans. Am. Math. Soc. 242 (1978) 73-121.

15. a) K.R. Parthasarathy, Probability measures on metric spaces, Academic Press (1967).
b) C. Berg, G. Forst, Potential theory on locally compact abelian groups, Springer Verlag, Berlin (1975).
16. J. Bellissard, Lecture on lattice gauge theory, Lecture given at the Lisboa Autumn School in Theoretical Physics (1980) CPT-81/PE. 1327.
17. G.F. De Angelis, D. De Falco, F. Guerra, R. Marra, Gauge fields on a lattice (Selected Topics), Proc. Intern. Universitätswoche Kernphysik, Schladming 1978, Acta Phys. Austr. Suppl. XIX, 205 (1978).
18. H.G. Dorsch, V.F. Müller, Lattice gauge theory in two space-time dimensions, Fortschritte der Physik 27 (1979) 547-559.
19. B. Durhuus, J. Fröhlich, T. Jonsson, Self-avoiding and planar random surfaces on the lattice, Nordita Preprint (1983).
20. B. Durhuus, J. Fröhlich, T. Jonsson, Critical properties of a model of planar random surfaces, E.T.H. Preprint (1983).
21. R. Balian, J.M. Drouffe, C. Itzykson, Gauge fields on a lattice, Phys. Rev. D 10 (1974) 3376-3395.
22. J. Fröhlich, K. Osterwalder, E. Seiler, Annals of Math., and paper in preparation (quoted in [4]).
23. J. Fröhlich, Some results and comments on quantized gauge fields, in Recent Developments in Gauge Theory, Cargese Summer Inst., 1979, Edts. G. t'Hooft et al, Plenum (1980).
24. I.M. Gelfand, N.Y. Vilenkin, Generalized functions, Vol. 4, Acad. Press, New York (1964).
25. M.M. Rao, Local functionals and generalized random fields with independent values, Theory Prob. Appl. 16 (1971) 457-473 (transl.).
26. B. Simon, The $P(\phi)_2$ Euclidean (Quantum) Field Theory, Princeton Univ. Press, Princeton, N.J. (1974).
27. E. Nelson, Construction of quantum fields from Markov fields, J. Funct. Anal. 12 (1973) 97-112.
28. K. Osterwalder, R. Schrader, Axioms for Euclidean Green's

function I, II, Comm. Math. Phys. 31 (1973) 83-112; 42 (1975) 281-305.

29. M. Göpfert, G. Mack, Proof of confinement of static quarks in 3-dimensional U(1) lattice gauge theory for all values of the coupling constant, Comm. Math. Phys. 82 (1982) 545-606.
30. J. Challifour, A path space formula for non-abelian gauge theories, Ann. of Phys. 136 (1981) 317-339.
31. K. Ito, Letts. Math. Phys. 2 (1978) 351-365.
32. T. Baaban, Recent results in constructing gauge fields, Physica 124A (1984) 79-90.

Acta Physica Austriaca, Suppl. XXVI, 233–254 (1984)

NON-STANDARD ANALYSIS; POLYMER MODELS, QUANTUM FIELDS+

by

S. ALBEVERIO
Mathematisches Institut
Ruhr-Universität
D-4630 Bochum 1
(Fed. Rep. Germany)

ABSTRACT

We give an elementary introduction to non-standard analysis and its applications to the theory of stochastic processes. This is based on a joint book with J.E. Fenstad, R. Høegh-Krohn and T. Lindstrøm. In particular we give a discussion of an hyperfinite theory of Dirichlet forms with applications to the study of the Hamiltonian for a quantum mechanical particle in the potential created by a polymer. We also discuss new results on the existence of attractive polymer measures in dimension $d \leq 5$, with applications to the $(\phi_1^2\ \phi_2^2)_d$-model of interacting quantum fields.

+Lecture given at the XXIII. Internationale Universitätswochen für Kernphysik, Schladming, Austria, February 20-March 1, 1984.

1. WHAT IS NON STANDARD ANALYSIS?

Non standard analysis is an extension of analysis to a richer structure, based on a larger number system $^*R$ than R, containing infinitesimal and infinite members. It has of course its historical roots way back in the origins of the infinitesimal calculus (Archimedes, Cavalieri, Leibniz, Euler), its modern systematic mathematical version has its roots in work by Skolem (1934), Schmieden and Laugwitz (1958), and most particularly Robinson (1959). For the history of the subject see e.g. [1], [2], for introductory books and articles see e.g. [3] - [7]. We shall here try to give an idea of non standard analysis by using a particular model of it, built with suitable sequences of real numbers, made into equivalence classes in a suitable way. The suitable was is here the use of an ultrafilter U containing the filter of all cofinite subsets of natural numbers (an ultrafilter being a filter with the ultrafilter property, i.e. for any set $A \subset N$ one has either $A \in U$ or $N-A \in U$; filter F of confinite subsets of N means of course the family of subsets of N of the form $N -\{k_1,\dots,k_n\}$, for some $k_i \in N$, $n \in N$; the existence of at least an ultrafilter U containing F is assured by e.g. some version of the axiom of choice).

Since for any set $A \subset N$ we have either $A \in U$ or $N - A \in U$, it is convenient to introduce a finitely additive measure p_U on N by $p_U(A) = 1$ iff $A \in U$, $p_U(A) = 0$ iff $A \notin U$. Let R^N be the set of all real-valued sequences a_n, $n \in N$. We introduce in R^N $\sim$-equivalence classes by setting $a \sim b$ for $a,b \in R^N$ iff

$\{i \in N | a_i = b_i\} \in U$, i.e. if $a_i = b_i$ p_U - a.s. (i.e. for p_U-a.a.i).

Let $^*R \equiv R^N/U$ be the set of all equivalence classes modulo U. Equality in $^*R$ is thus defined by $a = b$ as elements in $^*R \leftrightarrow a_i = b_i$ p_U - a.s..

Similarly one defines the arithmetical operations a+b,

$a-b$, $a.b$, a/b for $b \neq 0$, and the order relation $a < b$ in ${}^*R$, in the natural way "modulo U", i.e. e.g. $a + b = c \leftrightarrow a_i + b_i = c_i$ p_U - a.s.,

$a < b \leftrightarrow a_i < b_i$ p_U - a.s..

Let us define, for any $r \in R$, ${}^*r$ as the sequence $r_i = r$ for all i, taken modulo U, so that ${}^*r \in {}^*R$. In this way R is imbedded in ${}^*R$, in fact r can be looked upon as the element ${}^*r = (r,r,\dots,r)/\mathrm{mod}\ U$ of ${}^*R$. Let ${}^*0 \equiv 0$, ${}^*1 \equiv 1$, then ${}^*R$ with the sum + and $\cdot$ defined is a field with zero 0 and unit 1. It is easy to show that ${}^*R$ contains infinite and infinitesimal elements. In fact the element of ${}^*R$ defined as $\omega = (1,2,\dots,n,\dots)/U$ is certainly strictly larger than any element $r \in R$ (identified with ${}^*r$), since $\omega \leq {}^*r$ means by definition $\omega_i \leq r$

p_U -a.s., i.e. $\{i \in N | \omega_i \leq r\} \in U$.

But the latter is a finite set, by the definition of the ω_i, and U cannot contain any finite subset of U, by definition. Thus we must have $\omega > {}^*r$ for all $r \in R$, which means ω is infinite positive. Using the arithmetics of ${}^*R$, we then also get negative infinite numbers and positive and negative infinitesimals (by definition either 0 or smaller in modulus than any positive real).

All numbers which are not infinite are called finite. Given a point $r \in {}^*R$ we define the monad $\mu(r)$ of r as the set of all numbers $s \in {}^*R$ which differ from r only by some infinitesimal, i.e. $\mu(r) = \{s \in {}^*R | \exists\, \varepsilon$ infinitesimal, $s = r + \varepsilon\}$. One writes for $r, s \in {}^*R$, $r \approx s$ iff $r-s$ is infinitesimal. For $s \in {}^*R$ finite one defines the <u>standard part</u> of s, denoted by sts or ${}^os$, as the unique number $r \in R$ s.t. $s-r$ is infinitesimal. I.e. ${}^os$ is the unique real number in $\mu(s)$. Of course when operating with infinite numbers in ${}^*R$ one should take care of the fact that they satisfy the relations of an ordered algebraic field, e.g. if $\omega > 0$ is infinite, $1/\omega$ and $1/\omega^2$ are infinitesimal and $1/\omega^2 < 1/\omega$, since $\omega \cdot 1/\omega^2 = 1/\omega \not\approx 0$. For many illustrations of the arithmetics in $^*R$ see e.g. [8], [9].

Of course to do analysis more than simply extending R to $^*R$ is needed, e.g. one needs to extend n-ary real-valued functions on R to $^*R$-valued n-ary functions on $^*R$. Again the extension is done by operating modulo the ultrafilter, thus if F is such an n-ary function, with arguments $s_1, \ldots, s_n \in R$, then the extension $^*F$ of F to $^*R \times \ldots \times {}^*R$ is defined s.t..

$$^*F(s_1, \ldots, s_n) = t \ , \quad t, s_i \in {}^*R \leftrightarrow \{i \in N \,|\, F(s_1(i), \ldots, s_n(i)) =$$

$$= t(i)\} \in U \ .$$

In particular we then have the natural extension of belonging to a set $E \subset R$ as the set $^*E$ defined by

$$^*E \equiv \{s \in {}^*R \,|\, \{i \in N \,|\, s(i) \in E\} \in U\}.$$

It is clear that not all subsets of $^*R$ are of the form $^*E$ for some $E \subset R$: e.g. $\{n \in {}^*N \,|\, n \geq \omega\}$, where ω is infinite, is not of the form $^*E$ (since if it were of this form, then we would have $\{i \in N \,|\, \nu(i) \in E\} \in U$ for exactly all sequences ν s.t. $\{i \in N \,|\, \nu(i) \geq \omega(i)\} \in U$. But $E \subseteq N$ has a least element, say n_0. Thus $\{i \in N \,|\, (n_0, n_0, \ldots, n_0, \ldots)(i) \in E\} \in U$, but $^*n_0 < \omega$).

The operation * on sets has interesting topological properties, e.g., for $E = (0,1]$, $^*E$ contains all strictly positive infinitesimals as well as the monads of all numbers α, $0 < \alpha < 1$, as well as 1 and all numbers $1 - \varepsilon$, $\varepsilon > 0$ infinitesimal.

* is a Boolean homomorphism of sets, but is not σ-additive. The <u>elementary transfer principle</u> says roughly speaking the following. All sentences of the 1.order language (allowing quantifiers on numbers, <u>not</u> sets) using variables for numbers or sets in R and constants out of R, the elementary symbols ($\wedge, \vee, \neg, \rightarrow$) and quantifiers ($\forall, \exists$) (acting only on number variables) valid in R, remain valid when one replaces the values A_i of the set variables X_i by ${}^*A_i$ and the values r_i of the free number variables by ${}^*r_i$. E.g. the true statement (of 1. order logic) $n \in N \rightarrow \quad k \in N,\ k > n$ goes by transfer over to the true statement $n \in {}^*N \rightarrow \quad k \in {}^*N$, $k > n$. For more details about this elementary transfer principle see e.g. [3], [5], [7].

In analysis one encounters of course more complicated statements involving bounded quantifiers also on set variables, not only on numerical variables. Moreover one needs spaces of functions, functions on spaces of functions etc., in short one needs what is called the superstructure

$$V(R) \equiv \cup V_n(R), \quad V_{n+1}(R) \equiv V_n(R) \cup \{X \mid X \subset V_n(R)\}, \quad V_o(R) \equiv R ,$$

over R. The transfer principle extends to the above statement in the case where the A_i stand for arbitrary elements on V(R). What are the sets ${}^*A_i$ then entering by transfer a valid statement over ${}^*R$? Their elements are by definition the so called <u>internal sets</u> (and sets which are not internal are called external). Roughly speaking, internal sets "behave nicely" under transfer. It is important to get some intuition about internal sets. The internal non void subsets of ${}^*N$ have a least element, any internal subset of ${}^*N$ containing N contains some infinite number, internal non void sets with an upper bound have a least upper bound. ${}^*R$, ${}^*N$ are internal, but R,N are external. Any internal subset of R is finite, etc. For discussion of internal sets see e.g. [3], [5], [7].

We have by now very shortly summarized the basic structures of non standard analysis. Some developments of non standard analysis, close to "classical" reasonings in "infinitesimal calculus" (before (but also after!) Weierstrass' ε-δ method) have been worked out very early, in fact already in Robinson's book (e.g. $\lim a_n = a$, $a, a_n \in R \leftrightarrow a_\omega \approx a$,

where ω is a positive integer, so that $a_1,\ldots,a_n,\ldots a_\omega$ extends beyond indices in $\mathbb{N}$ the given sequence a_n; similarly e.g. for the concepts of continuity, uniform continuity of functions, integral ...).
(Why is $f(x) = \frac{1}{x}$ continuous but not uniformly continuous on $(0,1)$? For $\omega \in {}^*\mathbb{N} - \mathbb{N}$ we have $f(1/\omega) = \omega \not\approx f(1/\omega^2) = \omega^2$, and uniform continuity means $f(x) \approx f(y)$ $\forall$ $x,y \in {}^*(0,1)$, $x \approx y$, whereas continuity at $y \in (0,1)$ means $f(x) \approx f(y)$ $\forall$ $x \in (0,1)^*$, $x \approx y$.)

For applications of non standard analysis to calculus and analysis, including differential equations, see e.g. [1] - [10]. Some of these applications are well known. We shall rather spend the rest of the time in this lecture discussing some relatively new applications in probability theory. For a long time non standard analysis was lacking a good tool for handling measure theoretical and probabilistic questions. The new tool was found by P. Loeb in 1975 [11]. The natural object to consider in non standard analytic probability theory is an internal probability space (X,A,ν), consisting of an internal set X, an algebra A of subsets of X, and a finitely additive measure ν, with values in ${}^*[0,1]$, with $\nu(X) = 1$.

Defining for any $A \in A$, ${}^{o}\nu(A) \equiv \mathrm{st}\,\nu(A)$, one gets a number in $[0,1]$, and clearly ${}^{o}\nu$ is an additive $[0,1]$-valued set function on A, with ${}^{o}\nu(X) = 1$.

The basic question which arises is the following: Does there exist an extension $L(\nu)$ of ν which is σ-additive on the σ-algebra $\sigma(A)$ generated by A? The answer is yes (since in our model of non standard analysis we have "saturation", in the sense that $\bigcap(A_n,\ n \in \mathbb{N}) \neq \emptyset$ if $A_n \downarrow$, A_n internal $\neq \emptyset$, $\forall n \in \mathbb{N}$, and this holds not only for finite measures ν). Moreover $L(\nu)$ is "well approximated" by ν in the sense that for any $A \in \sigma\,(A)$ there exists an internal $B \in A$ s.t. $L(\nu)\ (A \Delta B) = 0$, when Δ means symmetric difference.

$L(\nu)$ is called the Loeb measure associated with ν. It is a standard σ-additive measure living on a non standard space X with a σ-algebra of subsets.

In applications of standard analysis the so called hyperfinite sets X are very useful. These are by definition the internal sets E s.t. there exists an internal 1-1 map of a hyperfinite subset $\{n \in {}^*N \mid n \leq \omega$, some $\omega \in {}^*N\}$, for some $\omega \in {}^*N$, into E.

Hyperfinite sets behave in a sense as finite sets, but can be taken to be "sufficiently dense" to become substitutes of the standard continuum. E.g. the standard interval [0,1] can be well modellized by the hyperfinite set $T = \{0, \Delta t, 2\Delta t, \ldots, \eta\Delta t = 1\}$, where $\Delta t = 1/\eta$, η being some fixed infinite integer, i.e. $\eta \in {}^*N - N$, so that Δt is infinitesimal.

One shows that Lebesgue measure λ on [0,1] is well modellized by the Loeb measure L(P), with P the counting measure on X = T defined by $P(A) \equiv |A|/|T|$, where $|A|$ for any internal $A \subset T$ means the number of elements in A. One has $\lambda(B) = L(P)\,(st^{-1}(B))$, i.e. the Lebegue measure of a subset $B \subset [0,1]$ is the Loeb measure of the subset of T whose standard partis B. Similarly Lebesgue measurable functions f on [0,1] have a corresponding object in internal functions F on T which are s.t. st F(t) = f(stt) (such F is called a lifting of f).
One has

$$\int_0^1 f dx = st\int {}^{o}F\, dL(P) \approx \int F\, dP = \sum_{t \in T} F(t) \frac{1}{|T|} ,$$

if F is "S-integrable" in the sense that $\int F dP$ is finite and $\int_A F dP \approx 0$ if $P(A) \approx 0$, thus the Lebesgue intgral is reduced to an hyperfinite sum ("fine discretization"). In [12] Anderson used Loeb measures to give a nice direct realization of Brownian motion as a hyperfinite random walk. Let B the hyperfinite random walk on T, defined by

$$B(\omega,t) \equiv \sum_{0}^{t-\Delta t} \omega(s)\sqrt{\Delta t} \quad , \qquad \omega \in \Omega \equiv \{-1,+1\}^T ,$$

where the values 1, -1 of $\omega(s)$, $s \in T$, are given equal probability. One proves that $B(\omega,t)$ has a standard part. Defining $b(\omega,{}^{o}t) = \text{st } B(\omega,t)$ one gets then a standard Brownian motion with "unusual" underlying probability space $\Omega = \{-1,1\}^T$, σ-algebra $L(A)$ (A internal subsets of T), and with "Wiener measure" $L(P)$, P being the counting measure on T. (This is a direct realization of Donsker approximation of Brownian motion by random walks.)

This characterization of Brownian motion (and other diffusion processes) has found many applications, e.g. in the study of stochastic differential equations and stochastic integrals (see e.g. [6], [7], [13], [14] and references therein).

A very nice application to the study of local times of Brownian motion has been obtained by E. Perkins [15]. The (standard) local time of Brownian motion is defined as the density $\ell(t,x)$ with respect to Lebesgue measure of

$$\int_0^t \delta(x-b(s))ds \quad .$$

Perkins has shown that $\ell(T,x) = {}^{o}L(t,x)$, with

$$L(t,x) \equiv \sum_{s<t} \chi_{\{x\}}(B(s))\sqrt{\Delta t} \ .$$

Related results in the higher dimensional case are contained in recent work by Stoll [16]. We also point out that above characterization of local times has permitted to E. Perkins to solve in particular the old problem of proving that Levy's formula for the local times

$$2\sqrt{\frac{2}{\pi}}\,\Omega(t,x) = \lim_{\delta \downarrow 0} \delta^{1/2}\, n(t,x,\delta) \ ,$$

for all $t \in [0,\infty)$ holds as uniform limit in x, a.s., where $n(t,x,\delta)$ is the number of excursions of the Brownian motion away from x that are greater than δ in length and are completed by time t.

Other recent applications of hyperfinite models are in the study of random fields [50], [51]. Another nice application of hyperfinite methods is in the domain of Dirichlet forms. In fact T. Lindstrøm has worked out a theory of hyperfinite Dirichlet forms, see [7]. Let us give a brief idea of what this is about. Let H be an internal real linear space, with dimension $n \in N$. A hyperfinite quadratic form on H is by definition a symmetric, non negative, bilinear form $\mathcal{E}(u,v)$ on H, $u,v \in H$.
One has $\mathcal{E}(u,v) = (u,Av)$, with A linear symmetric nonnegative. $\mathcal{E}$ defines, through A, a semigroup $Q^t \equiv (1-A\Delta t)^k$, $t \equiv k\Delta t$, $k \in {}^*N$, Δt infinitesimal with $0 < \Delta t \leq 1/||A||$. A basic result [7] says that every standard closed, densely defined quadratic form E on a (standard) Hilbert space K can be obtained from a hyperfinite one $\mathcal{E}$ on $H \subset {}^*K$ by taking the standard part of $\mathcal{E}$ restricted to K. Also the semigroup associated with $\mathcal{E}$ gives by taking standard parts the one associated with E.
In the standard theory of (symmetric) Markov processes the theory of special quadratic forms, called "Dirichlet forms", plays an important role, see [17]. In view of the applications of this theory to quantum theory, see e.g. [18] - [22], let us spend a few words on this subset. A standard Dirichlet form is a closed densily defined quadratic form E on $K \equiv L^2(X,m)$, with X Hausdorff and m a Random measure, with the contraction property $E(u^+, u^+) \leq E(u,u)$, $u^+ \equiv (u \vee o) \wedge 1$. Dirichlet forms are in 1-1 correspondence with symmetric Markov semigroups $T_t = e^{-tA}$ on $L^2(X,m)$, with $E(u,v) = (A^{1/2}u, A^{1/2}v)$, (,) being the scalar product in K.

Hyperfinite Dirichlet forms are defined correspondingly with X the hyperfinite space $\{s_1,\dots,s_N\}$, $N \in {}^*N$, $s_i \in R$, and K al internal maps $u: X \to {}^*R$ with scalar product $(u,v) \equiv \sum u(s_i)v(s_i)m(s_i)$, some $m \in ({}^*R)^{{}^*N}$. A hyperfinite quadratic form $\mathcal{E}$ on K is a Dirichlet form iff it is a discrete version of $\int \partial_i u \partial_j u \, dv_{ij} + \int u^2 k(dx)$; in fact

$$\mathcal{E}(u,u) = \frac{1}{\Delta t} \sum_1^N [u(i)-u(j)]^2 q_{ij} m(i) + \sum_1^N u(i)u(i) q_{io} m(i),$$

with q_{ij} $(N+1)\times(N+1)$ stochastic matrix.
Using this characterization, a potential theoretic and probabilistic theory of hyperfinite Dirichlet forms can be developed, correspondingly to the one of standard Dirichlet forms. In particular there is a nice control on the associated Markov processes and stochastic calculus. Natural conditions are known which guarantee that the standard part of the associated Markov process (chain!) is a strong Markov process. One interesting feature, that should be further exploited in the future,is that there makes no difference, in this framework, whether the state space of the process is finite or infinite dimensional, for the standard theory of the latter, which is rather different from the finite dimensional case, see [19], [20], [23] .

We shall see below how this theory comes in as a natural tool in the study of a problem of great interest in such disparate fields like quantum mechanics, polymer theory, and in the study of quantum fields. The starting point is the study of perturbations of Markov generators by potentials V with support on zero measure sets, e.g. $-\Delta+V$, with V of support with Lebesgue measure zero, two typical examples being :

a) $V(x) = -\sum_i \lambda_i \, \delta(x-x_i)$: this model (Schrödinger operator, with point interactions) has been studied quite extensively in recent years in connection with the study of the low energy limit of quantum mechanics, in solid state models, and for models of nuclear and atomic physics, see e.g. the surveys [24] - [25]. (The case of N centers x_i was first studied by methods of non standard analysis [26].)

b) $V(x) = -\lambda \int_0^t \delta(x-b(s))ds$, with b(s) a Brownian motion

(Schrödinger operator for a particle moving in a field of sources sitting along a path of Brownian motion, which can be looked upon as a "polymer").

On this model there has been recent work by Ph. Blanchard, J.E. Fenstad, R. Høegh-Krohn, W. Karwowski, S. Kusuoka, T. Lindstrøm and myself, see [7], [27]. The basic question is: when does

$$E'(u,v) \equiv E(u,v) - \int_C \lambda uv d\rho$$

gives a s.a. operator $H \neq H_o$ in $L^2(X,m)$, with H_o the self-adjoint operator associated with the known form E(e.g. $H_o = -\Delta$) and C a set of Lebesgue measure zero, ρ some probability measure concentrated on C?

The basic technique of non standard analysis which can be used here, cf. [7], is to find hyperfinite representations $\mathcal{E}'$, $\mathcal{E}$ on $L^2(Y,\mu)$, Y,μ hyperfinite s.t.

$$\mathcal{E}'(u,v) = (u,v) - \int_B \lambda_Y \, uv \, d\rho_Y$$

with $B \subset Y$ internal s.t. $stB = C$, λ_Y, ρ_Y internal function resp. probability measure on B s.t. $\rho = L(\rho_Y) \circ st^{-1}$.

One observes that the generator of $\mathcal{E}'$ is simply given by $(L'u)(i) = Lu(i)-\lambda(i)u(i)\frac{\rho(i)}{\mu(i)}$ with L the generator of $\mathcal{E}$.

The non-triviality of the perturbation $(H \neq H_o)$ amounts to s.t. $\mathcal{E}' = E' \neq$ st $\mathcal{E} = E$. There is a special intest in cases where the function λ <u>cannot</u> be taken to be standard, hence $\int_C \lambda uvd\rho$ has no standard meaning, i.e. the perturbation is genuinely nonstandard.

We shall now shortly sketch the technique used to handle these perturbation theorems. One computes $(L'-z)^{-1} =$ $= G_z(1-\lambda \frac{\rho}{\mu}G_z)^{-1} = G_z\sum_\ell (\lambda\frac{\rho}{\mu}G_z)^\ell$, with $G_z \equiv (L-z)^{-1}$, which is well defined for all $z < z_o$, for some $z_o \in {}^*R_+$ (z_o might be infinite). But the r.h.s. is also equal to $G_z + \hat{G}_z{}^*(\frac{1}{\lambda} - G_z')^{-1}\hat{G}_z$, with $\hat{G}_z$, $\hat{G}_z^*$, G_z' the operators given by the kernel of G_z as an operator $L^2(Y) \to L^2(B)$, resp.

$L^2(B) \to L^2(Y)$, resp. $L^2(B) \to L^2(B)$. One then tries to show that one can choose λ s.t.

1) st $(\frac{1}{\lambda} - G'_z)^{-1}$ exists,

2) $H \neq H_o$.

It turns out ([7], [27]) that this is possible if there is a $z_o \in {}^*R$ s.t.

$$\int G_{z_o} G_{z_o}(\cdot,y)d\rho(y) \in L^2(R^d,d\rho).$$

In this case H is a family parametrized by internal functions λ_Y s.t.

$$\frac{1}{\lambda_Y(x)} - \int G_{z_o}(x,y)d\rho(y)$$

has a standard part.

Also one gets criteria for when λ can be chosen bounded, standard. An application of this method to the problem a) with $V(x) = -\lambda\delta(x)$, $x \in R^3$, yields back the result that had been discussed before by other techniques of nonstandard analysis (using smooth non standard realization of the δ-function) in [26], [7], [30], [31] to the effect that $-\Delta-\lambda_\varepsilon(\alpha)\varepsilon^{-3}\chi_1(|x|/\varepsilon)$, with ε a positive infinitesimal and χ_1 the characteristic function of the unit sphere centered at the origin, has a standard part H (heuristically of the form $-\Delta-\lambda\delta(x)$), different from $-\Delta$, provided

$$\lambda_\varepsilon(\alpha) = (k + \tfrac{1}{2}) \frac{\pi^2}{2} \varepsilon + 8\pi\alpha\varepsilon^2 + \gamma\varepsilon^3,$$

with $\alpha,\gamma \in R$, $k \in N$. H is independent of k,γ, but depends on α. It turns out that the operator defined by setting $\alpha = \infty$ in the resolvent kernel of H is H_o, for all $\alpha \in R$, however $H \neq H_o$ (non-triviality of the perturbation). One should also note that any other choice of $\lambda_\varepsilon(\alpha)$ leads either to the non-existence of H (as standard part) or to its triviality. Then in short λ has to be chosen infinitesimal positive (= $\lambda_\varepsilon(\alpha)$) in order to get a non trivial perturbation of $-\Delta$ supported by the point 0, in 3 dimensions.

Extensions of these results to the case of infinitely many discrete sources, i.e. to $V(x) = -\sum_i \lambda_i \delta(x-x_i)$, are discussed in [26], [7].

Let us now shortly describe some applications of the method to the study of perturbations of $-\Delta$ (or more general Markov generators) by local time functionals of the form $H_\omega = -\Delta - \int_0^t \lambda_\omega(x)\delta(x-b(s)(\omega))ds$, in $L^2(R^d,dx)$. Physically this is the Hamiltonian for a quantum mechanical particle moving under the influence of a polymer modellized by a path of Brownian motion (developing from time 0 to t). Such models have been discussed in the physical literature, see e.g. [32]. Of course as a model of polymers, the physical dimensions are $d \leq 3$. Mathematically, we have here the situation described above with $C = C_\omega \equiv \{b,$

$$s \in [0,t]\}, \quad \rho_\omega(A) \equiv \int_A \delta(x-b(t)(\omega))dx \, .$$

The above method works for $d \leq 5$ and yields the following <u>Theorem[7]</u>: There are choices of λ_ω s.t. for Wiener a.a.ω, H_ω exists and is non trivial ($\neq -\Delta$) for $d \leq 5$. The support of the perturbation is the Brownian path C_ω. For $d \leq 3$, $\lambda_\omega(x)$ can be chosen as any standard, bounded function (in particular it can be chosen independent of ω). For $d = 4,5$, $\lambda_\omega(x)$ has to be chosen as a positive function taking only infinitesimal values.

Proof: The proof uses the above mentioned general hyperfinite technique, a basic estimate being

$$E[\,||\int G_z G_z(\cdot,y)d\rho_\omega(y)||_{L^2(\rho_\omega)}] = E\{\int_0^t\int_0^t\int_0^t \frac{e^{-z|b(s_3)-b(s_2)|}}{|b(s_3)-b(s_2)|^{d-4}} \cdot$$

$$e^{-z|b(s_3)-b(s_1)|}\,|b(s_3)-b(s_1)|^{-(d-4)}ds_1 ds_2 ds_3\} < \infty \text{ for } d \leq 5 \, .$$

The basic estimate for the possible choice of λ independent of ω for $d \leq 3$ is to control $X_z(t) \equiv \int G_z(b(t),y)\,d\rho_\omega(y)$ as $z \to -\infty$, uniformly in t, through Proklorov-Kolmogorov theorem and the estimate

$$E[|X_z(t) - X_z(s)|^3] \leq c_\delta |t-s|^{3/2-\delta}$$

for some $\delta > 0$.

Remark: It would be very intersting to have results on the spectrum of H, besides the lower bound. In particular, the well known questions of localizations (point spectrum versus absolutely continuous spectrum), discussed in other models (e.g. [33], [34]), could be studied in this model.

Besides having an intrinsic value as a quantum mechanical model, we shall now show how above model has applications to the study of classical polymer models of the type studied by Edwards, see e.g. [35], [37]. These models involve measures of the heuristic form

$$Z^{-1}\, e^{+\lambda(b,\tilde{b})\int_0^t\int_0^t \delta(b(s)-\tilde{b}(\tilde{s}))\,ds\,d\tilde{s}}\,dP\,d\tilde{P} \;,$$

where b, $\tilde{b}$ are two independent Brownian motions in R^d, with laws P resp. $\tilde{P}$, with Z a normalization constant.
Such a measure describes the interaction between b, $\tilde{b}$ obtained by penalizing ($\lambda>0$) resp. favorizing ($\lambda<0$) crossings ($b(s) = \tilde{b}(\tilde{s})$) of $b,\tilde{b}$. If we look upon b as a realization of a polymer, then the model describes two species of polymers in interaction (such models have been introduced in the description of rubber, e.g.[35]).
It turns out that we can give a meaning to such measures for $d \leq 5$.
Theorem[7]:For $d \leq 5$ there are choices of $\lambda(b,\tilde{b})$ s.t. the polymer measure

$$Z^{-1}\exp\{\lambda(b,\tilde{b})\int_0^t\int_0^t \delta(b(s)-\tilde{b}(\tilde{s}))\,ds\,d\tilde{s}\}\,dP\,d\tilde{P}$$

exists and is non trivial (i.e. different for $dPd\tilde{P}$). For $d = 4,5$ one has to choose λ positive infinitesimal. In this case λ can be chosen independent of b or $\tilde{b}$. For $d \le 3$ λ has to be chosen real valued and can be chosen independent of both b and $\tilde{b}$ [27].

Proof: The proof uses a hyperfinite realization of $T_\omega^t \equiv$ $\equiv e^{-tH_\omega}$, $t \ge 0$, with $H_\omega = -\Delta-\lambda_\omega\int_0^t \delta(x-b(s)(\omega))ds$, together with the hyperfinite Feynman-Kac formula

$$(u,T_\omega^t v)_{L^2(dx)} \approx E_{\tilde{P}}[\tilde{u}(B(t))\tilde{v}(\tilde{B}(0))\exp\{\sum_{s=0}^{t}\sum_{r=0}^{t}(\Delta t)^2$$

$$\cdot\, \lambda_\omega(\tilde{B}(s))\delta(B(r)-\tilde{B}(s))\}]$$

where $\tilde{u},\tilde{v}$ are liftings of u,v and $B,\tilde{B}$ are two independent Anderson hyperfinite random walks, with laws P, $\tilde{P}$, s.t. $st\,B = b$, $st\,\tilde{B} = \tilde{b}$ (in the sense explained before). δ is a hyperfinite realization of the δ-function so that $\exp\{\ \}$ is a hyperfinite realization of

$$\exp[\lambda\int_0^t\int_0^t \delta(b(s)-\tilde{b}(s'))ds\, ds'] .$$

Remark. It is an open question whether λ can be chosen independent of both b and $\tilde{b}$ for $d = 4,5$. Some related questions have been studied by Kusuoka [36].

Remark. The above seems to be the first mathematical result about existence of non trivial polymer measures for $d = 4,5$. We should note that the result needs λ positive infinitesimal, hence the model describes attracting polymers. For $d \le 3$ and λ negative, the above polymer measures constitute a basic approximation to the "self-avoiding" polymer measure

$$Z^{-1}\exp\{\lambda\int_0^t\int_0^t\delta(b(s)-b(s'))dsds'\}dPd\tilde{P}$$

studied by Westwater [37], [52] (see also [38], [39], [16],[49]).

We conjecture that also this one-Brownian motion polymer measure exists for d = 4 and is non trivial for λ infinitesimal positive.

We close our exposition with a remark concerning a model of quantum fields in space time dimension d of the ϕ^4_d-type, namely the model $(\phi_1^2\ \phi_2^2)_d$, with ϕ_1, ϕ_2 independent scalar fields over R^d.

By using an hyperfinite realization of Symanzik's [40] representation of this model in terms of a polymer gas we get an expression of the Schwinger function of this model, with classical Euclidean action

$$\sum_{i=1}^{2} \int (\nabla\phi_i)^2 dx + m_i^2 \int \phi_i^2 dx - \lambda \int \phi_1^2\ \phi_2^2 dx ,$$

in terms of measures of the above type

$$\exp\{\lambda \int_0^t \int_0^t \delta(b(s)-\tilde{b}(s'))ds ds'\} dP d\tilde{P} .$$

By the above result for the physical space-time dimension d = 4 we get existence for λ positive infinitesimal. This corresponds to an "attractive" ϕ^4_4-model (with the opposite sign of the coupling constant as usually taken).
We pointed out already in [24], [28], [29] the possibility of a non trivial ϕ^4_4-model in this sense. Recently similar possibilities have been discussed from other points of view, e.g. [41], [42]. The above result for the "zero components $(\phi_1^2\ \phi_2^2)_d$-model" might be looked upon as a sign for a possible way out of the inpasse of triviality results for scalar theories (with $\lambda > 0$) defined as limits from lattice models ([43], [48]).

Remark. In this lecture we have only discussed applications of non standard analysis to a few problems in the area of stochastic analysis, polymer models and quantum fields. Even within these areas there are many other recent applications on non standard methods we did not touch , e.g.

hyperfinite versions of scalar and gauge fields, see [7] (for an early paper in the area of quantum fields see [58]). In the related area of statistical mechanics recent applications of non standard analysis have been to the problem of the description of Gibbs states and of the global Markov property of lattice Gibbs fields in statistical mechanics [62] and to the study of the Boltzmann equation [59]. For these and many other applications of non standard methods we refer to [7].

ACKNOWLEDGEMENTS

It is a special pleasure to thank the organizers for their kind invitation and for making so pleasant our stay at the school. I am very happy to thank Jens-Erik Fenstad, Raphael Høegh-Krohn and Tom Lindstrøm for the joy of collaboration on our joint project of a book on non standard methods, on which these lectures are based.
I am also grateful to Witold Karwowski, Shigeo Kusuoka and John Westwater for the collaboration and very useful discussions, and to Christoph Kessler and Andreas Stoll for reading through the paper and for stimulating discussions. Several stays at the Centre de Physique Théorique and the Université d'Aix-Marseille II, Luminy, at ZiF, University of Bielefeld, and at the Mathematics Institute, University of Oslo, are gratefully acknowleged.

REFERENCES

1. A. Robinson, Non-standard analysis, North-Holland, Amsterdam (1970).
2. D. Laugwitz, Infinitesimalkalkül, Bibl. Institut, Mannheim (1978).
3. M. Davis, Applied non standard analysis, Wiley, New York (1977).

4. K.D. Stroyan, W.A.J. Luxemburg, Introduction to the theory of infinitesimals, Academic Press, New York (1976).
5. H.J. Keisler, Foundations of infinitesimal calculus, Prindle, Weber & Schmidt, Boston (1976).
6. N. Cutland, Non standard measure theory and its applications, Bull. London Math. Soc. 15 (1983) 525-589.
7. S. Albeverio, J.E. Fenstad, R. Høegh-Krohn, T. Lindstrøm, Non standard methods in stochastic analysis and mathematical physics, Acad. Press.
8. H.J. Keisler, Elementary Calculus, Prindle, Weber & Schmidt, Boston (1976).
9. J.M. Henle, E.M. Kleinberg, Infinitesimal Calculus, MIT, Cambridge (1979).
10. R. Lutz, M. Goze, Non-standard analysis: a practical guide with applications, Lect. Notes Maths. 881 (1981) Springer.
11. P. Loeb, Conversion from nonstandard to standard measure spaces and applications in probability theory. Trans. Amer. Math. Soc. 211, (1975) 113-122.
12. R.M. Anderson, A non standard representation for Brownian motion and Ito integration, Isr. J. Math. 25 (1976) 15-46.
13. J. Keisler, An infinitesimal approach to stochastic analysis, Mem. Am. Math. Soc.
14. T. Lindstrøm, Hyperfinite stochastic integration I-III, Math. Scand. 46 (1980).
15. E. Perkins, A global intrinsic characterization of Brownian local time, Ann. of Prob. 9 (1981) 800-817.
16. A. Stoll, A self-repelling random walk and polymer measures in two dimension, in preparation.
17. M. Fukushima, Dirichlet forms and Markov processes, North-Holland (1980).
18. S. Albeverio, R. Høegh-Krohn, L. Streit, Energy forms, Hamiltonians, and distorted Brownian paths, J. Math. Phys. 18 (1977) 907-917.
19. S. Albeverio, R. Høegh-Krohn, Dirichlet forms and diffusion processes on rigged Hilbert spaces, Z. Wahrscheinlichk. verwe. Geb. 40 (1977) 1-57.

20. S. Kusuoka, Dirichlet forms and diffusion processes on Banach spaces, J. Fac. Sci. Univ. Tokyo 1A 29 (1982) 79-95.

21. Contribution by S. Albeverio, R. Høegh-Krohn, M. Fukushima, L. Streit in New Stochastic Methods in Physics, Ed. C. De Witt-Morette, K.D. Elworthy, Phys. Repts. 77 (1982) 121-382.

22. S. Albeverio, R. Høegh-Krohn, Hunt processes and analytic potential theory on rigged Hilbert spaces, Ann. Inst. H. Poincaré B13 (1977) 269-291.

23. S. Albeverio, R. Høegh-Krohn, Diffusion fields, quantum fields, fields with values in Lie groups, to appear in Adv. in Prob., Stochastic analysis and application, Ed. M. Pinsky, Dekker (1984).

24. S. Albeverio, R. Høegh-Krohn, Schrödinger operators with point interactions and short range expansions, pp.11-27,in Proc. VII Int. Congress Math. Physics, Boulder, Eds. W.E. Brittin, K.E. Gustavson, W. Wyss, North Holland (1984).

25. S. Albeverio, F. Gesztesy, R. Høegh-Krohn, H. Holden, Some exactly solvable models in quantum mechanics and the low energy expansions, to appear in Proc. Leipzig Conf. Operator Algebras, 1983, Teubner 1984,and book in preparation.

26. S. Albeverio, J.E. Fenstad, R. Høegh-Krohn, Singular perturbations and non standard analysis, Trans. Am. Math. Soc. 252 (1979) 275-295.

27. S. Albeverio, J.E. Fenstad, R. Høegh-Krohn, W. Karwowski, T. Lindstrøm, Perturbation of the Laplacian supported by null sets, with applications to polymer measures and quantum fields, Bochum Preprint (1984).

28. S. Albeverio, Ph. Blanchard, R. Høegh-Krohn, Some applications of functional integration, Proc. Int. AMP. Conf., Berlin, 1981, Ed. R. Schrader, R. Seiler, D.A. Uhlenbrock, Lect. Notes Phys. 153 (1982) Springer,Berlin.

29. S. Albeverio, Ph. Blanchard, R. Høegh-Krohn, Newtonian diffusions and planets, with a remark on non-standard Dirichlet forms and polymers, to appear in Proc. LMS,

Symposium Stoch. Analysis, Swansea, 1983, Ed. A. Truman, D. Williams, Lect. Notes, Maths..

30. E. Nelson, Internal set theory: a new approach to non standard analysis, Bull. Am. Math. Soc. 83 (1977) 1165-1198.
31. Alonso y Coria, Skrinking potentials in the Schrödinger equation, Ph. D. Thesis, Princeton Univ. (1978).
32. S. Edwards, Y.B. Gulyaev, Proc. Phys. Soc. 83 (1964) 495.
33. Kirsch, Martinelli, On the spectrum of Schrödinger operators with a random potential, Comm. Math. Phys. 85 (1982) 329-350.
34. H. Holden, F. Martinelli, On absence of diffusion near the bottom of spectrum for a random Schrödinger operator on $L^2(R^\nu)$, Bochum Preprint 1983, to appear in Comm. Math. Phys.
35. S.F. Edwards, The theory of polymer solutions at intermediate concentration, Proc. Phys. Soc. 88 (1966) 265-280; Functional problems in the theory of polymers, pp. 53-59, in Functional Integration, Ed. A.M. Arthurs, Clarendon Press, Oxford (1975); The mathematics of a rubber band, New Scientist, Febr. 1981, pp. 480-482.
36. S. Kusuoka, unpublished. See [24], [29].
37. J. Westwater, On Edward's model for long polymer chains I, Comm. Math. Phys. 72 (1980) 131-174.
38. J. Rosen, A local time approach to the self-intersections of Brownian paths in space, Comm. Math. Phys. 88 (1983) 327-338.
39. S. Kusuoka, Asymptotics of polymer measures in one dimension; and On the path property of Edward's model for long polymer chains in three dimensions, to appear in Proc. Bielefeld Conf. Infinite dimensional analysis and stochastic processes., Ed. S. Albeverio.
40. K. Symanzik, Euclidean quantum field theory, in Local Quantum Theory, Ed. R. Jost, Academic Press, New York (1969).
41. G. Gallavotti, V. Rivasseau, ϕ^4 field theory in dimension 4: a modern approach to its unsolved problems, to appear;

A comment on ϕ_4^4 Euclidean field theory, Phys. Lett. 122B (1983) 268-270.

42. P.M. Stevenson, On the physics of ϕ^4 in 3 + 1 dimensions, Madison Preprint, 1983.
43. S. Albeverio, G. Gallavotti, R. Høegh-Krohn, The exponential interaction in R^n, Phys. Letts. 83B (1979) 177; Some results for the exponential interaction in two or more dimensions, Comm. Math. Phys. 70 (1979) 187-192.
44. M. Aizeman, Proof of the triviality of ϕ_d^4 field theory and some mean field features of Ising models for d> 4, Phys. Rev. Letts. 47 (1981) 1-4; Geometric analysis of ϕ_d^4 fields and Ising models, Comm. Math. Phys. 86 (1982) 1-48.
45. J. Fröhlich, On the triviality of $\lambda\phi_d^4$ theories and the approach to the critical point in $d \underset{(-)}{>} 4$ dimensions, Nucl. Phys. B 200 FS4 (1982) 281-296, and Lecture at this school.
46. T. Hattori, A generalization of the proof of triviality of scalar field theories, Tokyo Preprint (1983).
47. Y. Matsubara, T. Suzuki, I. Yotsuyanagi, On a possible inconsistency of the ϕ^4 and the Yukawa theories in four dimensions, Kanazawa Prepr. (1983).
48. K.R. Ito, Trajectories of the ϕ_4^4-model by the block spin transformations, ZiF-Preprint (1984).
49. A. Bover, G. Felder, J. Fröhlich, On the critical properties of the Edwards and the self-avoiding walk model of polymer chains, ETH Preprint (1983).
50. A. Stoll, A non standard construction of Levy Brownian motion, Ruhr-University Bochum, Preprint (1984).
51. C. Kessler, Hyperfinite representation of generalized random fields, Bochum Preprint (1984).
52. J. Westwater, On Edward's model for polymer chains, to appear in Proc. Bielefeld Encounters Math. and Phys. IV, Edts. S. Albeverio, Ph. Blanchard, World Scient. Publ. (1984).
53. G.F. Lawler, A self-avoiding random walk, Duke Math. J. 47 (1980) 655-692.

54. A.D. Sokal, An alternate constructive approach to the ϕ_3^4 quantum field theory, and a possible destructive approach to ϕ_4^4, Ann. I.H. Poincaré A 37 (1982) 317.
55. R.A. Brandt, Asymptotically free ϕ^4 theory, Phys. Rev. D14 (1976) 3381-3394.
56. D. Brydges, T. Spencer, Self-avoiding walk in 5 or more dimensions, Viriginia Courant Preprint (1984).
57. G. Felder, J. Fröhlich, Intersection properties of simple random walks: a renormalization group approach, ETH-Preprint (1984).
58. Ph. Blanchard, J. Tarski, Renormalizable interactions in two dimensions and sharp-time fields, Acta Phys. Austr. 49 (1978) 129-152.
59. L. Arkeryd, Asymptotic behaviour of the Boltzmann equation with infinite range forces, Comm. Math. Phys. 86 (1982) 475-484.
60. J.E. Fenstad, Is non standard analysis relevant for the philosophy of mathematics? Synthese (1984).
61. Ch. Kessler, Non standard treatment of the global Markov property of lattice fields, in preparation.
62. D.N. Hoover, H.J. Perkins, Non standard construction of the stochastic integral and applications to stochastic differential equations, I, II, Trans. Amer. Math. Soc. 275 (1983) 1-58.
63. M.M. Richter, Ideale Punkte, Monaden und Nichtstandard-Methoden , Friedr. Vieweg & Sohn, Braunschweig/Wiesbaden (1982).
64. J. Rosen, A local time approach to the self-intersections of Brownian paths in space, Commun. Math. Phys. 88 (1983) 327-338.
65. K. Gawedzki, A. Kupiainen, Triviality of ϕ_4^4 and all that in a hierarchical model approximation, IHES Preprint (1982).
66. J.E. Fenstad, Non standard methods in stochastic methods in stochastic analysis and mathematical physics, Jber. d. Dt. Math.-Verein 82 (1980) 167-180.
67. K. Symanzik, Lett. Nuovo Cim. 6 (1973) 77.

Acta Physica Austriaca, Suppl. XXVI, 255–257 (1984)

STATISTICAL MECHANICS OF RANDOM SURFACES+

by

J. FRÖHLICH
Theoretical Physics
ETH - Hönggerberg
CH-8093 Zürich

SUMMARY

The statistical mechanics of random surfaces has proven to be of growing importance in quantum field theory (string theory, random surface representations of gauge theory) and condensed matter physics (domain walls and interfaces, incommensurate phases, spin glasses, crystal surfaces, surface phenomena such as wetting, etc.). During the past three years, the author has been actively involved in developping a suitable mathematical framework for a statistical mechanics of random surfaces. Various applications to quantum field theory and condensed matter physics have been outlined. This work has been carried out in collaboration with M. Aizenman, D. Brydges, J. and L. Chayes, B. Durhuus, T. Jonsson, and T. Spencer. Research in various directions of random surface physics is presently

+Lecture given at the XXIII. Internationale Universitätswochen für Kernphysik,Schladming,Austria,February 20-March 1,1984

continuing. Several original papers and reviews have appeared in print. Therefore, and because of other obligations, the author has found himself unable to write a detailed account of his lectures. May the following references be useful.

1. Random surface representations of lattice gauge theories

B. Durhuus and J. Fröhlich, Commun. Math. Phys. 75 (1980) 103.
J. Fröhlich, Physics Reports 67 (1980) 137.
D. Brydges, J. Fröhlich and A. Sokal, to appear.
See also Ref.6.

2. Surface roughening

J. Fröhlich and T. Spencer, Commun. Math. Phys. 81 (1981) 527-602; (in particular §7 of this paper).

3. General aspects of the statistical mechanics of surfaces

J. Fröhlich, C.-E. Pfister and T. Spencer, in "Stochastic Processes in Quantum Field Theory and Statistical Physics", Lecture Notes in Physics 173 (Berlin-Heidelberg-New York, Springer-Verlag,1982).
J. Fröhlich and T. Spencer, in "Scaling and Self-Similarity in Physics", J. Fröhlich (ed.), Progress in Physics (Basel and Boston, Birkhäuser, 1983).

4. Lattice gauge theory and plaquette percolation

M. Aizenman, J.T. Chayes, L. Chayes, J. Fröhlich and L. Russo, Commun. Math. Phys. 92 (1983) 19.
M. Aizenman and J. Fröhlich, Nucl. Phys. B [FS](1984),in press.

5. Critical Properties of random surfaces

B. Durhuus, J. Fröhlich and T. Jonsson, Nucl. Phys. B225 [FS9] (1983) 185-203.

B. Durhuus, J. Fröhlich and T. Jonsson, Physics Letters 137 B (1984) 93-97.
Perhaps the most important paper in this series is:
B. Durhuus, J. Fröhlich and T. Jonsson, "Critical Behaviour in a Model of Planar Random Surfaces", preprint, submitted to Nuclear Physics B [FS]. (In this paper the triviality of the naive Nambu-Goto string theory is established.)

6. Review of "random surfaces in quantum field theory"

J. Fröhlich, "Quantum Field Theory in Terms of Random Walks and Random Surfaces", to appear in the proceedings of the Cargese summer school (1983), G't Hooft et al.(eds.).
A further review is presently in preparation.

Acta Physica Austriaca, Suppl. XXVI, 259–308 (1984)

STOCHASTIC QUANTIZATION AND GAUGE FIXING IN GAUGE THEORIES+

by

E. SEILER
Max-Planck-Institut für Physik und Astrophysik
Werner-Heisenberg-Institut für Physik
Postfach 401212
Munich, Fed. Rep. Germany

CONTENTS

+Lectures given at the XXIII. Internationale Universitätswochen für Kernphysik, Schladming, Austria, February 20-March 1, 1984.

INTRODUCTION

These lectures were supposed to focus on a specific non-compact lattice gauge model with a peculiar "stochastic" gauge fixing invented by Zwanziger. But to put this model into perspective I find it appropriate to widen the scope of these lectures somewhat.

So in the first part I will review briefly the connections between Quantum Mechanics and Stochastic Processes. These fall into three classes: Nelson's stochastic mechanics describing real time quantum mechanics, stochastic processes describing imaginary time ("Euclidean") quantum mechanics - typica ly leading to Feynman-Kac type formulae, and finally stochastic quantization in the sense of Parisi and Wu which treats Euclidean quantum mechanics with an additional auxiliary time parameter roughly corresponding to computer time in Monte Carlo simulations.

There is clearly some overlap between this first part and some other lectures given at this school, such as the ones of Ph. Blanchard [1], L. Streit [2] and P. Zoller [3]. But I think this need not be a drawback since it may be advantageous to see the material presented from different points of view.

In the second part I will discuss some applications of these concepts to gauge theories. I will concentrate on Zwanziger's modification of the Parisi-Wu idea ("stochastic gauge fixing") and on numerical studies of the non-compact lattice gauge model mentioned in the beginning which implement Zwanziger's scheme.

These last sections are based on a collaboration with Nucu Stamatescu and Dan Zwanziger. Clearly that part could not have been written without their help. I also take this opportunity to thank them for many enjoyable and fruitful discussions as well as correspondence.

Finally I should like to apologize for a certain

lack of mathematical rigor that will be perceptible throughout these lectures. This is so partly for pedagogical reasons (in the first part) because I want to concentrate on the intuitive ideas, and the mathematics has been presented in many other places, even at this school. More fundamentally, the lack of rigor in the second part is largely due to a lack of precise mathematical understanding; most of the second part is in fact "experimental mathematics". I hope that this presentation will stimulate further thought that is certainly necessary.

I. GENERALITIES

1. Stochastic Mechanics

In 1966 Edward Nelson [4] proposed an interesting derivation of the Schrödinger equation from a classical but stochastic dynamical law resembling Newtonian mechanics. Whatever one may think of the hidden variable interpretation of quantum mechanics suggested by this approach, the structure is interesting in itself and deserves to be known; one could even imagine that useful numerical schemes could be derived from it.

The idea can be described as follows (cf.[5]): The probability density

$$\rho(x,t) \equiv |\psi(x,t)|^2 \quad , \tag{1}$$

where ψ is a Schrödinger wave function in R^n satisfying

$$i\hbar\partial_t\psi = -\frac{\hbar^2}{2m}\Delta\psi + V\psi \quad , \tag{2}$$

obeys the continuity equation

$$\dot{\rho} = -\nabla J \tag{3}$$

where the current J is

$$J \equiv -\frac{i\hbar}{2m}\,\bar{\psi}\,\overleftrightarrow{\nabla}\psi \;=\; -\frac{\hbar}{m}\,\rho\nabla\,\mathrm{Im}\;\ln\psi\;. \tag{4}$$

It is possible to associate a Markov process to ρ described by a transition probability density $P(x,x';t)$ such that

$$\rho(x,t) = \int P(x,x';t)\rho(x',t)dx' \tag{5}$$

will obey a Fokker-Planck equation

$$\dot{\rho} = \frac{\nu}{2}\Delta\rho - \nabla(b\rho)\;; \tag{6}$$

(4) and (6) are compatible if

$$\nu = \frac{\hbar}{m} \tag{7}$$

and

$$\Delta|\psi|^2 = -i\nabla(\bar{\psi}\overleftrightarrow{\nabla}\psi) + \frac{2m}{\hbar}\nabla(b|\psi|^2)\;. \tag{8}$$

Nelson furthermore requires

$$\nabla\wedge b = 0 \tag{9}$$

which leads to the solution

$$\begin{aligned} b &= \frac{\hbar}{m}\nabla\;\ln|\psi| + \frac{\hbar}{m}\nabla\;\arg\psi \\ &= \qquad u \qquad\quad + \qquad v \end{aligned} \;. \tag{10}$$

The Schrödinger equation leads to the following equations of motion for u and v:

$$\dot{u} = -\frac{\hbar}{2m}\nabla(\nabla v) - \nabla(uv)$$

$$\dot{v} = -\frac{1}{m}\nabla u + \frac{1}{2}\nabla(u^2-v^2) + \frac{\hbar}{2m}\Delta u\;. \tag{11}$$

An equivalent description of the stochastic process is given by the Langevin equation

$$dx = b\,dt + \sqrt{\frac{\hbar}{2m}}\,dw \tag{12}$$

(cf. [1,2,3,4,6]) where dw is the increment of the Wiener process described by

$$\langle dw_i \rangle = 0 \tag{13}$$

$$\langle dw_i \, dw_j \rangle = 2\delta_{ij}\, dt \tag{14}$$

and the specification that the dw_i are Gaussian r.v.'s which are independent for different times.

We can see the connection between (12) and (6) as follows: Let $f(x(t),t)$ be a smooth function. Then "to order dt"

$$\begin{aligned} df(x(t),t) &= (\partial_t f)dt + (\partial_i f)dx_i \\ &\quad + \frac{1}{2}(\partial_i \partial_j f)dx_i \, dx_j \\ &= (\partial_t f)dt + (b\cdot\nabla f)dt \\ &\quad + (\partial_i f)\sqrt{\frac{\hbar}{2m}}\,dw^i \\ &\quad + \frac{\hbar}{4m}\,(\partial_i \partial_j f)dw_i \, dw_j \quad . \end{aligned} \tag{15}$$

Taking the expectation value of the Wiener process (at time t) we obtain

$$\langle df \rangle = \langle \partial_t f \rangle dt + \langle b\cdot\nabla f \rangle dt + \frac{\hbar}{2m}\,\langle \Delta f \rangle dt \quad . \tag{16}$$

Since this is true for any f, ρ has to satisfy the adjoint equation which is (6). An important <u>special case</u> is obtained for $\psi = \psi_0$ where ψ_0 is the ground state of the

Hamiltonian

$$H = -\frac{\hbar^2}{2m}\Delta + V \quad .$$

Since ψ_0 has no nodes we see immediately that $v = 0$; the process is stationary (this actually turns out to be true for any eigenstate of the Hamiltonian [4]).

As a simple example let us consider the ground state of the harmonic oscillator [8]

$$\psi_0 = (2\pi\sigma)^{-1/4}\exp(-\frac{x^2}{4\sigma}) \qquad (\sigma = \frac{\hbar}{2m\omega}) \tag{17}$$

which leads to the "drift"

$$b = -\frac{\hbar}{2m\sigma}x = -\omega x \quad . \tag{18}$$

The Fokker-Planck equation becomes

$$\dot{\rho} = \frac{\hbar}{2m}\frac{d^2}{dx^2}\rho + \omega\rho + \omega x\frac{d}{dx}\rho \quad , \tag{19}$$

and it is solved by

$$\rho(x,t) = \int\rho(x',0)P(x,x';t)dx'$$

with

$$P(x,x';t) = (2\pi\bar{\sigma}(t))^{-1/2}\exp[-\frac{1}{2\bar{\sigma}(t)}(x-e^{-\omega t}x')^2] \tag{20}$$

where

$$\bar{\sigma}(t) \equiv \sigma(1 - e^{-2\omega t}) \quad . \tag{21}$$

We note that

$$\langle x(0)x(t)\rangle = \sigma\, e^{-\omega|t|} \tag{22}$$

which looks very "Euclidean".

More generally we observe that the ground state process leads to the Fokker-Planck equation

$$\dot{\rho} = \frac{\hbar}{2m} \Delta\rho - \frac{\hbar}{2m} \nabla((\nabla \ln \rho_o)\rho) \tag{23}$$

which has the stationary solution $\rho = \rho_o = |\psi_o|^2$. It turns out that quite generally

$$\langle x(0)x(t)\rangle = (\psi_o, x\, e^{-H|t|} x\, \psi_o)\ ; \tag{24}$$

(24) is a remarkable formula because it links a "real time" object on the left-hand-side to an "imaginary time" quantity on the right-hand-side.

This general connection was noted by Guerra and Ruggiero [7,8]: "Euclidean Quantum Mechanics (or Field Theory) is the Ground State Process of Stochastic Mechanics".

The strange equation (24) can be true only because the "real time" left-hand-side is not accessible to measurement.

The mathematical reason for (24) lies in the fact that the "Fokker-Planck operator" for the ground state process

$$h \equiv - \frac{\hbar}{2m} \Delta + \nabla b \tag{25}$$

$$b = 2\nabla \log |\psi_o| \tag{26}$$

is not only hermitian with respect to the measure $\rho_o dx$ +) but actually conjugate to H:

$$h = \psi_o\, H\, \psi_o^{-1} \tag{27}$$

+) This is the property of "detailed balance".

(we normalized H such that $H\psi_o = 0$).

So the Fokker-Planck equation has the formal solution

$$\rho(t) = e^{-ht}\rho(0)$$

$$= \psi_o\, e^{-Ht}\psi_o^{-1}\, \rho(0) \quad . \qquad (28)$$

(By (27) and (28) it is also obvious how to find a domain of essential self-adjointness for h once one has one for H.) (28) makes manifest that (6) describes a relaxation process; the relaxation times are directly related to the spectrum of H.

This suggests a possible application of stochastic mechanics: First obtain an approximate expression for ψ_o and thereby for b, use this to set up an approximate ground state process (by a Langevin or Monte Carlo algorithm) and from this extract information on the energy levels (cf. [5]).

Appendix:
Verification of Eq.(27)

For simplicity we set $\hbar = 2m = 1$. Then

$$h = -\Delta + \nabla b$$

$$= -\Delta + 2\nabla\ (\nabla \log \psi_o)$$

$$= -\Delta + 2\nabla\ \frac{\nabla\psi_o}{\psi_o}$$

$$= -\Delta + 2V - 2\left(\frac{\nabla\psi_o}{\psi_o}\right)^2 + 2\left(\frac{\nabla\psi_o}{\psi_o}\right)\nabla \quad .$$

Now consider

$$H_f \equiv e^{-f}\ H\ e^{f}$$

$$= -\Delta + V - (\Delta f) - (\nabla f)^2 - 2(\nabla f)\nabla \quad .$$

Setting $f = -\log \psi_o$ we obtain

$$\nabla f = - \frac{\nabla \psi_o}{\psi_o}$$

$$\Delta f = \left(\frac{\nabla \psi_o}{\psi_o}\right)^2 - V$$

and $H_f = h$.

2. Euclidean Functional Integrals and Feynman-Kac Formulae

The Euclidean approach to Quantum Mechanics (or Quantum Field Theory) uses the positivity of the Hamiltonian to continue analytically to imaginary time, i.e., it replaces the time evolution e^{iHt} by e^{-Ht}. The Schrödinger equation gets replaced by a diffusion equation and a probabilistic or stochastic interpretation is very suggestive. It should be noted, however, that in this approach the stochastic processes play a purely auxiliary role since they do not take place in "real time". Remember, however, the strange identification of real and imaginary times occurring in Eq.(24). We give only a brief sketch; for details see [6,9].

If as before (but with $\hbar = 2m = 1$) we study the Hamiltonian

$$H = -\Delta + V \quad , \tag{29}$$

the Feynman-Kac formula gives a representation for the integral kernel of (with respect to Lebesgue measure dx):

$$e^{-tH}(x,y) = \int dP_{xy}^{t}(w)\; e^{-\int_0^t V(w(\tau))d\tau} \quad . \tag{30}$$

Here $dP_{xy}^{t}(w)$ is the conditional Wiener measure for Brownian paths starting at x and ending in y after time t. The most straightforward way to derive (30) is based on the Trotter product formula

$$e^{-tH} = \lim_{n\to\infty} (e^{\frac{t}{n}\Delta} e^{-\frac{t}{n}V})^n \tag{31}$$

(the limit is in the strong sense but often holds in a much stronger, e.g., trace norm sense).

The form of the Feynman-Kac formula found most frequently gives instead the solution of the imaginary time Schrödinger equation

$$\dot{\psi} = (\Delta-V)\psi \equiv H\psi \tag{32}$$

by

$$\psi(x,t) = \int dP^t(w) e^{-\int_0^t V(x+w(\tau))d\tau} \psi(x+w(t),0) , \tag{33}$$

where $dP^t(w)$ is now the standard Wiener measure for paths starting at the origin.

It is of course also possible to obtain an expression for e^{-th} in terms of the Wiener process; this can either be done directly using the so-called Cameron-Martin formula [6,9] or it can be derived from (27):

By the Itô calculus we have (cf.[6,8,9] and Eq.(15))

$$\begin{aligned} d \log \psi_0(w(t)) &= \nabla \log \psi_0 \, dw + \Delta \log \psi_0 \, dt \\ &= \nabla \log \psi_0 \, dw + [V(w(t)) - (\nabla \log \psi_0)^2]dt \quad . \end{aligned} \tag{34}$$

Recalling

$$b = 2\nabla \log \psi_0 \tag{35}$$

this implies

$$V(w(t))dt = d \log \psi_o + \frac{1}{4} b^2 dt - \frac{1}{2} b\, dw \tag{36}$$

or, replacing w(t) by x + w(t),

$$\exp\left(-\int_o^t V(w(\tau))d\tau\right) = \psi_o(x+w(t))^{-1}\psi_o(x)\exp\left(-\frac{1}{4}\int_o^t b^2\, dt + \frac{1}{2}\int_o^t b\, dw\right) . \tag{37}$$

Inserting this in (33) we obtain

$$\psi(x,t) = \int dP^t(w)\psi_o(x+w(t))^{-1}\psi_o(x) \cdot \cdot \exp\left(-\frac{1}{4}\int_o^t b(x+w(\tau))^2 d\tau + \frac{1}{2}\int_o^t b(x+w(\tau))dw\right) . \tag{38}$$

Since

$$\psi(x,t) = (e^{-Ht}\psi)(x,t) = (\psi_o^{-1}\, e^{-ht}\, \psi_o\, \psi)(x,t) , \tag{39}$$

(38) can be read as

$$e^{-ht}(x,y) = \int dP_{xy}^t(w) \exp\left(-\frac{1}{4}\int_o^t b^2\, d\tau + \frac{1}{2}\int_o^t b\, dw\right) . \tag{40}$$

From the Feynman-Kac formula one obtains easily the standard Euclidean functional integral expressions for the Schwinger functions, for instance

$$<x(t_1) \;....\; x(t_n)> \equiv (\psi_o, x\, e^{(t_2-t_1)H} x \;...\; e^{(t_n-t_{n-1})H} \psi_o)$$

$$= \lim_{T\to\infty} \frac{1}{Z_T} \int dx \int dP^T_{xx}(w) \cdot$$

$$\cdot \exp(-\int_0^T V(w(\tau))d\tau) w(t_1) \;...\; w(t_n) \;, \tag{41}$$

where $T \geq t_1 \geq t_2 \geq ... \geq t_n \geq 0$ and

$$Z_T = \int dx \int dP^T_{xx}\, e^{-\int_0^T V(w(\tau))d\tau} \equiv Tr\, e^{-TH} \;. \tag{42}$$

Formally (41) can be read as

$$<x(t_1) \;....\; x(t_n)>$$

$$= \lim_{T\to\infty} \frac{1}{\tilde{Z}_T} \int e^{-S_T} x(t_1) \;...\; x(t_n) \prod_{t=0}^{T} dx(t) \;, \tag{43}$$

where S_T is the Euclidean action in time T and

$$\tilde{Z}_T \equiv \int e^{-S_T} \prod_{t=0}^{T} dx(t) \;.$$

This is so because the conditional Wiener measure corresponds to the formal expression

$$\frac{1}{N_T} \exp(-\int_0^T \dot{w}(\tau)^2 d\tau) \prod_{t=0}^{T} dw(\tau) \;. \tag{44}$$

3. Stochastic Quantization [10]

This approach, as the previous one, uses stochasticity as an auxiliary tool, but this time the stochastic processes take place in an additional time that is analogous to computer time in Monte Carlo algorithms.

The Feynman-Kac formulae of the last section lead to Euclidean functional integrals; the idea is now to set up an ergodic stochastic process possessing the Euclidean functional measure as its unique equilibrium measure. Clearly this does not specify the process uniquely (as can be seen from the various Monte Carlo algorithms). But the simplest choice is based on a Langevin equation and a corresponding Fokker-Planck equation resembling closely those discussed in Section 1.

We only have to make the identifications (formally)

$$|\psi_0|^2 = \frac{1}{Z} \exp(- \frac{1}{\hbar} S) \tag{45}$$

where Z is a normalizing factor,

$$b = - \frac{1}{\hbar} \nabla S \tag{46}$$

where ∇ now really means a functional gradient. At this formal level there is no difference between ordinary Quantum Mechanics and Quantum Field Theory, so we assume that S is a functional of some Euclidean fields symbolized as ϕ.

The Fokker-Planck equation now reads (with a suitably rescaled time)

$$\dot{\rho} = \hbar^2 \Delta\rho + \nabla(\nabla S)\rho \quad , \tag{47}$$

and the Langevin equation

$$d\phi = \nabla S \, dt + \hbar \, dw \quad ; \tag{48}$$

dw is now a high dimensional Wiener process, one for each space-time point and field component. For a scalar field ϕ we may take

$$\langle dw(x)dw(y)\rangle = 2\delta(x-y)dt \quad . \tag{49}$$

We can also determine a potential $\tilde{V}$ belonging to the ground state wave function

$$\psi_o = Z^{-1/2} e^{-1/2\hbar \, S} \quad :$$

It is

$$\tilde{V} = \frac{\hbar^2 \Delta\psi_o}{\psi_o} = - \hbar \, \Delta \, S + (\nabla S)^2 \quad . \tag{50}$$

(Δ and ∇ have to be interpreted as functional derivatives, of course.) So one can write again a Feynman-Kac formula leading to a functional integral with one extra dimension. Formally its density is given by

$$e^{-\frac{1}{\hbar}\tilde{S}}$$

where

$$\tilde{S} = \frac{1}{4} \int \dot{\phi}^2 \, d\tau + \int \tilde{V} \, d\tau \quad . \tag{51}$$

The "supereuclidean" models arising in this way have interesting properties, such as a certain supersymmetry, analyzed for instance by E. Gozzi and R. Kirschner [11], following the work of Parisi and Sourlas [12].

Stochastic quantization can be used to construct a perturbation expansion [10]. It is also possible to use the auxiliary time for a regularization by replacing the Wiener process by a suitable non-Markovian process: this amounts to replacing $\int\dot{\phi}^2 d\tau$ in (51) by $(\phi, C^{-1}\phi)$ where C^{-1} is a suitable operator; in view of applications to gauge theory it is useful to choose C^{-1} "local in space-time".

Let me close this section with a remark about the ergodicity of the stochastic process described by (47), (48): Using a Feynman-Kac formula it is not difficult to see that - if we approximate the space of fields by a finite dimensional space - the transition probability density from any configuration to any other is strictly positive. A classic theorem due to Perron and Frobenius (see [9,12]) then assures us that there is at most one ground state; since we have already found the ground state ψ_o we have a unique equilibrium distribution and we can compute averages with respect to that by taking time averages.

For infinitely many degrees of freedom this argument cannot work; it has to fail because of the possible occurrence of phase transitions.

II. APPLICATION TO GAUGE THEORIES

1. Some Possible Constructions

a) Gauge field theories formally fit into the framework of Schrödinger Quantum Mechanics used in the first part, if we work in the temporal gauge $A_o = 0$. Formally their Hamiltonian is of the form

$$H = -\frac{g^2}{2}\Delta + \frac{1}{2g^2}V \tag{52}$$

where

$$V = \int \vec{B}^2 \, d\tau \quad . \tag{53}$$

It has to be supplemented by the Gauss law constraint on the physical subspace that expresses the absence of external (infinitely heavy) charges by the requirement that the physical states must be gauge invariant (under time independent gauge transformations).

To be on firm ground one should work on a spatial lattice; H is then the Kogut-Susskind lattice Hamiltonian and has again the form (52). Δ is now the sum of "link Laplaceans" Δ_{xy}; Δ_{xy} is the Laplace-Beltrami operator with respect to the link variable U_{xy}. U_{xy} is an element of a compact Lie group G but may be taken as a unitary matrix associated to the link <xy> of the lattice; the Hilbert space $\mathcal{H}$ consists of all square-integrable functions of all these link variables. On the lattice V has to be interpreted, for instance, as

$$V = \sum_{\substack{\text{plaquettes} \\ P}} \operatorname{Re} \operatorname{tr} U_{\partial P} \ . \tag{54}$$

If we knew the ground state ψ_o of H we could now set up the ground state process in the form (cf. [14])

$$U_{xy}^{-1} \, dU_{xy} = \frac{1}{g^2} \, b_{xy} \, dt + g \, dw_{xy} \tag{55}$$

where now b_{xy} and dw_{xy} take values in the Lie algebra of G; dw_{xy} is the increment of the Wiener process on the group G [29], and

$$b_{xy} = b_{xy}^a \, L^a \tag{56}$$

where L^a $(a = 1,2,\ldots,c)$ runs through an orthonormal basis of the Lie algebra of G (with respect to the positive definite Killing metric), and

$$b_{xy}^a = \frac{d}{d\varepsilon} \, 2 \log \psi_o \, (e^{\varepsilon L^a} U_{xy}) \Big|_{\varepsilon=0} \ . \tag{57}$$

(In the last expression we only displayed the dependence of ψ_o on the selected link variable U_{xy} .)

By the Gauss law constraint ψ_o is gauge invariant. Therefore the drift b will always be perpendicular to the orbits of the gauge group which acts by changing U_{xy} to $V_x U_{xy} V_y^{-1}$, and it will be "gauge covariant" (equivariant in mathematical terms), i.e.,

$$b_{xy}(V_x U_{xy} V_y^{-1}) = V_x b_{xy}(U_{xy}) V_x^{-1} \quad . \tag{58}$$

(Remember that b_{xy} corresponds to an infinitesimal left translation of U_{xy} .)

This implies that b_{xy}, considered as a vector field = derivation, commutes with gauge transformations.

We are interested in the expectation values of gauge invariant observables F. Zwanziger observed [15] (in a slightly different context, see below) that we may modify the diffusion process by adding an arbitrary drift $b_{||}$ tangential to the orbits of G without affecting those expectation values. This follows from the following more general fact:

<u>Proposition</u>: Let M be a Riemannian manifold on which a Lie group G acts isometrically. Let Y be a vector field tangential to the orbits of G, i.e.,

$$Y = \lambda X$$

where X is an infinitesimal group translation. Let b be a vector field commuting with X. Consider the following two Fokker-Planck operators:

$$h \equiv -\Delta + b^* \tag{59a}$$

$$\tilde{h} \equiv -\Delta + b^* + Y^* \tag{59b}$$

where the adjoint is taken with respect to the invariant

$L^2(M)$, and the vector fields b, Y are considered as derivations. Let ρ_t and $\tilde{\rho}_t$ be solutions to the corresponding Fokker-Planck equations $\dot{\rho}_t = -h\rho_t$ and $\dot{\tilde{\rho}}_t = -\tilde{h}\tilde{\rho}_t$, respectively, with the same initial conditon ρ_o. Let finally F be a G-invariant function on M.+)

Then

$$(\rho_t, F)_{L^2(M)} = (\tilde{\rho}_t, F)_{L^2(M)} \quad .$$

Proof:

$$\rho_t = \exp(-th)\rho_o \quad , \quad \tilde{\rho}_t = \exp(-t\tilde{h})\rho_o \ ;$$

therefore

$$(\tilde{\rho}_t, F) = (\rho_o, e^{-t\tilde{h}^*} F) \ .$$

The proposition follows from

$$e^{-t\tilde{h}^*} F = e^{-th^*} F \quad . \tag{60}$$

To see this we use the Duhamel formula

$$e^{-t\tilde{h}^*} = e^{-th^*} + \int_o^t ds\ e^{-s(h^*-Y)}\, Y\, e^{-(t-s)h^*} = e^{-th^*} + \int_o^t ds\ e^{-s(h^*-Y)}\, \lambda X\, e^{-(t-s)h^*} \ . \tag{61}$$

Applying this to F we may drop the second term because X can be pushed through $e^{-(t-s)h^*}$ where it will hit and annihilate F (cf.[16]).

This principle will be used later in the context

+) There are some additional technical conditions to insure existence of the relevant integrals etc., but we will not bother to write them down.

of Parisi-Wu stochastic quantization.

But let us first make a few remarks on a possible computational scheme:
In general it is hard to find a good candidate for a ground state wave function. Formally for the free electromagnetic field the ground state is proportional to $\exp[-1/2(\vec{B},|\Delta|^{-1/2}\vec{B})]$ which shows long range correlations. It may be expected that in a nonabelian theory these long range correlations are not there; this kind of disorder should be connected to confinement. One might try something like $\exp \gamma \sum_{\langle xy\rangle} \mathrm{tr} U_{xy}$, but this is not gauge invariant. A better candidate might be the gauge average of that function, but averaging over the gauge group destroys the local structure again and it is hard to obtain an explicit expression for the drift.

Another possibility would be to use approximate ground state wave functions obtained by some variational method (cf. [17]). But it is certain that the problem of obtaining a good approximate drift for the ground state process requires much more study.

b) It is also possible to write Feynman-Kac formulae for Hamiltonian lattice gauge theories. In analogy with the case of ordinary Quantum Mechanics one obtains:

$$e^{-Ht}(\underline{U},\underline{U}')$$

$$= \int dP^t_{\underline{U},\underline{U}'}(w) \exp\left(-\int_0^t V(w(\tau))\,d\tau\right) \tag{62}$$

(we put $g^2 = 2$ for simplicity), $dP^t_{\underline{U},\underline{U}'}$ is the conditional Wiener measure on the configuration space $= G^\dagger$ of links. To enforce Gauss's law one should actually study $P_o e^{-Ht}$ where P_o is the projector into the gauge invariant subspace. This leads to

$$(P_o e^{-Ht})(\underline{U},\underline{U}')$$

$$= \int d\underline{V} \int dP^t_{\underline{U}^{\underline{V}},\underline{U}'}(w)\exp(-\int_0^t V(w(\tau)d\tau) \quad , \tag{63}$$

i.e., we simply have to average over the gauge transforms $\underline{U}^{\underline{V}}$ of the starting point $\underline{U}$.

c) Parisi and Wu [10] originally proposed their scheme for gauge theories because it can be implemented - at least in perturbation theory for gauge invariant operators - without gauge fixing: Even though there is no equilibrium measure on the space of vector potentials because the system will diffuse forever along the gauge orbits, there should be an equilibrium measure on the space of these orbits.

2. Zwanziger's Scheme

Zwanziger [15] proposed a modification of the Parisi-Wu stochastic quantization by adding a drift tangential to the gauge orbits in complete analogy to our discussion given in the previous section. By the same arguments as given there it is seen that (at least for a Euclidean lattice version preserving gauge invariance, such as Wilson's) the expectation values of gauge invariant observables are unaffected by this modification. Formally these arguments carry over to the continuum because formally the gauge transformations still act isometrically with respect to the (weak) Riemannian metric $g(A,A)=\sum_{a,\mu}\int (A^a_\mu)^2$ (cf.[19]). A formal argument for the equivalence of Zwanziger's scheme with the usual Faddeev-Popov prescription was also given by Baulieu and Zwanziger [18].

We will now describe Zwanziger's scheme in a more detailed (but still formal)way: The "classical" drift is given by $-\delta S/\delta A^a_\mu(x)$ where $S = -1/2\|F\|^2$ in the continuum.

We want to invent a convenient extra drift term $b_{||} \equiv K_{GF}$ is tangential to the gauge orbits. Any vector field (considered as a derivation) tangential to the gauge orbits has the form

$$\int (D_\mu(A)v)^b(x) \frac{\delta}{\delta A_\mu^b(x)} dx \quad , \tag{64}$$

where $D_\mu(A)$ is the covariant derivative and $v(x)$ is a Lie algebra valued function (for each x) on the space of connections. (64) follows from the fact that an infinitesimal gauge transformation is of the form

$$\delta A_\mu^a(x) = (D_\mu(A)v)^a(x) \ . \tag{65}$$

For practical reasons we want to choose a local $v(x)$ depending only on the vector potential $A(x)$ and its derivatives at x. We also want v to correspond to some standard gauge condition like the Landau or back ground gauge. So we choose

$$v(x)^a \equiv D_\mu(\bar{A})^{ab}(A_\mu - \bar{A}_\mu)^b(x) \tag{66}$$

where $\bar{A}$ is a background field which we will put equal to zero from now on. The components of the gauge fixing drift are then

$$K_{GF\mu}^a(x) = \alpha \, D_\mu^{ab}(A)(\partial\cdot A)^b(x) \ . \tag{67}$$

This can be split into a gradient of a gauge fixing function and a "pure curl" (i.e. divergenceless term):

$$K_{GF\mu}^a(x) = -\frac{\delta S_{GF}}{\delta A_\mu^a(x)} + g \, f^{abc} \, A_\mu^b(\partial\cdot A)^c(x) \ ,$$

$$S_{GF} = \frac{\alpha}{2} \int (\partial\cdot A)^2 dx \quad , \tag{68}$$

where f^{abc} are the structure constants of G and we used the usual continuum normalization for the vector potential that produces the factor g in the nonlinear term.

The appearance of a drift whose curl does not vanish is a novel feature; it leads to some differences with respect to the situation discussed in the first part where $\nabla\wedge b = 0$ was assumed.

First of all it should be noted that this feature cannot be avoided: There does not exist any gauge fixing surface that is orthogonal to the gauge orbits (even locally), and hence a drift tangential to the gauge orbits cannot be written as a gradient [20,21].

The reason is that the Frobenius ingegrability condition [22] is not fulfilled for a vector field tangential to the gauge orbits. A simple analogue to that situation is given by the following example in R^3: The vector field

$$X \equiv \frac{\partial}{\partial z} + \alpha(x\frac{\partial}{\partial y} - y\frac{\partial}{\partial x}) \tag{69}$$

is tangential to the orbits of the group $\mathcal{R}$ acting on R^3 in a "screwy" way. X is dual to the 1-form

$$\omega = dz + \alpha(xdy - ydx) \tag{70}$$

(using the Euclidean metric). The Frobenius integrability condition for the (local) existence of a 2-surface orthogonal to X or ω requires that there is a 1-form σ such that

$$d\omega = \sigma \wedge \omega$$

and therefore

$$\omega \wedge d\omega = 0 \quad .$$

But $d\omega = 2\alpha\ dx\wedge dy$ and

$$\omega \wedge d\omega = 2\alpha \, dx \wedge dy \wedge dz \neq 0 \quad . \tag{71}$$

A few general remarks are in order about diffusion processes with $\nabla \wedge b \neq 0$. First it should be noted that $h = -\Delta + \nabla b$ will not be symmetric with respect to the equilibrium measure $\rho_0 dx$; the process is not reversible, detailed balance does not hold. This also means that the spectrum of $\exp(-th)$ will in general be complex; the relaxation will go through damped oscillations (if at all). More specifically, the operator

$$H \equiv \rho_0^{-1/2} \, h \, \rho_0^{1/2}$$

has a positive "real part", i.e.,

$$H + H^* \geq 0 \quad . \tag{72}$$

(In the terminology of [23] this means that iH is a "dissipative operator"; its spectrum lies in the upper half plane.)

To see this let $\rho_0 = \exp(-2f)$. Then

$$\begin{aligned} H &= e^f(-\Delta + \nabla b)e^{-f} \\ &= -\Delta + (\Delta f) - (\nabla f)^2 + 2(\nabla f)\nabla + (\nabla b) \\ &\quad + b\nabla - (b\nabla f) \quad , \end{aligned}$$

$$\begin{aligned} H^* &= -\Delta - (\Delta f) - (\nabla f)^2 - 2(\nabla f)\nabla \\ &\quad + b\nabla - (b\nabla f) \quad , \end{aligned}$$

$$\frac{1}{2}(H + H^*) = -\Delta - (\nabla f)^2 + \frac{1}{2}(\nabla b) - (b\nabla f) \quad .$$

By the stationary Fokker-Planck equation $(-\Delta + \nabla b)e^{-2f} = 0$, hence

$$2\Delta f - 4(\nabla f)^2 + (\nabla b) - 2(b\nabla f) = 0$$

and thus

$$\frac{1}{2}(H + H^*) = -\Delta - (\Delta f) + (\nabla f)^2$$
$$= (\nabla - (\nabla f))^*(\nabla - (\nabla f)) \geq 0 \quad . \tag{73}$$

(Note that $1/2(H + H^*)$ again has the "supersymmetric" form that occurred in Section I.3, Eq. (50) .)

A further phenomenon characteristic for diffusion processes with $\nabla \wedge b \neq 0$ is the presence of steady currents in the equilibrium measure: The Fokker-Planck equation has the form

$$\dot{\rho} = -\nabla J \tag{74}$$

with the current

$$J = b\rho - \nabla\rho \quad . \tag{75}$$

If we define the velocity field v by $\rho^{-1}J$ we see

$$v = b - \nabla \log\rho \quad . \tag{76}$$

Since $\nabla \wedge b \neq 0$ we also obtain $\nabla \wedge v \neq 0$ and hence

$$v \neq 0 \quad , \qquad J = v\rho \neq 0 \tag{77}$$

even in equilibrium ($\dot{\rho} = 0$).

These phenomena will be seen in the numerical simultions to be discussed in the next section. It should be remarked, however, that they will not be felt as long as we restrict ourselves to gauge invariant observables (for instance in a Wilson lattice gauge theory with a stochastic gauge fixing analogous to (67)). Put differently, the process induced on the <u>space of orbits</u> will be <u>re-</u>

versible.

3. Properties of the Stochastic Gauge Fixing Force

In this section we analyze the gauge fixing drift (67) in more detail. In particular we will investigate in which sense it actually "fixes" the gauge; in this context we will encounter a remarkable fact which we call "spontaneous compactification". Our arguments will mostly be presented in a formal way in the continuum; to make them rigorous it will often be necessary to introduce a discretization (non-compact lattice approximation).

We will first study the properties of K_{GF} by considering the dynamical system

$$\dot{A} = K_{GF}(A) \quad . \tag{78}$$

The first observation is very simple:

Proposition: If A evolves according to Eq. (78) we have

$$\frac{d}{dt}||A||^2 = -2\alpha||\partial A||^2 \leq 0 \quad . \tag{79}$$

In (79) $||A||^2$ stands for $\sum_{a\mu} \int A_\mu^a(x)^2 dx$, etc. The proof of this proposition is easy:

$$\begin{aligned} \frac{d}{dt}||A||^2 &= 2\alpha \int A_\mu^a(x) D_\mu(A)^{ab}(\partial A)^b dx \\ &= -2\alpha \int |\partial A|^2 dx \quad . \end{aligned}$$

This means that K_{GF} has the character of a restoring force. It is also obvious that the subspace given by $\partial A = 0$ consists of fixed points under (78). There are "essentially" no others because $D(A)f = 0$ implies $f = 0$ whenever A is irreducible. So the manifold $\{D(A)\partial A = 0$ but $\partial A \neq 0\}$ is of lower dimension than the manifold $\{\partial A = 0\}$.

Of course one is interested in the stability properties of the fixed points. Zwanziger [15] uncovered the following remarkable fact that becomes a theorem in a finite dimensional lattice approximation (see[24]):

Fact: On the surface given by $\partial A = 0$ there is an open convex set Ω, bounded along each ray such that the points in Ω are stable fixed points and the points outside $\bar{\Omega}$ are unstable fixed points of the force K_{GF}. The boundary $\partial\Omega$ is the Gribov horizon.

Remark: In a finite dimensional lattice approximation the set Ω is actually bounded in the ordinary sense.

We will give the proof for this latter situation. This means that all derivatives have to be interpreted as finite differences and one has to take care in transcribing for instance exterior products of forms to the lattice. The required formalism is described in detail in [24].

We will linearize the evolution $\dot{A} = K_{GF}$ around the surface $C \equiv \{\partial A = 0\}$. In particular we are interested in

$$\frac{d}{dt}(\partial A) \equiv \frac{d}{dt} v \tag{80}$$

in linear approximation, which gives

$$\dot{v} = hv \quad , \tag{81}$$

where the linear operator h can be written as $h = \delta_{\overset{\circ}{A}} d$ - the Faddeev-Popov operator - ($\overset{\circ}{A}$ is the reference point on C around which we are linearizing). In components we have

$$(hv)^a = D_\mu(\overset{\circ}{A})^{ab}\partial_\mu v^b \; ; \tag{82}$$

h is symmetric because $\partial A = 0$. It has a null space L_0 of constants, but for appropriate (e.g., periodic) boundary conditions L_0 is orthogonal to the space of v's of the form $v = \partial A$. On $L_0^\perp$, $\partial_\mu\partial_\mu \equiv \Delta$ is strictly negative. Ω is now the subset of C on which $h < 0$. Because h depends linearly on $\overset{\circ}{A}$, Ω is convex:

$$(v, h(\alpha \overset{o}{A} + (1\ \alpha)\overset{o}{A}')v)$$

$$= \alpha(v, h(\overset{o}{A})v) + (1-\alpha)(v, h(\overset{o}{A}')v) < 0 \tag{83}$$

if both terms are <0 and $\alpha \in [0,1]$. To see that Ω is bounded along rays we consider

$$(v, h(\lambda \overset{o}{A})v) =$$

$$= -(\partial_\mu v, \partial_\mu v) + \lambda f^{abc}(\overset{o}{A}{}^b_\mu v^a, \partial_\mu v^c) \ . \tag{84}$$

We are done if we can show that the second term can be >0 for any $\overset{o}{A} \neq 0$ and suitable v.

Such a v is found as follows: Pick a lattice point m_o and indices μ_o, b_o such that $A^{b_o}_{\mu_o}(m_o) \neq 0$. Consider v's that are different from zero only at m_o and its neighbor in μ_o direction $m_o + e_{\mu_o}$. The second term of (84) then becomes

$$\lambda v^a(m_o) f^{ab_oc} \overset{o}{A}{}^{b_o}_{\mu_o}(m_o) v^c(m_o + e_{\mu_o}) \ . \tag{85}$$

We pick $v(m_o + e_{\mu_o})$ arbitrarily and choose

$$v^a(m_o) \equiv \mathrm{sgn}\lambda\ f^{ab_oc} A^{b_o}_{\mu_o}(m_o) v^c(m_o + e_{\mu_o}) \ .$$

This choice clearly makes (85) positive.

It can be shown [24] that on a unit lattice the diameter of Ω is less than 32 (in 4 space-time dimensions) and that the boundary $\partial\Omega$ comes close to the origin at constant A^a_μ: There are boundary points with $A \sim 2\pi/N$ where N is the number of lattice sites along the edge of a cubic lattice.

To understand the properties of stochastic gauge fixing it is instructive to study a simple two-dimensional example. (A slightly different example was suggested to me by Dan Zwanziger to whom I am also indebted for correspondence about it; Nucu Stamatescu did the numerical analysis presented below.) We choose

$$b \equiv K_{GF} = -\alpha \begin{bmatrix} 2xy^2 \\ (1-x^2)y \end{bmatrix} \tag{86}$$

($\alpha > 0$). The role of the submanifold C is played by the axis $\{y = 0\}$. b is "restoring": Under $\binom{\dot{x}}{\dot{y}} = b$ we have

$$\frac{d}{dt}(x^2 + y^2) = -2\alpha x^2(1+y^2) \leq 0 \; ; \tag{87}$$

h is replaced by the one-dimensional operator $-\alpha(1 - x^2)$ and

$$\Omega = \{\binom{x}{y} \,|\, y = 0, \quad |x| < 1\} \; . \tag{88}$$

The points $\binom{\pm 1}{0}$ form the "Gribov horizon" $\partial\Omega$. The orbits of b are

$$u \equiv x \exp(-\frac{x^2}{2} - y^2) = \text{const} \tag{89}$$

and are orthogonal to the level surfaces of $y/(1-x^2)$, i.e., the parabolae

$$y = v(1 - x^2) \; . \tag{90}$$

A "portrait" of the force is given in Fig.1.

Obviously the orbits u = const are <u>not</u> the orbits of an isometry of R^2, i.e., an Euclidean motion. This is impossible if we want a compact $\bar{\Omega}$. But at least near the horizon the orbits become approximate circles:
Putting $x = 1+\xi$, $y = \eta$, u = const becomes to second order in (ξ,η)

$$\xi^2 + \eta^2 = \text{const} \; . \tag{91}$$

So the behavior of our diffusion will at least be approximately α independent near the horizon, if we average over the orbits.

Let us try to understand some properties of the equilibrium measure $\rho_o dxdy$. It obeys the stationary Fokker-Planck equation

$$\nabla(\nabla - b)\rho_o = 0 \quad . \tag{92}$$

Proving the existence of a normalizable ρ_o is not completely trivial and I will not attempt to do it here. It is, however, intuitively plausible that ρ_o will be mainly concentrated around Ω.

We are lucky to have an explicit solution for $\alpha = 9$: It is

$$\rho_o = \text{const} \exp(-\tfrac{3}{2}x^2 - 3y^2) \quad . \tag{93}$$

Because the current $J = (\nabla - b)\rho_o$ is divergenceless it obeys $\oint_L (n \cdot J) d\ell = 0$ over any closed curve L (n is the outer normal to L). Specifically if we integrate over an orbit we obtain

$$\oint_{u=\text{const}} (n \cdot \nabla)\rho_o d\ell = 0 \quad . \tag{94}$$

This tells us something about the density in the space of orbits. In particular for $\alpha \to \infty$ the density ρ_o will have the form

$$\rho_o(\alpha = \infty) = \sigma(x)\theta(1-x^2)\delta(y) \tag{95}$$

and (94) allows to determine σ: The normal to the orbit is

$$n = \begin{bmatrix} 1-x^2 \\ -2xy \end{bmatrix} ((1-x^2)^2 + 4x^2y^2)^{-1/2} \quad , \tag{96}$$

and

$(n.\nabla\rho)$

$$= \frac{1}{1-x^2}\,((1-x^2)\sigma'(x)\delta(y)-2xy\sigma(x)\delta'(y))$$

$$= \frac{1}{1-x^2}\,((1-x^2)\sigma'(x)+ 2x\sigma(x))\delta(\underline{y}) \quad . \tag{97}$$

Inserting this in (94) we obtain

$$(1-x^2)\sigma' + 2x\sigma = 0 \tag{98}$$

which is solved by

$$\sigma = \text{const}\ (1-x^2) \quad . \tag{99}$$

The density $\sigma(x)$ corresponds to the Faddeev-Popov determinant (det h = const$(1-x^2)$).

The linear vanishing of the density on orbit space near the horizon is a more general fact: If we are in the situation of an isometric action of the group (like in the true gauge models) we may put $\alpha=0$ without changing the distribution in orbit space. But the free diffusion will produce a flat distribution; if we integrate over the orbits the integrated distribution will go to zero like the orbit volume near the horizon (on the horizon the gauge orbits have lower than generic dimension). Of course the linear vanishing near the horizon follows also from the linear vanishing of the Faddeev-Popov determinant at generic points of the horizon.

To give an illustration of these facts I present a numerical simulation of the model (courtesy of I.O. Stamatescu). The Langevin equation

$$d\binom{x}{y} = b\,dt + \binom{dw_x}{dw_y}$$

was simulated with a random walk algorithm (see the next

section) for α = 20,000 and step size 0.003. Figs. 2a,b,c show histograms of the distribution in x after 2×10^6, 10×10^6 and 24×10^6 steps, respectively, together with the const $\cdot$ $(1-x^2)$ prediction. Convergence to that paraboly is clearly visible. The spread in y is very small (about 0.05 near the horizon where it is widest).

4. Numerical Study of Zwanziger's Scheme

The numerical investigation is done by discretizing

- → Space-time
- → Langevin time
- → Field space .

This leads to a random walk algorithm in field space. We introduce a step size η in the field space. We may think of the resulting set of allowed field configurations as a simple cubic lattice (of very high dimension). We define our algorithm by giving a probability p_0 for not moving and a hopping probability to go from A to a neighboring point $\vec{A} \pm \eta \vec{e}_i$,

$$p(\vec{A} \pm \eta \vec{e}_i, \vec{A}) = \frac{1-p_0}{2N} (1 \pm \tanh \frac{\eta}{2} b_i(\vec{A})) ,$$
$$\frac{1-p_0}{2N} = \frac{\varepsilon}{\eta^2} . \tag{100}$$

The tanh is introduced "by hand" to avoid negative probabilities; it should be noted, however, that the algorithm can be trusted to give a good approximation to the Langevin process only if $1/2\eta b_i << 1$ so that the tanh can be replaced by its argument. N is the number of dimensions of field space; N = (number of colors) $\times$ (number of space-time dimensions) $\times$ (number of space-time lattice points).

To justify (100) we make a finite difference approximation of the Fokker-Planck equation:

$$\frac{1}{\varepsilon}\,(\rho(\vec{A},t+\varepsilon)\ -\rho(\vec{A},t))$$

$$= \frac{1}{\eta^2}\sum_{i=1}^{N}(\rho(\vec{A}+\eta\vec{e}_i,t)\ +\ \rho(\vec{A}-\eta\vec{e}_i,t)$$

$$-\ 2\rho(\vec{A},t))$$

$$-\ \frac{1}{2\eta}\sum_{i=1}^{N}(b_i(\vec{A}+\eta\vec{e}_i)\rho(\vec{A}+\eta\vec{e}_i,t)$$

$$-\ b_i(\vec{A}-\eta\vec{e}_i)\rho(\vec{A}-\eta\vec{e}_i,t))\ . \tag{101}$$

This equation can be read as

$$\rho(\vec{A},t+\varepsilon)\ =\ p_0\rho(\vec{A},t)$$

$$+\ \sum_{i,\pm} p(\vec{A},\vec{A}\pm\eta\vec{e}_i)\rho(\vec{A}\pm\eta\vec{e}_i,t) \tag{102}$$

with

$$p_0\ =\ 1\ -\ \frac{2N\varepsilon}{\eta^2}$$

which has to be ≥ 0, and the transition probabilities given by (100) without the tanh.

The random walk algorithm obtained this way is an alternative to the usual Monte Carlo algorithms and comparable in speed [26].

The lattice gauge model studied by this method is of the "non-compact" variety, i.e., the action is discretized naively, without forcing the lattice gauge fields to live in the group G instead of its Lie algebra. For instance the field strenght has the following form:

$$F^a_{\mu\nu}(m)\ =\ \Delta_\mu\ A^a_\nu(m)\ -\ \Delta_\nu\ A^a_\mu(m)$$

$$+\ f^{abc}\ s_\nu\ A^b_\mu(m)\ s_\mu\ A^c_\nu(m) \tag{103}$$

where Δ_μ is the forward lattice derivative and s_ν is a forward symmetrizer in ν direction, i.e., $s_\nu A_\mu^b(m)=1/2(A_\mu^b(m)+A_\mu^b(m+\eta e_\nu))$.

The classical action is

$$S \equiv \frac{1}{4} \sum_{\substack{m,a \\ \mu,\nu}} (F_{\mu\nu}^a(m))^2 \quad , \tag{104}$$

and the total drift used in the simulation is

$$\vec{b} = -\beta \vec{\nabla} S + \vec{K}_{GF} \tag{105}$$

where $\vec{K}_{GF}$ is as before given by

$$K_{GF\mu}^a(x) = D_\mu^{ab}(A)(\partial A)^b(x) \quad ; \tag{106}$$

D_μ^{ab} is a forward covariant lattice derivative:

$$[D(A)v]_\mu^a(m) \equiv \Delta_\mu v^a(m) + f^{abc} A_\mu^b(m) s_\mu v^c(m) \quad . \tag{107}$$

Clearly this naive procedure is problematic since it does not respect the gauge structure; it introduces a non-gauge-invariant regularization, and to recover gauge invariance in the continuum one has to allow in general non-gauge-invariant counterterms. This has not been done, but something can be said about a possible mass counterterm for the gauge field.

The simulations (without such counterterms and for the group G = SU(2)) gave a qualitatively very different picture from the one obtained with compact lattice gauge models: There does not seem to be quark confinement. The theory at weak coupling behaves essentially like QED; at stronger coupling it looks a lot like a system with Higgs fields in the fundamental representation or a system with a mass term for the gauge fields (which is essentially the same thing).

But it is not possible to cancel this effect by a negative mass counterterm: The system would immediately become unstable, i.e., it would drift out to infinite A's along the abelian directions.

On the other hand our naive system still seems to show asymptotic freedom (an non-renormalizable mass counterterm does not seem compatible with that).

Furthermore the system shows typcial features normally associated with continuum gauge fields such as the Gribov horizon.

Most of our simulations have been done for $\alpha\sim\beta$ because that is convenient; the step size η hs to be adjusted to the strength of the drift (cf. (100)) and it is therefore good to have all forces of about equal strength.

The dependence of expectation values on α is a measure of the violation of gauge invariance; this becomes weak for weak coupling ($\beta\to\infty$).

To test for confinement we mainly relied on Polyakov loops. We measured both the expectations of single Polyakov loops $P_{1/2}$ in the fundamental representation (Fig. 3a) which for $\alpha=\beta$ are remarkably well fitted by the Coulomb self-energy even for g^2 as large as 10. They stay significantly non-zero wherever they can be measured, and their square agrees within errors with the "asymptotic" value of Polyakov loop correlations ("∞" separation, see Fig.4).

In Fig.3b the expectation values of the adjoint Polyakov loop P_1 are shown. They deviate earlier (at smaller g^2) from the Coulomb law prediction than the fundamental ones and show clearly convergence to the ∞ coupling prediction as g^2 increases, which is $\neq 0$ for P_1.

To test for asymptotic freedom we followed Creutz's idea and measured the following two functions:

$$F(\frac{1}{g^2}) \equiv \ln \frac{\langle W(4,4)\rangle\langle W(2,2)\rangle}{\langle W(4,2)\rangle^2} \tag{108}$$

on an 8^4 lattice, and

$$G(\frac{1}{g^2}) \equiv \ln \frac{\langle W(2,2)\rangle\langle W(1,1)\rangle}{\langle W(2,1)\rangle^2} \tag{109}$$

on a 4^4 lattice ($W(n,m)$ stands for a $n\times m$ Wilson loop). (108) and (109) represent the same quantity on lattices of different mesh a; according to the renormalization group they should approximately agree if we move $g^2(a)$ appropriately, namely we should have

$$F(\frac{1}{g^2}) \cong G(\frac{1}{g^2} - 0.064) \ . \tag{110}$$

The results are shown in Fig.5; they are not smashingly precise but at least hint at asymptotic freedom.

To study the gauge fixing force we also ran simulations at $\beta = 0$. The system remains stable ("self-compactifying") and shows the typical damped slow oscillations that should be expected (see Section II.1). In Fig. 6 we show a run starting at the origin. To test stability further, we also started with configurations outside the Gribov horizon; such configurations are easy to construct with constant A. The system can be seen to perform a sort of cycle: First it is pushed off the gauge fixing surface by the unstable modes, this makes $\int(\partial\cdot A)^2$ grow. Once this is large, the system behaves essentially classically (the randomness is unimportant) and returns inside the Gribov horizon according to the equation (79),

$$\frac{d}{dt}||A||^2 = -2\alpha||\partial\cdot A||^2 \ .$$

This can be clearly seen in Figs. 7, 8. (Fig. 8 actually

represents a run with $\beta \neq 0$ but $\alpha >> \beta$.) For more details see Ref.[25].

The non-compact lattice gauge model discussed here raises many questions, some of which were already raised by earlier work on a non-compact lattice model [27] and which would require hard analytic work to be answered; for instance a perturbative study should be done to attack the question of counterterms.

But the stochastic gauge fixing certainly remains an interesting alternative to the standard Faddeev-Popov procedure. Together with a regularization in the auxiliary time (instead of a lattice regularization) which is compatible with the gauge structure it might also open new possibilities for the rigorous construction of gauge theories.

Finally it might be worthwhile to think harder about possible computational uses of stochastic mechanics, possibly along the lines of Section II.1. In this context it should be noted that F. Guerra and R. Marra [28] have given a completely general version of stochastic mechanics that also covers, at least in principle, field theories with fermions.

FIGURE CAPTIONS

Fig. 1: Portrait of the "gauge fixing" force (86). The "circles" are the lines of force given by $x \exp(-x^2/2-y^2) = \text{const}$. The parabolae $y = \text{const} \cdot (1-x^2)$ are always perpendicular to the force.

Fig. 2: Numerical simulation of the diffusion under the drift force (86). The histograms give the number of points falling into the corresponding intervals in x; the parabola is the theoretical prediction for $\alpha = \infty$.
(a): 2×10^6 steps
(b): 10×10^6 steps

(c): 24×10^6 steps .

Fig. 3: Average values of single Polyakov loops:

(a) Fundamental loops $\langle P_{1/2} \rangle$ on 8^4 and 4^4 lattices

(b) Adjoint loops $\langle P_1 \rangle$ on 4^4 lattice .

The dashed line is the Coulomb self-energy prediction.

Fig. 4: Polyakov loop correlation functions $\langle P_{1/2}(0) P_{1/2}(R) \rangle$:

(a) $g^{-2} = 0.2$

(b) $g^{-2} = 0.25$

(c) $g^{-2} = 0.3$

(d) $g^{-2} = 0.5$.

The arrow is the corresponding value of $\langle P_{1/2} \rangle^2$ (cf. Fig. 3a).

Fig. 5: Asymptotic Freedom Test: Note that the large full circles and the empty circles seem to fall on the same curve.

Fig. 6: Diffusion under the gauge fixing force (106) for $\alpha=0.15$ on a 4^4 lattice. The run was started at the origin. The crosses represent the action $||F||^2$, the triangles $||\partial \cdot A||^2$, the circles $||A||^2$.

Fig. 7: Same as before, but with $\alpha= 5$ on a 6^4 lattice and a start outside the Gribov horizon.

Fig. 8: Diffusion under both the gauge fixing and the classical force ($\alpha = 1$, $\beta = 0.25$) on a 4^4 lattice. The system was started outside the Gribov horizon.

REFERENCES

1. Ph. Blanchard, these proceedings.
2. L. Streit, these proceedings.
3. P. Zoller, these proceedings.
4. E. Nelson, Phys. Rev. 150 (1966) 1079; Dynamical Theories of Brownian Motion (Princeton University Press, 1967); "Connection Between Brownian Motion and Quantum Mechanics", in Lecture Notes in Physics, vol. 100 (Springer-Verlag, Berlin-Heidelberg-New York, 1979) (Einstein Symposium Berlin).

5. G. Jona-Lasinio, "Stochastic Processes and Quantum Mechanics", talk given at the Colloque en l'Honneur de L. Schwartz, Ecole Polytechnique, June 1983 , to appear in Asterisque.
6. C. de Witt-Morette and D. Elworthy, Phys. Rep. 77 (1981).
7. F. Guerra and P. Ruggiero, Lett. Nuov. Cim. 31 (1973) 1022.
8. F. Guerra, Phys. Rep. 77 (1981) 263.
9. B. Simon, Functional Integration in Quantum Physics (Academic Press, New York-San Francisco-London, 1979).
10. G. Parisi and Y.S. Wu, Scientia Sinica 24 (1981) 483.
11. E. Gozzi, Phys. Lett. 129B (1983) 432, (Err. 134B (1983) 477); Phys. Lett. 130B (1983) 83; Phys. Rev. D28 (1983) 1922; "The Onsager's Principle of Microscopic Reversibility and Supersymmetry", CCNY-HEP-83/16;
R. Kirschner, "Quantization by Stochastic Relaxation Processes and Supersymmetry", KMU-HEP 84-01.
12. G. Parisi and N. Sourlas, Nucl. Phys. B206 (1982) 321; Phys. Rev. Lett. 43 (1979) 744.
13. P. Walters, Introduction to Ergodic Theory (Springer-Verlag, Berlin-Heidelberg-New York, 1982).
14. A. Guha and S.C. Lee, Phys. Rev. D27 (1982) 2412; Phys. Lett. 134B (1984) 216.
15. D. Zwanziger, Nucl. Phys. B192 (1981) 259; Phys. Lett. 114B (1982) 337; Nucl. Phys. B209 (1982) 336.
16. S. Helgasson, Differential Geometry and Symmetric Spaces (Academic Press, New York-San Francisco-London, 1962), Chapter X, Prop. 2.1.
17. N.D. Hari Dass, P.G. Lauwers and A. Patkos, Phys. Lett. 124B (1983) 387; Phys. Lett. 130B (1983) 292.
18. L. Baulieu and D. Zwanziger, Nucl. Phys. B193 (1981) 163.
19. I.M. Singer, Comm. Math. Phys. 60 (1978) 7;
O. Babelon and C.-M. Viallet, Phys. Lett. 85B (1979) 246.
20. M. Creutz, I. Muzinich, T. Tudron, Phys. Rev. D19 (1979) 531.
21. A. Chodos and V. Moncrief, J. Math. Phys. 21 (1980) 364.
22. H. Flanders, Differential Forms with Applications to the Physical Sciences (Academic Press, New York-London, 1963).

23. I.C. Gohberg and M.G. Krein, Introduction to the Theory of Non-selfadjoint Operators (American Mathematical Society Translations, Providence, R.I., 1969).

24. E. Seiler, I.O. Stamatescu and D. Zwanziger, "Monte Carlo Simulations of Non-Compact QCD with Stochastic Gauge Fixing", Nucl. Phys. B, in print.

25. E. Seiler, I.O. Stamatescu and D. Zwanziger, "Numerical Evidence for a Barrier at the Gribov Horizon", Nucl. Phys. B, in print.

26. I.O. Stamatescu, U. Wolff and D. Zwanziger, Nucl. Phys. B225[FS9] (1983) 377.

27. A. Patrascioiu, E. Seiler and I.O. Stamatescu, Phys. Lett. 107B (1981) 364.

28. F. Guerra and R. Marra, "Discrete Stochastic Variational Principles and Quantum Mechanics", preprint Università di Roma, 1983.

29. H.P. McKean, Stochastic Integrals (Academic Press, New York-San Francisco-London 1969);
N. Ikeda, S. Watanabe, Stochastic Differential Equations and Diffusion Processes (North Holland Publishing Co., Amsterdam-Oxford-New York, 1981).

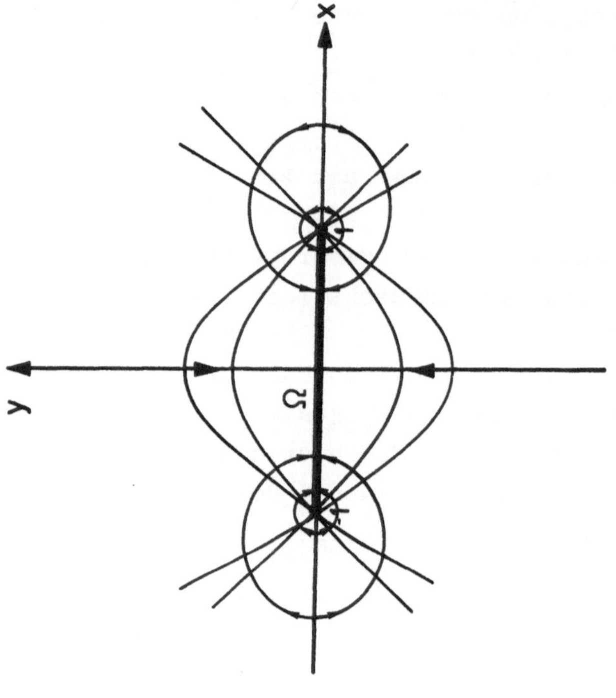

Fig. 1

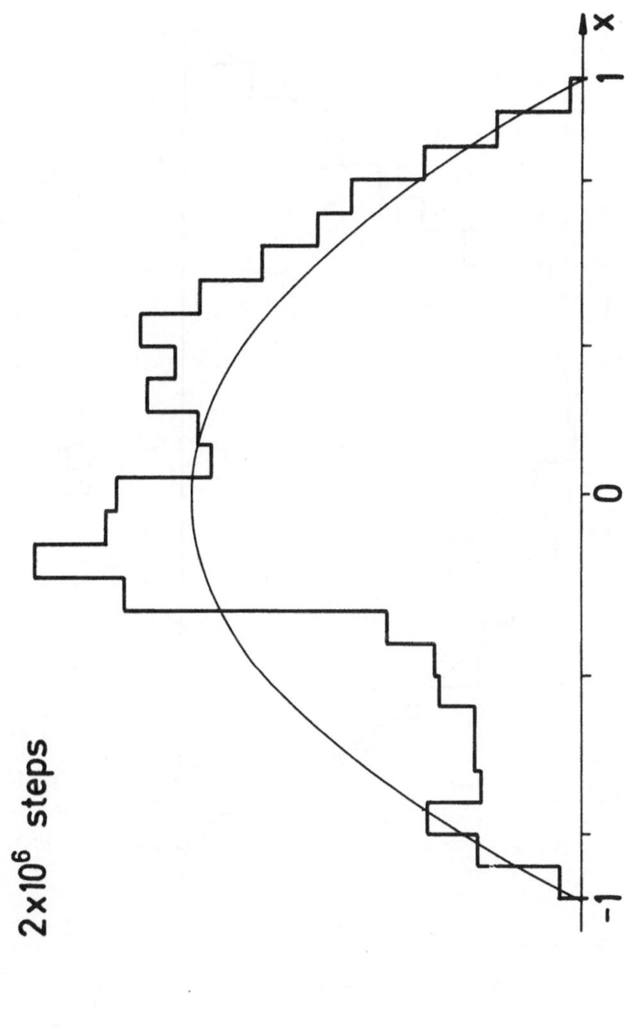

Fig. 2a

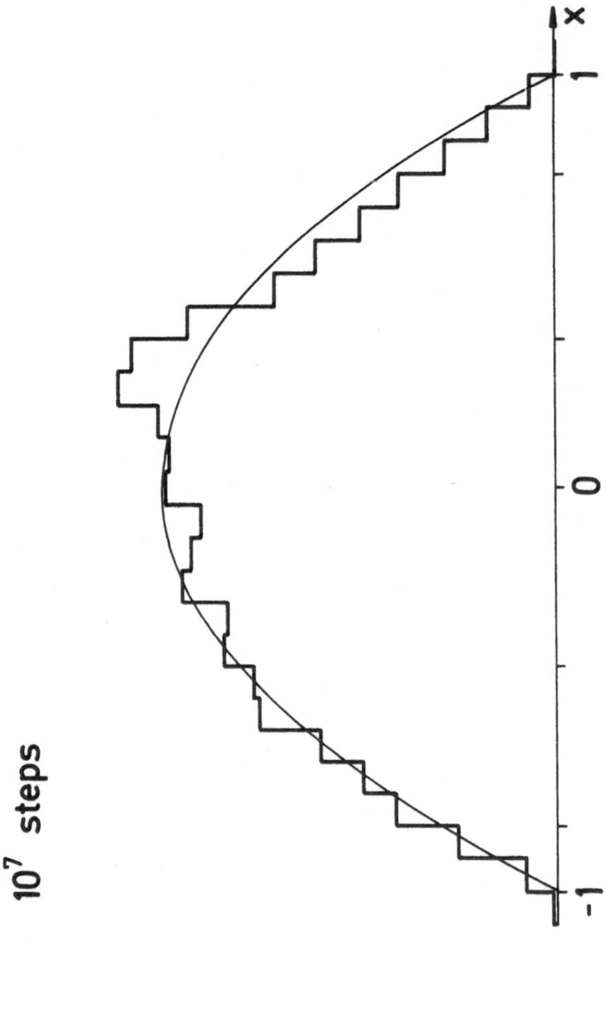

Fig.2b

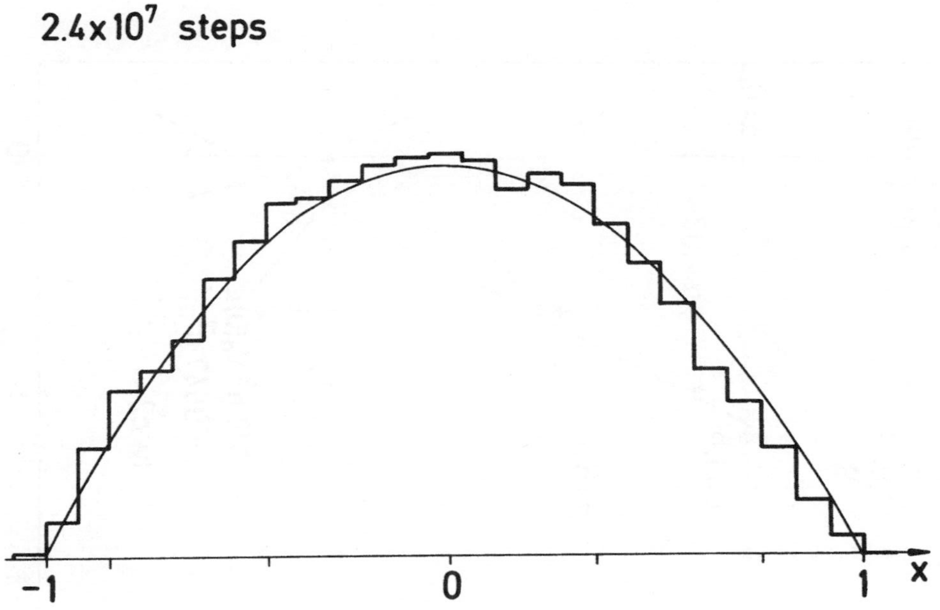

Fig. 2c

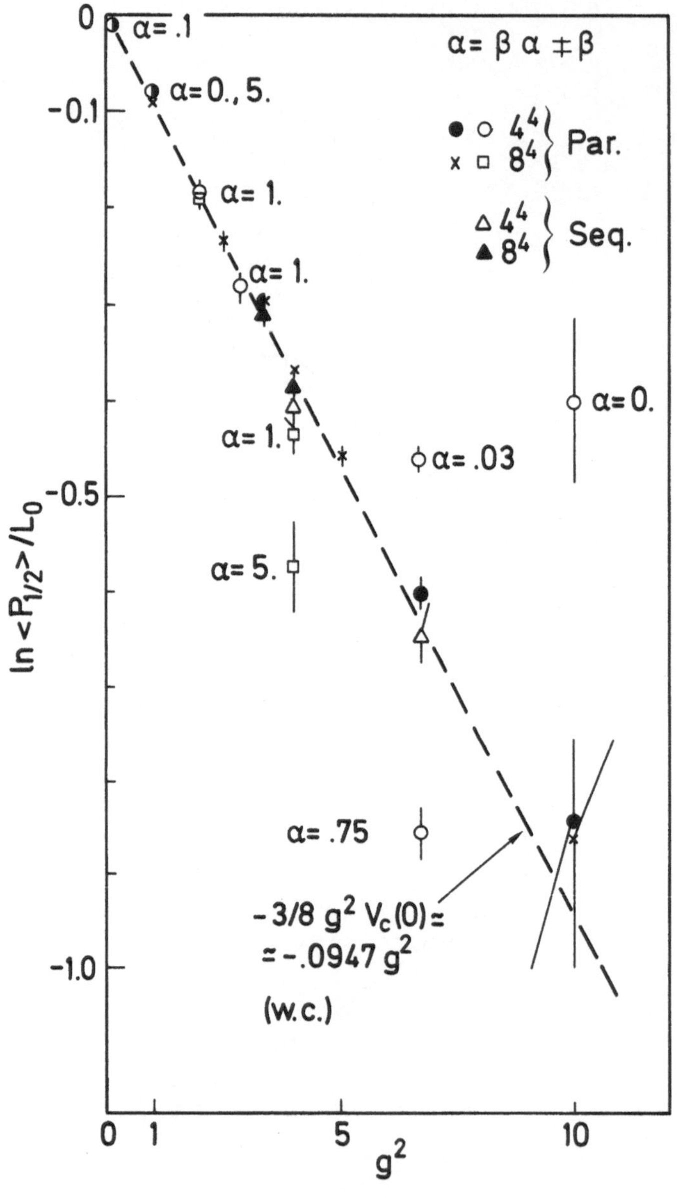

α= β α ∓ β
4⁴ Par.
8⁴
4⁴ Seq.
8⁴
α= .1
α=0., 5.
α= 1.
α= 1.
α= 1.
α= 5.
α= .03
α=0.
α= .75
-3/8 g² V_c(0) ≃
≃ -.0947 g²
(w.c.)
0
-0.1
-0.5
-1.0
ln <P_{1/2}>/L_0
0
1
5
10
g²

Fig. 3a

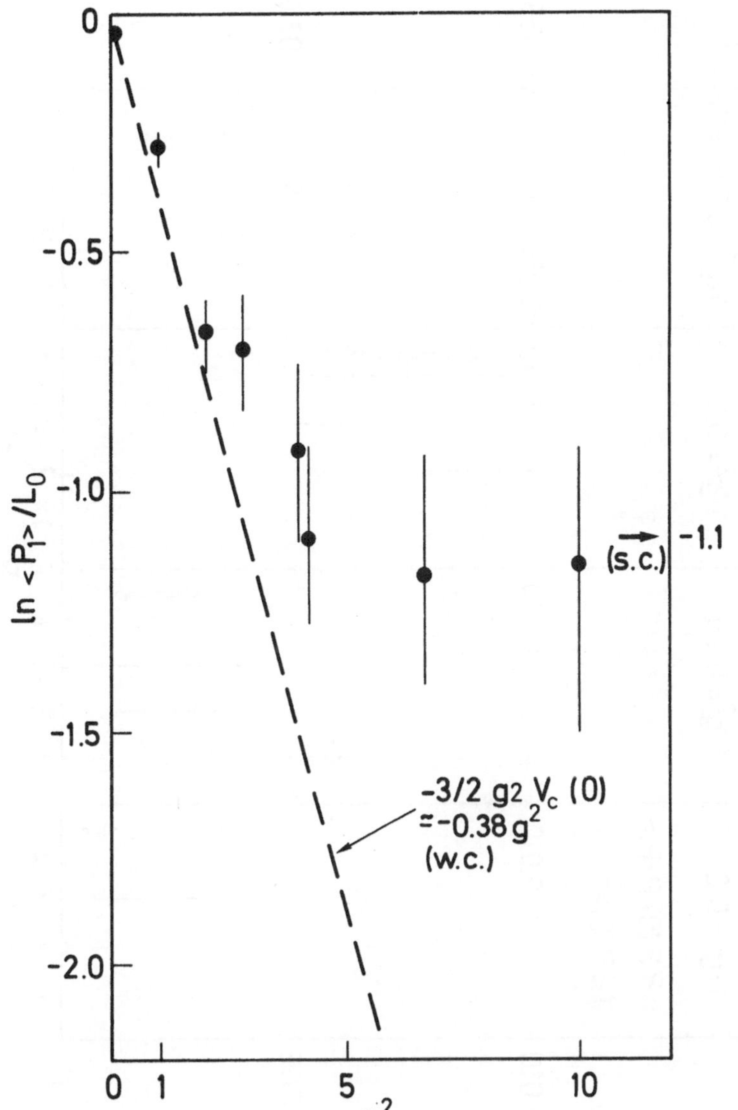

Fig. 3b

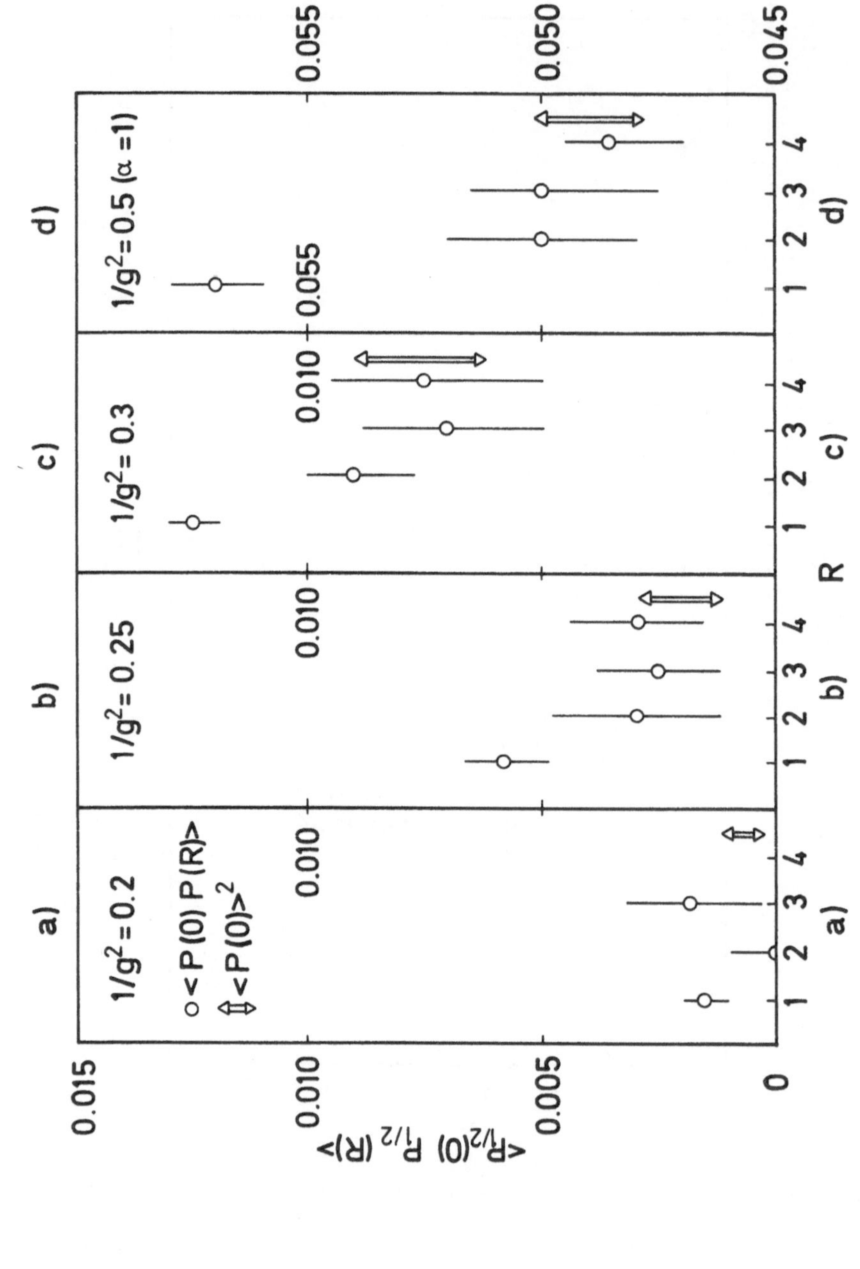
a)
b)
c)
d)
1/g²=0.2
1/g²= 0.25
1/g²= 0.3
1/g²=0.5 (α =1)
○ <P(0) P(R)>
⇕ <P(0)>²
<P_{1/2}(0) P_{1/2}(R)>
0.015
0.010
0.005
0
0.055
0.050
0.045
R
1 2 3 4

Fig. 4

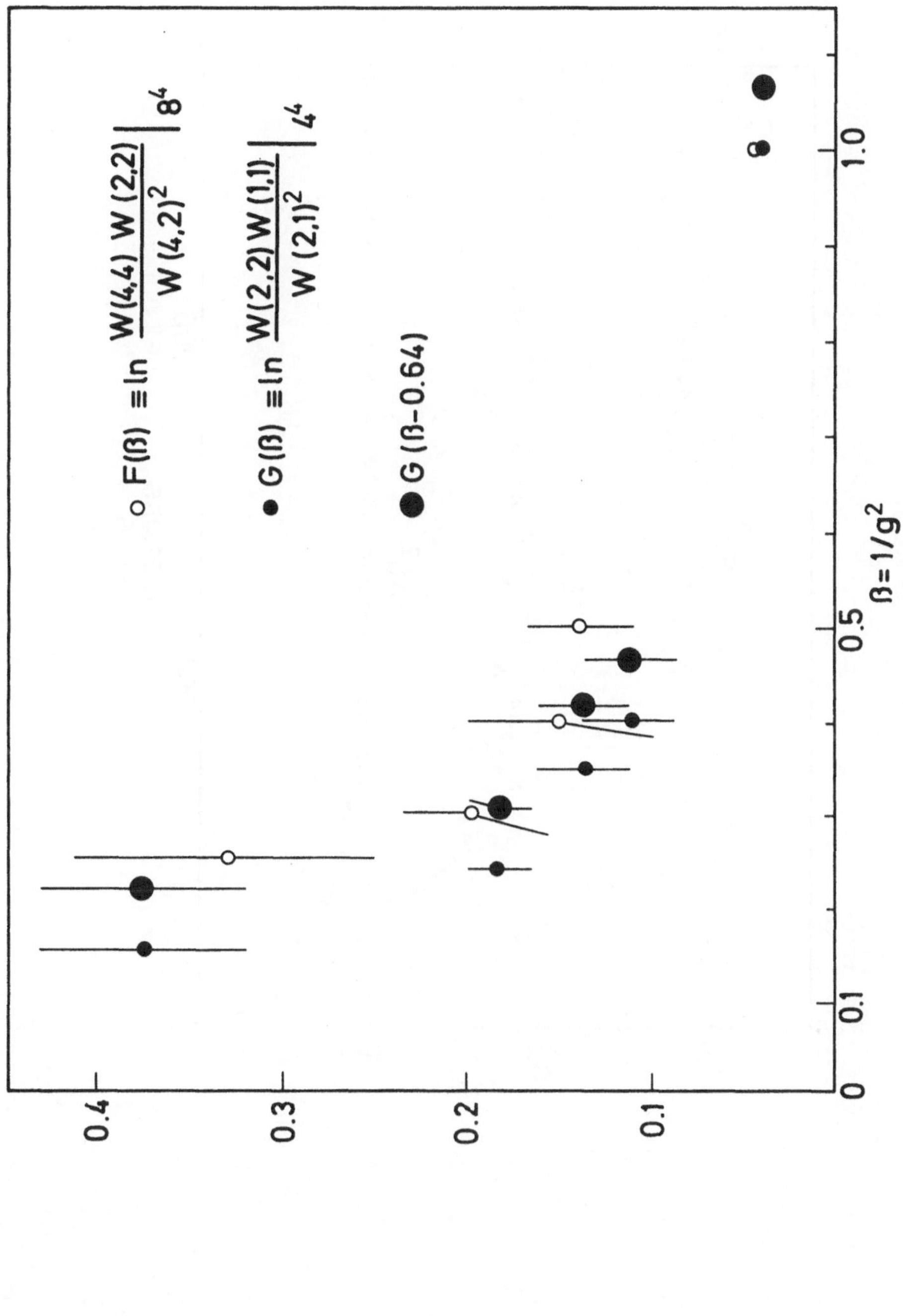

○ F(β) ≡ ln $\frac{W(4,4)\,W(2,2)}{W(4,2)^2}\Big|_{8^4}$
● G(β) ≡ ln $\frac{W(2,2)\,W(1,1)}{W(2,1)^2}\Big|_{4^4}$
● G (β−0.64)
0.4
0.3
0.2
0.1
0
0.1
0.5
1.0
β = 1/g²

Fig. 5

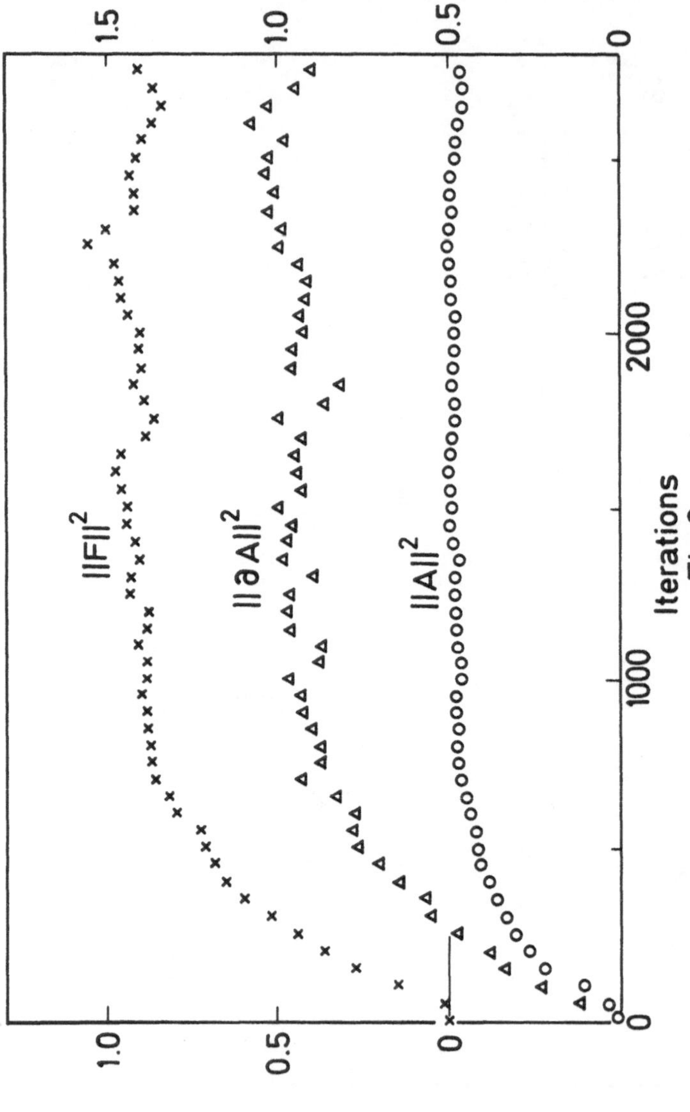

Fig.6

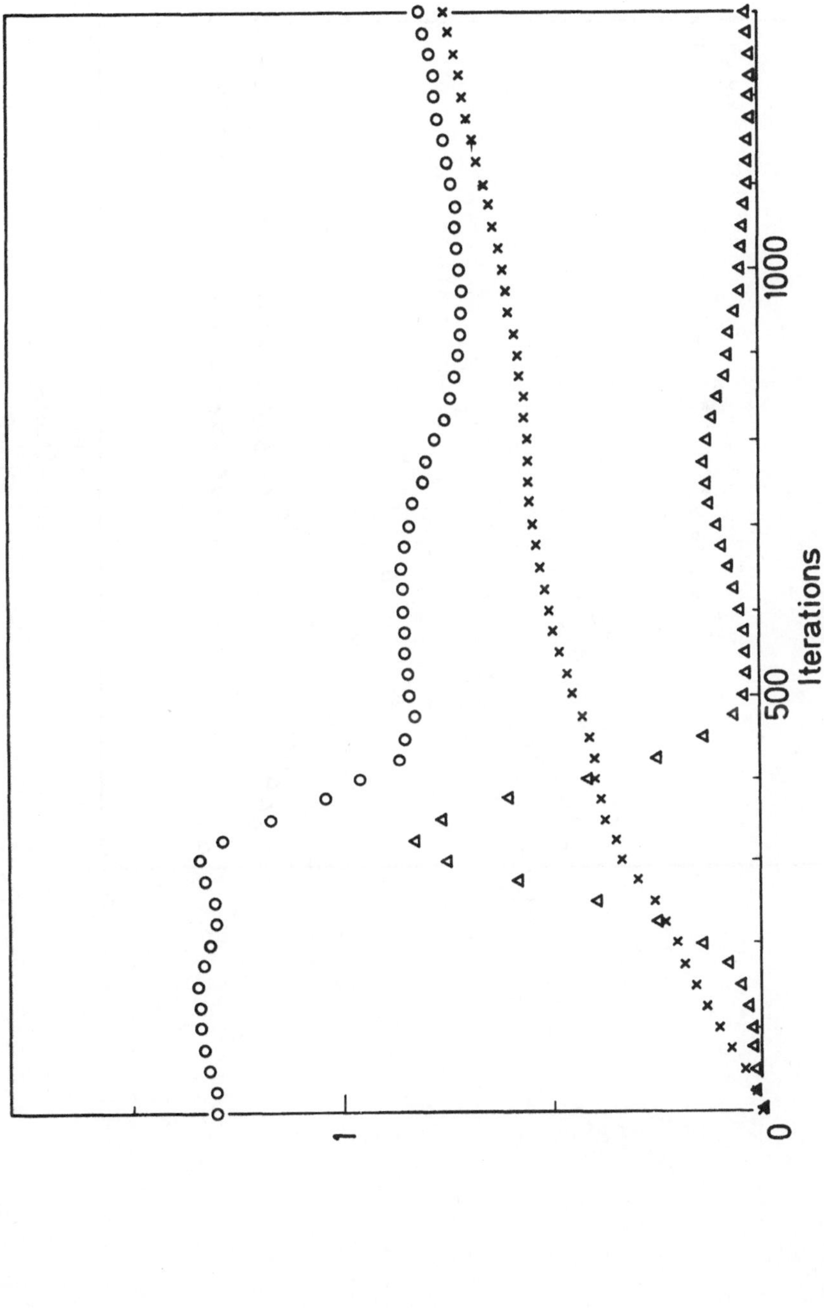

Fig. 7

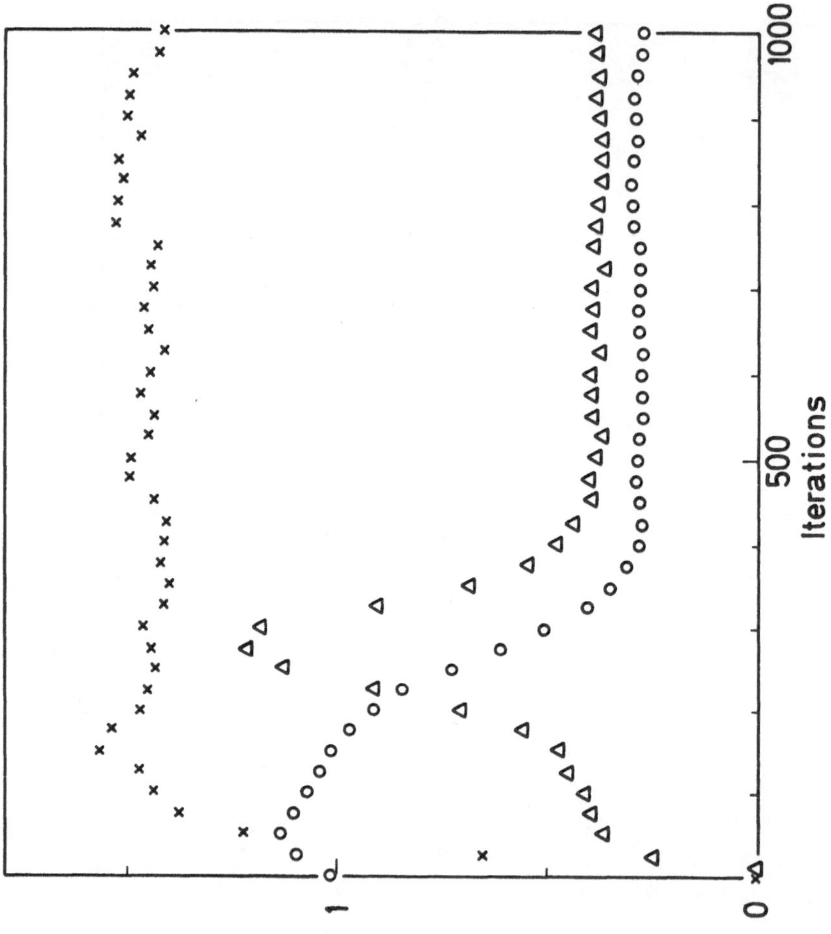

Fig. 8

Acta Physica Austriaca, Suppl. XXVI, 309–347 (1984)

NUMERICAL CALCULATIONS IN QUANTUM FIELD THEORIES+

by

C. REBBI
Brookhaven National Laboratory,
Upton,N.Y., U.S.A.

CONTENTS

+Lectures given at the XXIII. Internationale Universitätswochen für Kernphysik,Schladming,Austria,February 20-March 1,1984.

3. Monte Carlo studies of Quantum Chromo-Dynamics
   - Absence of a phase transition
   - Continuum limit - scaling behaviour
   - Results for: Force between static quarks and string tension
     - Mass gap
     - Deconfinement temperature

4. Systems with fermions
   - General formulation
   - The quenched approximation
   - Calculations of quark model spectrum
   - Simulations with dynamical fermions

FIRST LECTURE

Quantum field theories in continuous space-time.
Dynamical variables are matter fields, such as scalar fields $\phi_i(x)$, $\bar{\phi}_i(x)$ and spinor fields $\psi_{si}(x)$, $\bar{\psi}_{si}(x)$, and gauge fields $A_\mu(x)$; i denotes indices labelling internal degrees of freedom, s is a spinor index.
The gauge field strength is

$$F^i_{\mu\nu}(x) = \partial_\mu A^i_\nu(x) - \partial_\nu A^i_\mu(x) + g f_{ijk} A^j_\mu(x) A^k_\nu(x) \ .$$

If τ^i are infinitesimal generators of the algebra then f_{ijk} is defined by

$$[\tau^j, \tau^k] = -f_{ijk}\tau^i \ ;$$

g is the coupling constant.
Throughout the lectures we shall use units with $\hbar = c = 1$.
The gauge potential is used to define gauge invariant couplings between matter fields at different space-time locations.
These are given by expressions of the type

$$\bar{\psi}(y)\; P\, e^{ig \int_{x,\gamma}^{y} A_\mu^{\;i}\, \tau_i\, dx^\mu}\; \psi(x)\ ,$$

where the integration is along some path γ joining the points x and y, the exponential is path ordered and the internal indices of the matter fields have been omitted. The matter fields belong to the representation of the gauge group for which the τ_i are infinitesimal generators.
The action of the gauge field is given by

$$S = \frac{1}{4} \int d^4x\; F^i_{\mu\nu} F^{\mu\nu}_i \ .$$

The total action contains also a term involving the matter fields; it turns out however that some of the most important dynamical features of the systems we shall be considering are determined by the gauge part of the action and we shall concentrate therefore on the quantum mechanical properties of a gauge system per se. Considerations involving the matter field part of the action will be presented in fourth lecture. Vacuum expectation values (or quantum averages) of observables O are expressed through functional integrals in Euclidean space-time, after a Wick rotation $t \to it$ has been performed. (Relevant quantities are analytically continued back to Minkowski space-time, if neccessary, at the end of the computations.)

$$\langle O \rangle = \frac{\int DA\ O(A)\, e^{-S(A)}}{\int DA\ e^{-S(A)}} \ .$$

Numerator and denominator in the above formula must be given precise mathematical meaning. This is achieved by the process of regularization which relies on a perturbative expansion

$$\int DA\, e^{-S} = \int DA\, e^{-A^2-gA^3-g^2A^4} =$$

$$= \int DA\, e^{-A^2}(1-gA^3-g^2A^4 \ldots\ldots) \;.$$

The regularization requires the introduction of a cut-off parameter M ,

$$\langle O \rangle = \langle O \rangle\ (g,M) = O_o + gO_1 + g^2O_2 \ \ldots \;.$$

Eventually a renormalization is performed: the cut-off parameter M is sent to infinity with g = g(M) so that O tends to a well defined limit.

The whole procedure may be invalidated if one needs to study phenomena which are governed by intrinsically large values of the coupling constant or, at an even more fundamental level, if the observables under investigation are given by expressions with an essential singularity at g = O, e.g.

$$\langle O \rangle = M\, e^{-\frac{const}{g^2}}\ (\ 1 + O(g)) \quad .$$

The lattices formulation achieves a non-perturbative regularization[1,2]. The continuum of points in space-time is replaced by the discrete set of vertices of a generally hypercubical lattice[3]. We shall denote by x the integer valued coordinates of the lattice, by symbols $\hat{\mu}$, $\hat{\nu}$ unit displacements in the μ, ν direction and by a the length of the lattice spacing.

Matter fields ϕ_x, $\bar{\phi}_x$, ψ_x, $\bar{\psi}_x$ are associated with the vertices of the lattice. The gauge fields, which are now finite elements U_x^μ of the gauge group, are associated with the oriented links of the lattice. ($U_x{}^\mu$ is defined on the link with origin in x and end point in

$x + \hat{\mu}$.) They reduce to exponential potentials in the limit of very small lattice spacing ,

$$U_x^\mu \simeq e^{igA_\mu^i a\tau_i} \quad .$$

Of particular importance are plaquette variables, defined as follows:

$$U_x^{\mu\nu} (\text{or } U_\square) = U_x^{\nu +} U_{x+\nu}^{\mu +} U_{x+\mu}^{\nu} U_x^{\mu} ,$$

which reduce to exponentiated field strenghts in the continuum limit ,

$$U_x^{\mu\nu} \approx \exp(ig F_{\mu\nu}^i a^2 \tau_i) .$$

The action is defined through the plaquette variables. One constructs first an "internal energy"

$$E_\square = 1 - \text{Tr } U_\square ,$$

where the trace is suitably defined and normalized. $E_\square$ is always larger or equal to 0 and equals zero if the plaquette variable reduces to the identity. In the continuum limit

$$E_\square \approx g a^4 F^2 \quad ;$$

this motivates the definition of the lattice action (of the pure gauge field)

$$S = \frac{\text{const}}{g^2} \sum_\square E_\square \quad .$$

Because of an analogy with statistical mechanics, the quantity $\frac{\text{const}}{g^2}$ is very frequently denoted by β. We shall refer to g as the coupling constant, to β as the coupling parameter.

The expectation values of the observables are now given by multiple integrals over the gauge manifolds. If one restricts the system to a finite volume V, which is let to infinity only at the end of the calculation, the expressions for the quantum averages are mathematically well defined, independently of any perturbative expansion:

$$<O> = \frac{\int \prod_{x,\mu} dU_x^{\mu} O(U)\ e^{-S(U)}}{\int \prod_{x,\mu} d\,U_x^{\mu}\ e^{-S(U)}} \quad .$$

The following examplifies the definition of the lattice gauge system in the two cases where the gauge group is the discrete group Z_2 or the group U(1).

If the group is Z_2 the link variables U_x^{μ} can be represented by ±1, the group operation being ordinary multiplication. The plaquette variables are then also either +1 or -1. With a terminology taken from the theory of spin glasses, one says that the plaquettes for which $U_{\square} = -1$ are frustrated. As internal energy we may define

$$E_{\square} = 1 - U_{\square} \quad .$$

$E_{\square}$ takes value 0 for a non-frustrated plaquette, 2 for a frustrated one. The total action

$$S = \frac{1}{g^2} \sum_{\square} E_{\square} \equiv \beta \sum_{\square} E_{\square}$$

equals then $\frac{2}{g^2}$ (or 2β) times the number of frustrated plaquettes.

The link variables U_x^{μ} in a U(1)-gauge theory can be represented by complex numbers of unit modulus: $U_x{}^{\mu} = e^{i\theta}$ .We then have

$$U_{\square} = e^{i(\theta_x{}^{\mu} + \theta^{\nu}_{x+\hat{\mu}} - \theta^{\mu}_{x+\hat{\nu}} - \theta^{\nu}_{x})} \equiv e^{i\theta_{\square}} .$$

A frequently used (but not unique) definition of $E_{\square}$ and S is

$$E_{\square} = 1 - \text{Re}\, U_{\square} = 1 - \cos\theta_{\square} ,$$

$$S = \frac{1}{g}\sum_{\square} E_{\square} \equiv \beta \sum_{\square} E_{\square} .$$

The expressions for the quantum expectation value of the observables in the lattice theory

$$\langle O \rangle = \frac{\int \Pi dU\; O(U)\; e^{-\frac{c}{g^2}\sum_{\square} E_{\square}}}{\int dU\; e^{-\frac{c}{g^2}\sum_{\square} E_{\square}}}$$

are formally analogous to the expressions giving the thermal averages in the statistical formulation of thermodynamics, with the correspondence $c/g^2 \leftrightarrow 1/kT$ (hence the notation $\beta \equiv c/g^2$). The dimensionality (4) of the lattice system is however a reminder that we are dealing with a quantum field theory. Strong coupling calculations, corresponding to high temperature expansions in the above analogy, can be performed. Weak coupling expansions, corresponding to low temperature expansions, are also possible and are quite similar to the perturbative expansions of the continuum theory. The lack of rotational invariance introduces however substantial complications in the perturbative expansions, and the formalism based on the continuum is more convenient for such expansions.

The regularization provided by the lattice must be eventually removed to achieve a continuum limit. This is done by the process of renormalization. The lattice spacing a is sent to 0, but the coupling constant g must also be correspondingly modified.
Assume that one can calculate a definite physical quantity, e.g. a quantity r with dimension of a length. It will be given by an expression

$$r = a\, f(g) \ .$$

For the existence of a continuum limit there must exist a critical value of the coupling constant g_0 such that

$$f(g) \to \infty \quad \text{for } g \to g_0 \ .$$

One can then set

$$a = a(g) = r/f(g)$$

and obtain the continuum limit by letting $a \to 0$ while $g \to g_0$ with

$$a = a(g) \quad \text{or} \quad g = g(a) \ .$$

The relationship $r = a\, f(g)$ only sets a scale, i.e. allows one to assign a definite value to the lattice spacing in correspondence to definite values of g close to g_0. However if one can also calculate other quantities, e.g. a mass $m = a^{-1} f_1(g)$, then specific predictions follow for the continuum limit. One finds for instance

$$m\, r = \lim_{g \to g_0} f(g) f_1(g) \ .$$

Notice that the function $f_1(g)$, as well as all the corresponding functions appearing in the expressions for other observables, must behave for $g \to g_0$ in a way such that a finite continuum limit ensues for m. We express this property by saying that g_0 is a scaling critical point.

It is apparent from all of the above discussion that in the definition of a continuum limit g cannot be taken arbitrarily, but must approach a scaling critical point g_0. For non-Abelian models, relevant to the theory of strong interactions, $g_0 = 0$. Strong coupling results must therefore be extrapolated to weak coupling, and one faces the always difficult task of obtaining information on the system for intermediate values of the coupling constants. During the last few years it has been found that such information can be obtained by the use of numerical methods, which effectively simulate with high speed computers the quantum fluctuations of a physical system.

SECOND LECTURE

We shall illustrate methods of numerical simulation considering for definiteness lattice gauge theories where the gauge group is Z_N, i.e. the subgroup of U(1) consisting of rotations by multiples of $2\pi/N$.

The link variables U_x^{μ} can be represented by complex numbers (of unit modulus),

$$U_x^{\mu} = e^{\frac{2\pi i\, n_x^{\mu}}{N}} \quad , \quad n_x^{\mu} = 0, \ldots N-1 ,$$

or, simply, by the integers n_x^{μ}. The group operation is addition modulus N,

$$e^{\frac{2\pi i\, n_1}{N}} \; e^{\frac{2\pi i\, n_2}{N}} = e^{\frac{2\pi i\, n_3}{N}} \qquad \text{with } n_3 = n_1 + n_2 \pmod{N} .$$

Our considerations extend indeed very easily to any system with discrete gauge group. The group elements U can be labelled by integers n, and the composition law can be codified in a multiplication table ,

$$(n_1, n_2) \rightarrow n_3 = MT(n_1, n_2) \ .$$

Plaquette variables and vacuum expectation values are defined as follows:

$$n_\square = n_x^\mu + n_{x+\hat\mu}^\nu - n_{x+\hat\nu}^\mu - n_x^\nu \qquad (\mathrm{mod}\ N)$$

$$E_\square = 1 - \cos 2\pi \frac{n_\square}{N}$$

$$\langle O \rangle = \frac{\sum_{\{n_x^\mu\}} O(n)\ e^{-\beta \sum_\square E_\square}}{\sum_{\{n_x^\mu\}} e^{-\beta \sum_\square E_\square}} \ .$$

The sums appearing in the above equation are finite but, if the lattice extends for M sites in each dimension, the number of terms in the sums is N^{4M^4} and a direct summation, even with the highest speed computers, is impossible but for extremely small values of M.

One resorts to methods of numerical sampling, or Monte Carlo simulations (which apply also to spin systems). The collection of gauge variables is defined as a configuration C ,

$$C \equiv \{n_x^\mu\} \ .$$

We have

$$\langle O \rangle = \sum_C O(C)\ e^{-S(C)} \Big/ \sum_C e^{-S(C)} \qquad ;$$

the denominator in the above equation will be frequently denoted by Z. If we can define a stochastic sequence

$$C_1 \to C_2 \to \; \dots\dots \to C_i \to C_{i+1} \to \dots$$

with the property that the probability $P^{(i)}(C)$ of encountering a definite configuration C at the ith step approaches

$$P^{(i)}(C) \to \frac{e^{-S(C)}}{Z}$$

for $i\to\infty$, then it will be possible to approximate the quantum averages with averages taken of a larger number N of configurations occuring in the sequence

$$\langle O\rangle \approx \frac{1}{N} \sum_{i=i_o+1}^{i_o+N} O(C_i) \; .$$

Requirements are that all of relevant parts of phase space are explored and that N is large enough to guarantee a sufficiently accurate determination of $\langle O\rangle$.

The passage from one configuration to the next is effected according to a transition probability $p(C\to C')$, satisfying

$$p(C \to C') \geq 0 \quad , \qquad \sum_{C'} p(C\to C') = 1 \; .$$

If p obeys a property of detailed balance

$$p(C\to C')/p(C'\to C) = e^{-S(C')}/e^{-S(C)} \quad ,$$

then $e^{-S(C\;)}/Z$ is an eigenvector of the stochastic process. Indeed

$$P(C') = \sum_C \frac{e^{-S(C)}}{Z} \quad p(C\to C') = \sum_C \frac{e^{-S(C')}}{Z} p(C'\to C) =$$

$$= \frac{e^{-S(C')}}{Z} \quad .$$

Although detailed balance is not an absolute requirement for convergence to the appropriate distribution, it is satisfied by all the algorithms used in practical simulations. The most widely used algorthm is an application of the algorithm originally introduced by Metropolis et al.[4] It proceeds as follows.

Upgrade U_x^{μ} with all other $U_{x'}^{\mu}$ fixed:

select a new candidate $\hat{U}_x^{\mu}$ according to $p_0(U\to\hat{U})$;

$$p_0 \text{ obeys } p_0(U \to \hat{U}) = p_0(\hat{U} \to U) \;;$$

evaluate $\Delta S = S(\hat{U};\, U'\ldots) - S(U;U'\ldots)$ (this requires only a local computation) ;

if $\Delta S \leq 0$ accept and $U \to \hat{U}$;

if $\Delta S > 0$ accept (and $U \to \hat{U}$) with probability $e^{-\Delta S}$, otherwise reject.

Proceed to another link variable $U_{x'}^{\mu'}$.

The algorithm satisfies detailed balance:

$$\frac{p(U\to\hat{U})}{p(\hat{U}\to U)} = \underset{S(\hat{U})>S(U..)}{(\text{assuming} \quad)} = \frac{p_0(U\to\hat{U})\,e^{-\Delta S}}{p_0(\hat{U}\to U)} = e^{-\{S(\hat{U},\ldots)-S(U,\ldots)\}} =$$

$$= \frac{e^{-S(\hat{U},\ldots)}}{e^{-S(U,\ldots)}} \quad .$$

For the computer implementation one keeps all of the n_x^{μ} in memory. Then the steps are as follows:

→ select link variable n_x^{μ}, calculate contribution to S from all plaquettes involving n_x^{μ} (appropriate boundary conditions must be implemented);

extract new candidate $\tilde{n}_x^{\mu}$;

calculate same contribution to S with $\tilde{n}_x^\mu$ replacing n_x^μ ;
calculate $\exp(-\Delta S)$;
extract r, $0 \le r \le 1$;
if $r \le \exp(-\Delta S)$, $n_x^\mu \to \tilde{n}_x^\mu$.

Alternative algorithms which are frequently used are the modified Metropolis algorithm:

the step $U_x{}^\mu \to \hat{U}_x{}^\mu$ is performed m times at fixed x, μ before proceeding to new x', μ';

the heat bath algorithm:

if we let $m \to \infty$, then
$p(U \to \hat{U}) = p(\hat{U}) \propto e^{-S(\hat{U};\; U' \text{ fixed})}$;
the heat bath method may be convenient if a fast algorithm is available to extract $\hat{U}$ according to $p(\hat{U})$.

Results for four-dimensional Z_N lattice gauge models.
Monte Carlo simulations can be used to evaluate $E \equiv \langle E_\square \rangle$ as function of β (as a reminder:

$$e^{-S} = e^{-\beta \sum_\square E_\square} \quad , \; 0 \le E_\square \le 2 \; ;$$

$$\langle E_\square \rangle = 0 \quad \text{for} \quad \beta = \infty \quad , \qquad \langle E_\square \rangle = 1 \quad \text{for} \quad \beta = 0) \; .$$

Results obtained by varying β very slowly in the course of the simulation exhibit hysteresis loops which signal phase transitions. Figures 1, 2 and 3 illustrate MC results for the Z_2 Z_3 Z_4 Z_5 Z_6 and Z_8 models[5].
The Z_2 Z_3 and Z_4 systems exhibit a structure of two phases separated by a first order transition; the Z_N systems, for $N>4$, exhibit a 3 phase structure, with 2 continuous phase transitions at

$$\left.\begin{array}{lll} \beta'_{cr} & \to & \text{const} \\ \beta''_{cr} & \sim & N^2 \end{array}\right\} \quad \text{for } N \to \infty \; .$$

MC simulations where the initial configuration has a mixed phase structure can be used to investigate the nature of the phase transitions (see figure 4). Simulations done at several values of β in the neighbourhood of a phase transition show a rather linear drift of E as function of the number of iterations if the transition is of first order. This corresponds to the stabel phase expanding and overtaking the whole system. A markedly different behaviour is observed with continuous phase transitions.
The U(1) gauge system, which can also be considered to be the limit of a Z_N system for $N \to \infty$, has a 2 phase structure [5,6]:

	β_{cr}	$\longrightarrow \beta = \frac{1}{g^2}$.
strong coupling phase	weak coupling phase	

For $\beta < \beta_{cr}$ the system is in a strong coupling phase: the force between static charges tends to a constant value as the separation increases to infinity and electric charges are confined. For $\beta > \beta_{cr}$ the system is in a weak coupling phase: the force between static charges exhibits a Coulombic behaviour. The continuum limit (i.e. the free quantized photon field) is recovered for $\beta > \beta_{cr}$.

Simulations of systems with continuous gauge group can be done along the lines illustrated above, but require more memory to represent the gauge dynamical variables and much more arithmetic to implement the group multiplications needed for the evaluation of ΔS. By separating the dynamical variables into suitable subsets, the upgrading of all the variables belonging to the same subsets can be done simultaneously (red-black algorithms). This allows optimal utilization of vectorized, parallel or special purpose processors [7].

THIRD LECTURE

Quantum Chromo-Dynamics or QCD, the gauge theory of strong interactions, is based on quark matter fields $\psi_{x,c}, \bar{\psi}_{x,c'}$ which transform according to the fundamental representation of SU(3), and gauge fields $U_{x,cc'}$ which transform according to the adjoint representation of SU(3). (Spin and other internal indices of ψ and $\bar{\psi}$ are left implicit.)

A typical gauge invariant coupling would be given by

$$\sum_{cc'} \bar{\psi}_{x+\hat{\mu},c} \, U^{\mu}_{x,cc'} \, \psi_{x,c} \; .$$

Many important dynamical effects are already present at the level of the pure gauge system. We shall therefore consider a theory with gauge dynamical variables only, in this lecture, and postpone the treatment of systems with quark fields to the next one.

By arguments based on perturbative expansions [8] one can show that $g_{cr} = 0$ defines a possible continuum limit and that, if the continuum theory is indeed recovered for vanishing coupling constant, the lattice spacing must behave for small g as follows:

$$a(g) \propto \left(\frac{8\pi^2}{33} \frac{6}{g^2}\right)^{\frac{51}{121}} e^{-\frac{4\pi^2}{33} \frac{6}{g^2}}$$

$$\text{or} \quad \propto \left(\frac{8\pi^2}{33} \beta\right)^{\frac{51}{121}} e^{-\frac{4\pi^2}{33} \beta} \qquad \left(\beta = \frac{6}{g^2}\right) .$$

It is convenient to define a lattice scale parameter Λ by

$$a(\beta) = \frac{1}{\Lambda} \left(\frac{8\pi^2}{33}\beta\right)^{\frac{51}{121}} e^{-\frac{4\pi^2}{33}\beta} (1 + O(1/\beta)) .$$

If a physical quantity q has dimensions of $[\text{mass}]^d = [\text{length}]^{-d}$, it will be given by an expression

$$q(\beta) = f(\beta) a^{-d} .$$

One then expects a scaling behaviour

$$f(\beta) \sim c[\Lambda a(\beta)]^d$$

which implies that the continuum value of q is $c\Lambda^d$.

Very early MC simulations gave evidence for the absence of a phase transition in the SU(3) model (and also in the SU(2) system, which for many aspects behaves similarly to the SU(3) system itself[9]). This implies that the properties of the strong coupling phase extend all the way to the scaling domain and persist in the continuum limit. Then numerical simulations were used to produce evidence for the scaling behaviour of many observables as $\beta \rightarrow \infty$ and to calculate their continuum values.

We illustrate now the calculation of the force between static quarks and of the string tension, i.e. the limiting value of the force for very large separation, using results from a very recent MC study done with a Cyber 205 vectorized processor by D. Barkei, K. Moriarty and the author [10].

Let us define by U_{ij} the product of U_x^μ dynamical variables along a rectangular closed path, extending for i and j links, and by W_{ij} the quantity

$$W_{ij} = \frac{1}{3} \text{Re Tr } U_{ij} .$$

By expressing the vacuum expectation value of W_{ij} as the result of the propagation in time of a quark-antiquark pair created at a separation of j lattice links and by using an expansion into a complete set of physical states, one can prove

$$\langle W_{ij}\rangle \approx |\langle q\bar{q}|0\rangle| e^{-E_{q\bar{q}}t} \quad \text{for large t.}$$

It follows

$$E_{q\bar{q}}(r = ja) = \lim_{i\to\infty} -\frac{1}{a} \ln \frac{\langle W_{ij}\rangle}{\langle W_{i-1,j}\rangle}$$

and

$$F(\sqrt{j(j-1)}a) = \frac{E(ja)-E((j-1)a)}{a} =$$

$$= \lim_{i\to\infty} -\frac{1}{a^2} \ln \frac{\langle W_{ij}\rangle\langle W_{i-1,j-1}\rangle}{\langle W_{i-1,j}\rangle\langle W_{i,j-1}\rangle} \equiv \lim_{i\to\infty} -\frac{1}{a^2} X_{ij} .$$

In particular if $F(r) \underset{r\to\infty}{\to} \sigma$ (string tension), then

$E(r) \underset{r\to\infty}{\sim} \sigma r$ and, for $i,j \to \infty$,

$$\langle W_{ij}\rangle \sim e^{-\sigma rt} = e^{-\sigma A}$$

with A = area of the rectangle enclosed by the path. This generalizes to any closed path γ. One expects a behaviour

$$\langle W_\gamma\rangle \sim e^{-\sigma A}$$

as the area A enclosed by γ becomes larger and larger. A nonvanishing σ (area law) signals confinement.

The expectation values $\langle W_{ij}\rangle$ have been calculated on a $16^3 \times 32$ lattice for sizes up to i = j = 8 at the values

$\beta = 5.6$, 5.8, 6, 6.2, 6.4, 6.6. These values of the coupling parameter are known to span the domain where the transition toward the scaling regime takes place. Over 1100 MC iterations, i.e. upgradings per dynamical variable, have been performed at each of the values of β, and the expectation values $\langle W_{ij} \rangle$ have been calculated averaging over 100 configurations separated by 10 iterations each.
To evaluate the force one needs to extrapolate

$$X_{ij} \underset{i\to\infty}{\rightarrow} \frac{F(a\sqrt{j(j-1)})}{a^2}$$

to large values of i. A simple functional form

$$X_{ij} \approx b_j + \frac{c_j}{i(i-1)}$$

(motivated by the expected behaviour for small i) has been assumed and fit through the data points. Figure 5 illustrates this fitting procedure for $\beta = 6.2$.
The values thus found for b_j are in turn consistent, for all β, with an expression for the force

$$b_j \approx \frac{\alpha}{j(j-1)} + \tau a^2$$

or equivalently

$$F(r) \approx \frac{\alpha}{r^2} + \sigma,$$

consisting of the superposition of a constant and a Coulombic term. This behaviour is illustrated, always for $\beta = 6.2$, in the bottom graph of figure 5 and in figure 6. The values found for σa^2 appear to scale as

$$(\Lambda a)^2(\beta) = \left(\frac{8\pi^2}{33}\beta\right)^{\frac{51}{121}} e^{-\frac{4\pi^2\beta}{33}},$$

with no substantial $O(\frac{1}{\beta})$ corrections for $\beta \geq 6$. One can thus determine

$$\Lambda \approx 9.6 \times 10^{-3} \sqrt{\sigma} .$$

The scaling behaviour of σa^2 is illustrated in figure 7. (The fact that the last point, for $\beta = 6.6$, is above the scaling curve is attributed to an overestimate of the asymptotic value of the force, due to the smallness of the lattice spacing for such value of β.)

The above results allow one to express the lattice spacing a in function of β and σ. Then all the values found for the force (at all values of β) can be re-expressed in physical units of σ, for the force, and $\sigma^{-1/2}$, for the separation. The data so rescaled are shown in figure 8. All of the points lie, within statistical errors, on a universal curve, which conforms the scaling behaviour.

Perturbative arguments [11] can be used to determine the expected short distance behaviour of the force, with no free parameters once Λ is fixed. The line in figure 8 represents the theoretical prediction for the short distance behaviour,and one sees that theory and numerical results are in agreement.

In a confining pure gauge system the spectrum of states (also referred to as glueballs) begins with a lowest stable state of mass $m_g > 0$; m_g is the mass-gap of the theory. It can be evaluated from the numerical calculation of

$$G_{\vec{x},t} = \langle E^{\mu\nu}_{\vec{x}t}\, E^{\mu\nu}_{o} \rangle - \langle E_{\square} \rangle^2$$

where $E^{\mu\nu}_{\vec{x},t}$ represents the energy of a plaquette at $\vec{x},t$. By inserting of a complete set of physical states, $G_{\vec{x},t}$ can be re-expressed as follows:

$$G_{\vec{x},t} = \sum_{n\neq\emptyset} |\langle n|E_{\Box}|\emptyset\rangle|^2 \; e^{-E_n t + i\vec{p}_n\cdot\vec{x}} \quad .$$

It is even better to consider

$$\tilde{G}_t = \sum_{\vec{x}} G_{\vec{x},t} = \sum_{\substack{n\neq\emptyset \\ \vec{p}_n=0}} |\langle n|E_{\Box}|0\rangle|^2 e^{-m_n t} \quad .$$

Clearly information on the lower masses in the spectrum of pure gauge excitations can be obtained from a study of the rate of decay of $\tilde{G}_t$ for large t.
MC evaluations of $\tilde{G}_t$, or of similar quantities, related in any case to the response of the system to a perturbation and therefore to the correlation $G_{\vec{x},t}$, have allowed several investigators to estimate m_g [12].

Thermodynamical effects in a quantum field theory can be calculated by expressing the partition function

$$Z(T) = \sum_n \langle n|e^{-\frac{H}{T}}|n\rangle$$

as a functional integral

$$Z(T) = \iint_{\text{configurations}} e^{-S} \quad ,$$

where the propagation occurs for a finite extent $t = \frac{1}{T}$ in time, and periodic boundary conditions in time are imposed.
Thus one can study thermodynamical properties of lattice quantum field theories by considering a system extending for n_t spacings in time, n_s spacings along the spatial directions, and $n_t \ll n_s$. This describes a medium at a physical temperature

$$T = \frac{1}{n_t a} \quad .$$

(These considerations also imply a warning on the use of very small lattices.)
The free energy of an individual quark can be evaluated recalling that a static source at $\vec{x}$ implies a modification of the action

$$S \to S + ig\int A^{0}(\vec{x},t)dt$$

in the continuum case.
The free energy of an isolated quark relative to the vacuum is therefore formally given in the continuum theory by

$$e^{-\frac{\Delta F}{T}} = \frac{\sum_n \langle n|e^{-\frac{H}{T}} e^{i\int A_0 dt}|n\rangle}{\sum_n \langle n|e^{-\frac{H}{T}}|n\rangle} = \langle e^{i\int A_0 dt}\rangle \quad .$$

But the lattice analogy of $e^{i\int A_0 dt}$ is the product of the $U^{\mu=4}_{\vec{x},t}$ gauge variables along a line extending in the time direction (a path closed by virtue of the periodic boundary conditions). Let such product be denoted by $U_{\vec{x}}$ and let

$$W_{\vec{x}} = \frac{1}{3}\,\mathrm{Tr}\,U_{\vec{x}} \quad .$$

We then have

$$e^{-\frac{\Delta F}{T}} = \langle W_{\vec{x}}\rangle \quad .$$

If quarks are confined, $\Delta F = \infty$ and $\langle W_{\vec{x}}\rangle = 0$.
If thermal fluctuations deconfine quarks, ΔF = finite and $\langle W_{\vec{x}}\rangle > 0$. MC simulations have shown that $\langle W_{\vec{x}}\rangle = 0$ for large enough n_t and small enough $a(\beta)$ (remember $T = 1/n_t\, a(\beta)$), but a transition to $\langle W_{\vec{x}}\rangle > 0$ is observed when, modifying n_t or β, T is brought above a critical value T_c.
Thus the thermodynamical properties of the gauge medium and

in particular the presence of a confining phase transition can be and have been studied by MC numerical techniques [13].

FOURTH LECTURE

Numerical calculations involving fermionic matter fields. The complete action for QCD is the sum of the pure gauge action S_G and the action of the quark matter fields S_M. S_M will be of the form

$$S_M = \sum_{xy} \bar{\psi}_x \, (D + m)_{xy} \, \psi_y \quad ,$$

where color and possible spin indices of the quark fields ψ_x, $\bar{\psi}_x$ have been left implicit, and the matrix D+m represents the generalization to the lattice of Dirac's operator of the continuum theory:

$$\gamma^\mu (\partial_\mu + ig \, A^i_\mu \, \tau_i) + m \quad .$$

Such a generalization is not obvious. A straightforward lattice transcription of Dirac's operator, obtained by replacing first order derivatives by central differences, leads to a theory which exhibits in the continuum limit a 16-fold degeneracy of states (2^d for a d-dimensional system). A modification of the action due to Wilson[14] removes the degeneracy at the cost of breaking chiral symmetry when $m = 0$; in another formulation, originated by Kogut and Susskind [15], which preserves a continuum chiral symmetry for $m = 0$, the spin components of ψ_x are distributed among the vertices in a 2^d cell. Then the fermions are represented by single component, anticommuting dynamical variables ψ_x, $\bar{\psi}_x$ ("single component" refers only to spin - ψ and $\bar{\psi}$ are still 3-component complex vectors in SU(3) color space). The Kogut-Susskind formulation reduces the unwanted degeneracy to 4. Other generalizations are possible; we need not go into details. In any event S_M will contain nearest neighbor couplings of the type

$$\bar{\psi}_{x+\hat{\mu}} \, U_x^{\mu} \, \psi_x \quad .$$

The U_x^{μ} variables needed to make the couplings gauge invariant render D+m explicitly dependent on the gauge fields, thus inducing the coupling between matter and gauge degrees of freedom.

For Monte Carlo simulations one needs to approximate quantum averages taken over gauge and fermionic degrees of freedom, e.g.

$$G_{x,y} = \langle \bar{\psi}_x \, \Gamma \, \psi_x \, \bar{\psi}_y \, \Gamma \, \psi_y \rangle =$$

$$= Z^{-1} \int \Pi d \, U_x^{\mu} \Pi \, (d\bar{\psi}_x d\psi_x) \bar{\psi}_x \, \Gamma \, \psi_x \, \bar{\psi}_y \, \Gamma \, \psi_y \, e^{-S_G - S_M} \quad ,$$

where Γ represents some suitable generalization of the continuum-matrices and

$$Z = \int \Pi \, dU_x^{\mu} \Pi \, (d\bar{\psi}_x \, d \, \psi_x) e^{-S_G - S_M} \quad .$$

From the discussion of the mass gap (Lecture 3), it should be clear that information about the quark model spectrum can be derived from the analysis of the rate of decay in time of

$$\tilde{G}_t = \sum_{\vec{x}} G_{\vec{x},t,0} =$$

$$= \sum_{\substack{n \\ \vec{p}_n = 0}} |\langle n | \bar{\psi} \, \Gamma \, \psi | 0 \rangle|^2 e^{-m_n t} \quad .$$

It is however not obvious how one should sample integrals over elements of a Grassman algebra, i.e. over the anti-commuting variables $\bar{\psi}$ and ψ. A possibility is to exploit the fact that such integrals have really been introduced to represent sums over occupation numbers of quark and anti-quark states. Such states are finite in number, in a lattice of finite volume, and thus the quantum averages are, for the fermionic part, sums over a finite number of binary variables, 0 and 1. This reformulation has been successfully used for the numerical simulation of 2-dimensional (1 space - 1 time) systems [16]. In more than 1 spatial dimension, however, the terms appearing in the sums are no longer positive definite, and a probabilistic interpretation as well as a practical sampling procedure become impossible.

For 4-dimensional systems all numerical computations have exploited the fact that the action S_M is bilinear in the fermionic fields. The Gaussian integration over $\bar{\psi}_x$ and ψ_x can be done explicitely and one finds (always referring to $G_{x,y}$ for exemplification)

$$G_{x,y} = \tilde{Z}^{-1} \int \Pi \, dU_x^\mu \langle \bar{\psi}_x \Gamma \psi_x \bar{\psi}_y \Gamma \psi_y \rangle_u \cdot$$

$$\cdot \, e^{-S_G} \, \mathrm{Det} \, (D(U)+m) \, ,$$

where

$$\tilde{Z} = \int \Pi \, d \, U_x^\mu \, e^{-S_G} \, \mathrm{Det} \, (D(U)+m)$$

and

$$\langle \bar{\psi}_x \Gamma \psi_x \bar{\psi}_y \Gamma \psi_y \rangle_u$$

stands for the expectation value of the fermionic observable in the presence of a fixed gauge field configuration U_x^μ.

This expectation value can be expressed in terms of quark propagators. For instance, assuming for simplicity that Γ is such that only connected diagrams are present,

$$<\bar{\psi}_x \, \Gamma \, \psi_x \, \bar{\psi}_y \, \Gamma \, \psi_y>_u =$$

$$= \mathrm{Tr} \, \{\Gamma (D(U)+m)^{-1}_{xy} \, \Gamma (D(U)+m)^{-1}_{yx}\} \, .$$

Thus the quantum averages have been reduced to averages over ordinary bosonic variables only (integrals over the group manifolds, as in the pure gauge system). A numerical calculation of G_{xy} still faces two serious levels of computational complexity.

A) The measure is now of the type

$$e^{-S_{eff}}$$

with

$$S_{eff} = S_G - \log \mathrm{Det} \, (D(U)+m) = S_G - \mathrm{Tr} \ln \, (D(U)+m) \, .$$

For a Monte Carlo simulation one needs to evaluate the change Δ induced by a change $U^{\mu}_{x} \to \hat{U}^{\mu}_{x}$. Whereas S_G is local, the second term makes S_{eff} non-local. The values of all the dynamical variables $U^{\mu'}_{x'}$ enter in the calculation of the variation ΔS_{eff} induced by any one of them. An exact calculation of ΔS_{eff}, while feasible, becomes so time consuming to make the implementation of the MC algorithm for any lattice of realistic size impossible. The only way out appears to be, at present, in replacing the exact calculation of ΔS_{eff} with some fast, but only approximate computation. While we shall be mentioning more sophisticated scheme of approximation later, the simplest approximate calculation consists in neglecting the second term in S_{eff} altogether, and to sample the gauge field configurations according to the pure gauge measure $\exp(-S_G)$.

Such an approximation, which has been called the "quenched" [17] or "valence" [18] approximation, is not necessarily as drastic as it may seem. It corresponds to neglecting (in terms of a perturbative expansion) the effects of internal quark loops, and several arguments of phenomenological and theoretical nature have been brought to justify schemes making the same or even more restrictive (to planar diagrams) approximations.

B) Whether the gauge field configurations are sampled according to $\exp(-S_o)$ or some method taking into account the effects of internal quark loops (i.e., the effects of dynamical fermions), one still needs to evaluate

$$(D(U) + m)^{-1}_{xy} \quad .$$

This is another time consuming operation, which requires the solution of a large system of linear equations (not a full matrix inversion, since generally only a few elements of $(D+m)^{-1}$ are required, for instance all those where y is set at the origin). Relaxation methods, or methods based on the algorithm of conjugate gradients, are normally used to find the propagators. The fact remains that only few inversions (of the order of a few tens, or at most a few hundreds) can be done in a definite calculation, and then estimates of masses and similar fermionic observables are based on relatively limited samples.

Many calculations of the masses of lowest states in the quark model spectrum and of other fermionic observables, such as the chiral symmetry breaking condensate $\langle\bar{\psi}\psi\rangle$, have nevertheless been performed, in the quenched approximation [19]. The results of different calculations exhibit discrepancies larger than what can be justified by statistical fluctuations, which can however be easily understood considering the various kinds of computational limitations. Among the most important:

- The generally small extents in space and time of the lattices used for the calculations.
  Only recently lattices extending for as much as $10^3 \times 20$ (space and time sizes) have been considered, and, assuming that $\beta \geq G$ is required to have enough resolution in the lattice approximation to the continuum, and that the values quoted for the string tension in the previous lecture are correct, this makes $a \leq \sim .125$ fm, $10\,a \leq \sim 1.25$ fm, a size certainly not too large.
- The limited sampling.
- The possible distortions introduced by the lattice transcription of the Dirac equation. While all formulations ought to become equivalent in the continuum limit, the differences in the spectra determined within different formalisms may still be sizeable at the values considered for a.

Bearing the above limitations in mind, the results have been generally successful. All give evidence for a dynamical realization of chiral symmetry, with $m_\pi^2 \to 0$ as $m_q \to 0$. The other masses (m_ρ, m_p, etc.) appear to remain at finite values when $m_q \to 0$, values generally compatible with the scale set by the calculations of pure gauge observables and with the experimental mass splittings among various states. I shall not quote here specific numbers, because the situation is still too much in a state of flux to quote definite values, and a systematics of all investigations performed and all results obtained is beyond the scope of these lectures.

Finally, let me briefly mention just one [20] (among several interesting methods which have been proposed [21]) possibility to go beyond the quenched approximation.

If $\delta U_x^\mu = \hat{U}_x^\mu - U_x^\mu$ is sufficiently small (and one would think that the nature of the continuum limit is such that it should be possible to explore all of the relevant phase space with small fluctuations δU_x^μ from configuration to configuration, although this may show the simulation), one can linearize the variation of S_{eff}:

$$\Delta S_{eff} = \Delta S_G - \text{Tr } \Delta \ln (D(U)+m) \approx$$

$$\approx \Delta S_G - \text{Tr } (D(U)+m)^{-1} \delta D(U) =$$

$$= \Delta S_G - \text{Tr } (D(U)+m)^{-1} \frac{\delta D}{\delta U} \delta U .$$

The structure of the lattice Dirac operator is such that $\frac{\delta D}{\delta U}$ is local. Thus the linearized variation will be easily calculable if a few relevant matrix elements (involving neighbouring sites) of $(D(U)+m)^{-1}$ are known. These will be given by expectation values

$$<\psi_x \bar{\psi}_y>_u$$

where the sites x and y are neighbouring. Indeed, to the whole expression $(D(U)+m)^{-1} \frac{\delta D}{\delta U}$, which multiplies δU_x^μ in the linearized variation of S_{eff}, one can give the form of expectation value, in the given gauge field configuration, of a fermionic current j_x^μ, involving fermions at neighbouring sites.

One thus finds

$$\Delta S_{eff} \approx \Delta S_G - \text{Tr } \{<j_x^\mu>_u \delta U_x^\mu \},$$

an expression quite reminiscent of the formula for the continuum action. The physical significance of the above equation is that the dynamical reaction of the fermions manifests itself (to first order in δU_x^μ) via the current that the gauge field U_x^μ excites in the vacuum.

The crucial observation is now that the vacuum expectation value of the current is the same (given identical couplings to the gauge field) for a fermionic or for a bosonic system. It is only the sign with which

$$\mathrm{Tr}\ \{\langle j_x^\mu\rangle_u\ \delta U_x^\mu\}$$

enters in ΔS_{eff} which differentiates between the two cases (it would be + for bosonic fields). Thus one may introduce a parallel set of bosonic variables $\bar{\phi}_x \phi_x$, the pseudofermions, and calculate $\langle j_x^\mu\rangle_u$ by a Monte Carlo simulation over these variables. The method proceeds then as follows. For a given gauge field configuration a few MC iterations over the pseudofermions are used to evaluate, approximately, $\langle j_x^\mu\rangle_u$. The values thus found are input in the upgrading of all the gauge field variables, approximating ΔS_{eff} with

$$\Delta S_G - \mathrm{Tr}\ \{\langle j_x^\mu\rangle_u\ \delta U_x^\mu\}\ .$$

The method has been tested in some cases [17,22,23] and appears to produce reasonably accurate results in situations where the dynamical effects of the fermions can be computed by other reliable means; it also appears computationally implementable, in the sense that the CP-time requirements, although larger than for a pure gauge simulation, are contained enough to allow for practical calculations. The method is currently being applied to problems, where the dynamical effects of the fermions are expected to be particularly relevant.

REFERENCES

1. K. Wilson, Phys. Rev. D10 (1974) 2445; Phys. Reports 23 (1975) 331.
2. For recent reviews see e.g.: M. Creutz, L. Jacobs and C. Rebbi, Phys. Reports 95 (1983) 203; Lattice Gauge Theories and Monte Carlo Simulations, ed. C. Rebbi (World Scientific, Singapore, 1983).
3. For a formulation based on a random lattice see, N.H. Christ, R. Friedberg and T.D. Lee, Nucl. Phys. B210

[FS6] (1982) 310.
4. N. Metropolis, A.W. Rosenbluth, M.N. Rosenbluth, A.H. Teller and E. Teller, J. Chem. Phys. 21 (1953) 1087.
5. M. Creutz, L. Jacobs and C. Rebbi, Phys. Rev. Lett. 42 (1979) 1390; Phys. Rev. D20 (1979) 1915.
6. B. Lautrup and M. Nauenberg, Phys. Lett. 95B (1980) 63.
7. See e.g., D. Barkai, K. Moriarty and C. Rebbi, Comp. Phys. Communications, in press, 1984.
8. H.D. Politzer, Phys. Rev. Lett. 30 (1979) 1346; D. Gross and F. Wilczek, Phys. Rev. Lett. 30 (1973) 1343; Phys. Rev. D8 (1976) 3633.
9. M. Creutz, Phys. Rev. Lett. 43 (1979) 553; Phys. Rev. Lett. 45 (1980) 313.
10. D. Barkai, K. Moriarty and C. Rebbi, BNL preprint 1984.
11. A. Billoire, Phys. Lett. 104B (1981) 472.
12. B. Berg, A. Billoire and C. Rebbi, Ann. of Phys. 142 (1982) 185; M. Falcioni et al., Phys. Lett. 110B (1982) 295; K. Ishikawa, G. Schierholz and M. Teper, Phys. Lett. 110B (1982) 399; B. Berg and A. Billoire, Phys. Lett. 113B (1982) 65; K.H. Mütter and K. Schilling, Phys. Lett. 117B (1982) 75.
13. L. McLerran and B. Svetitsky, Phys. Lett. 98B (1981) 195; J. Kuti, J.J. Polonyi and K. Szlachanyi, Phys. Lett. 98B (1981) 199; K. Kajantie, C. Montonen and E. Pietarinen, Zeit. Phys. C9 (1981) 253; J. Engels, F. Karsch, H. Satz and I. Montvay, Phys. Lett. 101B (1981) 89; 102B (1981) 332; Nucl. Phys. B205 (1982) 545.
14. K. Wilson, in: New Phenomena in Subnuclear Physics, ed. A. Zichichi (Plenum Press, N.Y., 1977).
15. J. Kogut and L. Susskind, Phys. Rev. D11 (1975) 395.
16. R. Blankenbeckler, J. Hirsch, D. Scalapino and R. Sugar, Phys. Rev. Lett. 47 (1982) 1628.
17. E. Marinari, G. Parisi and C. Rebbi, Nucl. Phys. B190 (1981) 266.
18. D. Weingarten, Phys. Lett. 109B (1982) 57; Nucl. Phys. B215 [FS7] (1983) 1.

19. H. Hamber and G. Parisi, Phys. Rev. Lett. 47 (1982) 1792; Phys. Rev. D28 (1983) 247; F. Fucito, G. Martinelli, C. Omero, G. Parisi, R. Petronzio and F. Rapuano, Nucl. Phys. B210 FS6 (1982) 407; G. Martinelli, C. Omero, G. Parisi and R. Petronzio, Phys. Lett. 117B (1982) 434; A. Hasenfratz, P. Hasenfratz, C.B. Lang and Z. Kunszt, Phys. Lett. 117B (1982) 81; C. Bernard, T. Draper, K. Olynyk, Phys. Rev. D27 (1983) 227; R. Gupta and A. Patel, Caltech preprint CALT-68-966 (1982); H. Lipps, G. Martinelli, R. Petronzio and F. Rapuano, CERN preprint TH.3548 (1983); K.C. Bowler, G.S. Pawley, D.J. Wallace, E. Marinari and F. Rapuano, Nucl. Phys. B220 (1983) 137.

20. F. Fucito, E. Marinari, G. Parisi and C. Rebbi, Nucl. Phys. B180 (1981) 369.

21. D. Weingarten and D. Petcher, Phys. Lett. 99B (1981) 333; H. Hamber, Phys. Rev. D24 (1981) 951; D. Scalapino and R. Sugar, Phys. Rev. Lett. 46 (1981) 519; J. Kuti, Phys. Rev. Lett. 49 (1982) 183; A. Hasenfratz, P. Hasenfratz, Z. Kunszt and C.B. Lang, Phys. Lett. B110 (1982) 282.

22. H. Hamber, E. Marinari, G. Parisi and C. Rebbi, Phys. Lett. 124B (1983) 199.

23. S. Otto and M. Randeria, Nucl. Phys. B220 (1983) 479; A.N. Burkitt and R.D. Kenway, Edinburgh Univ. preprint (1983).

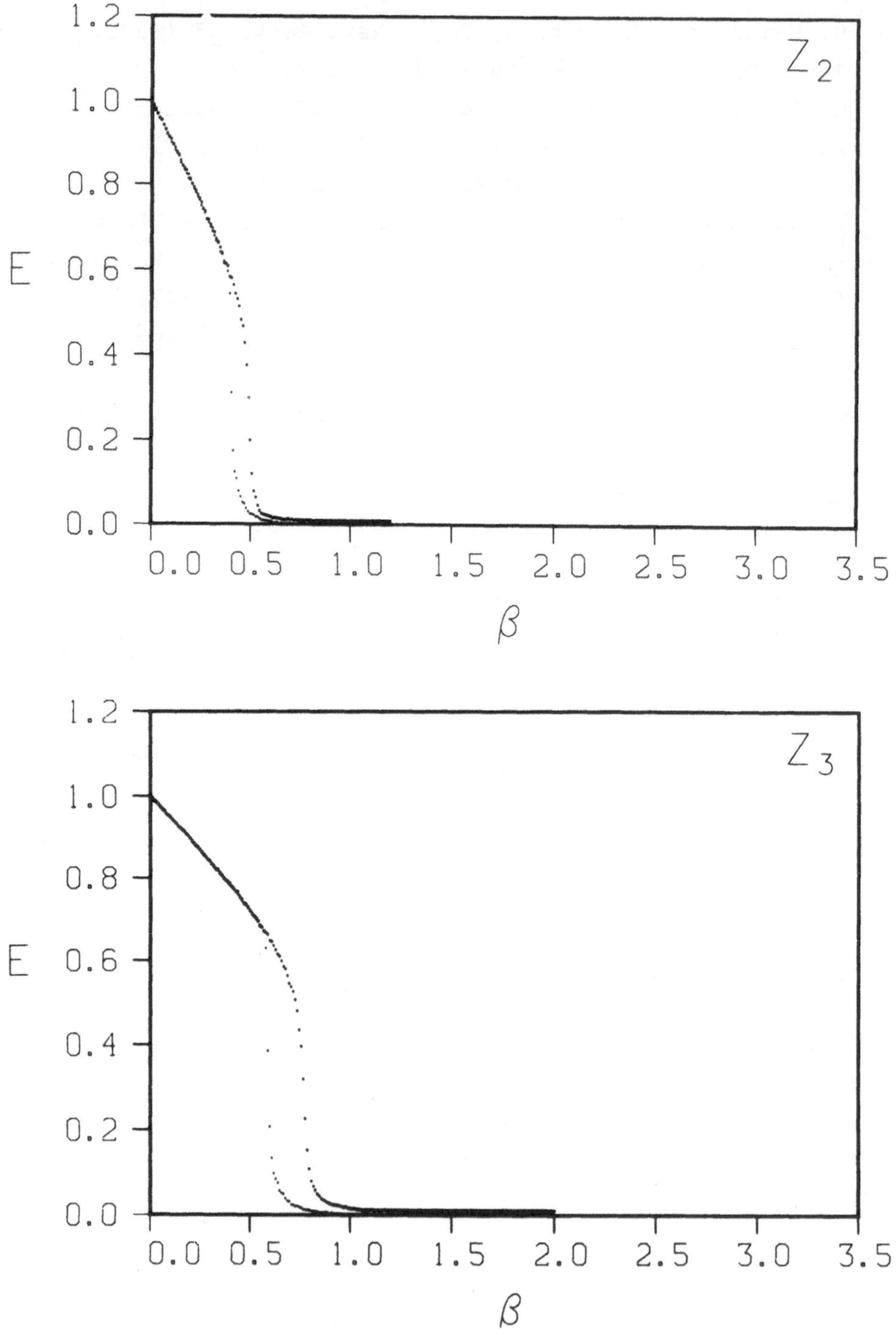

Z2
E
1.2
1.0
0.8
0.6
0.4
0.2
0.0
0.0
0.5
1.0
1.5
2.0
2.5
3.0
3.5
β
Z3
E
1.2
1.0
0.8
0.6
0.4
0.2
0.0
0.0
0.5
1.0
1.5
2.0
2.5
3.0
3.5
β

Fig.1.

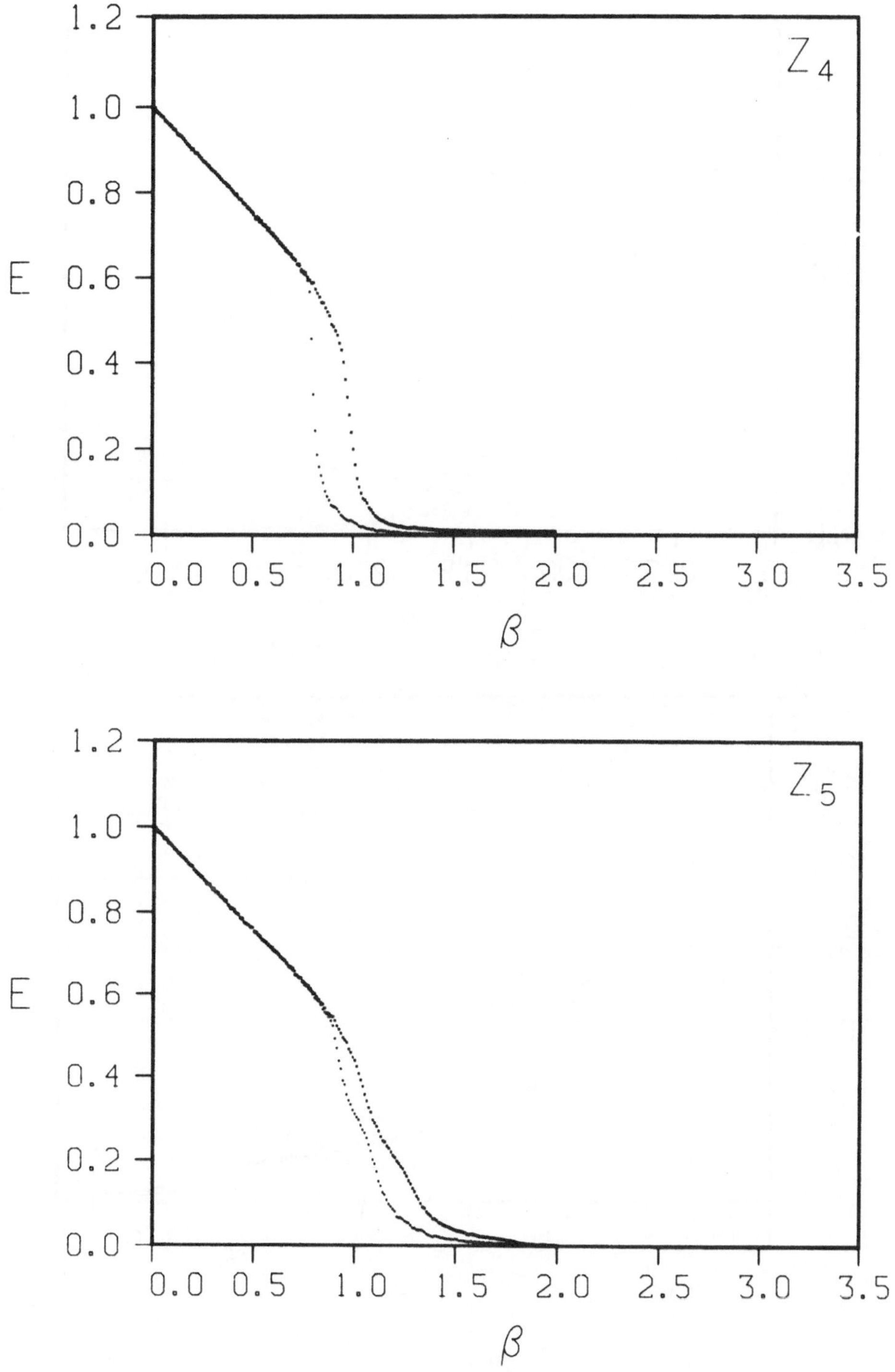
Z4
E
1.2
1.0
0.8
0.6
0.4
0.2
0.0
0.0 0.5 1.0 1.5 2.0 2.5 3.0 3.5
β
Z5
E
1.2
1.0
0.8
0.6
0.4
0.2
0.0
0.0 0.5 1.0 1.5 2.0 2.5 3.0 3.5
β

Fig.2.

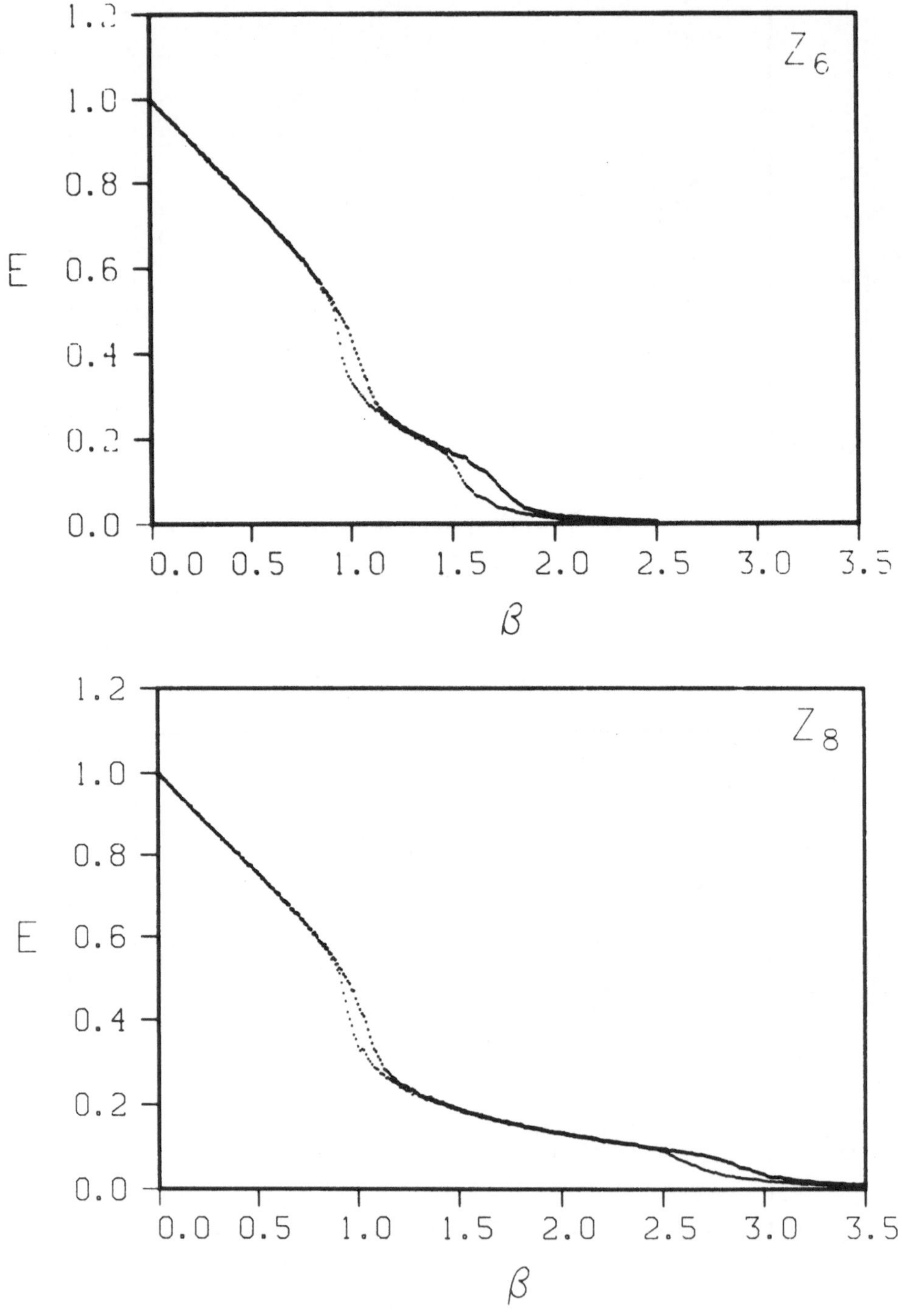

Z6
E
0.0
0.2
0.4
0.6
0.8
1.0
1.2
0.0 0.5 1.0 1.5 2.0 2.5 3.0 3.5
β
Z8
E
0.0
0.2
0.4
0.6
0.8
1.0
1.2
0.0 0.5 1.0 1.5 2.0 2.5 3.0 3.5
β

Fig.3.

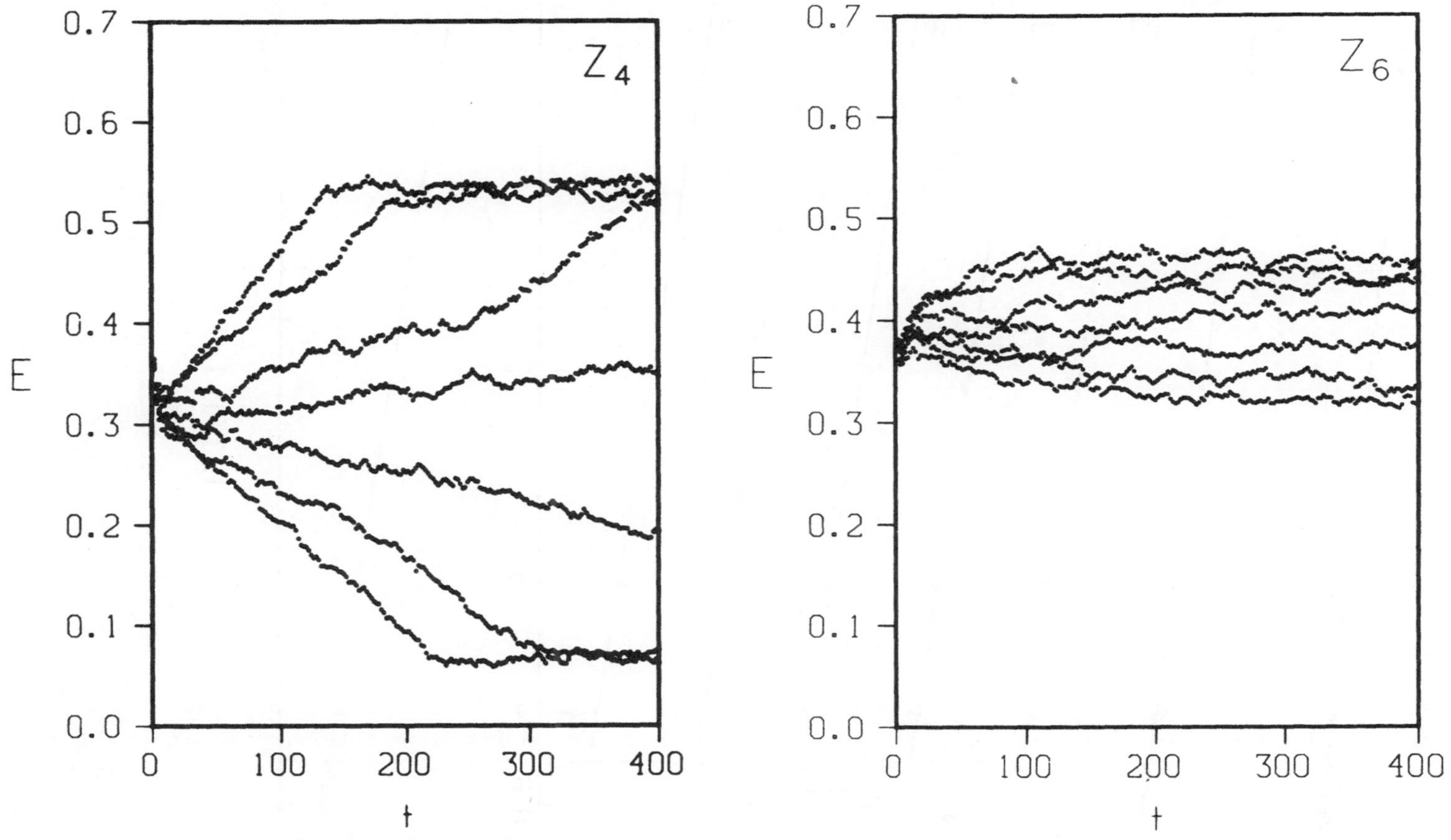

Fig.4.

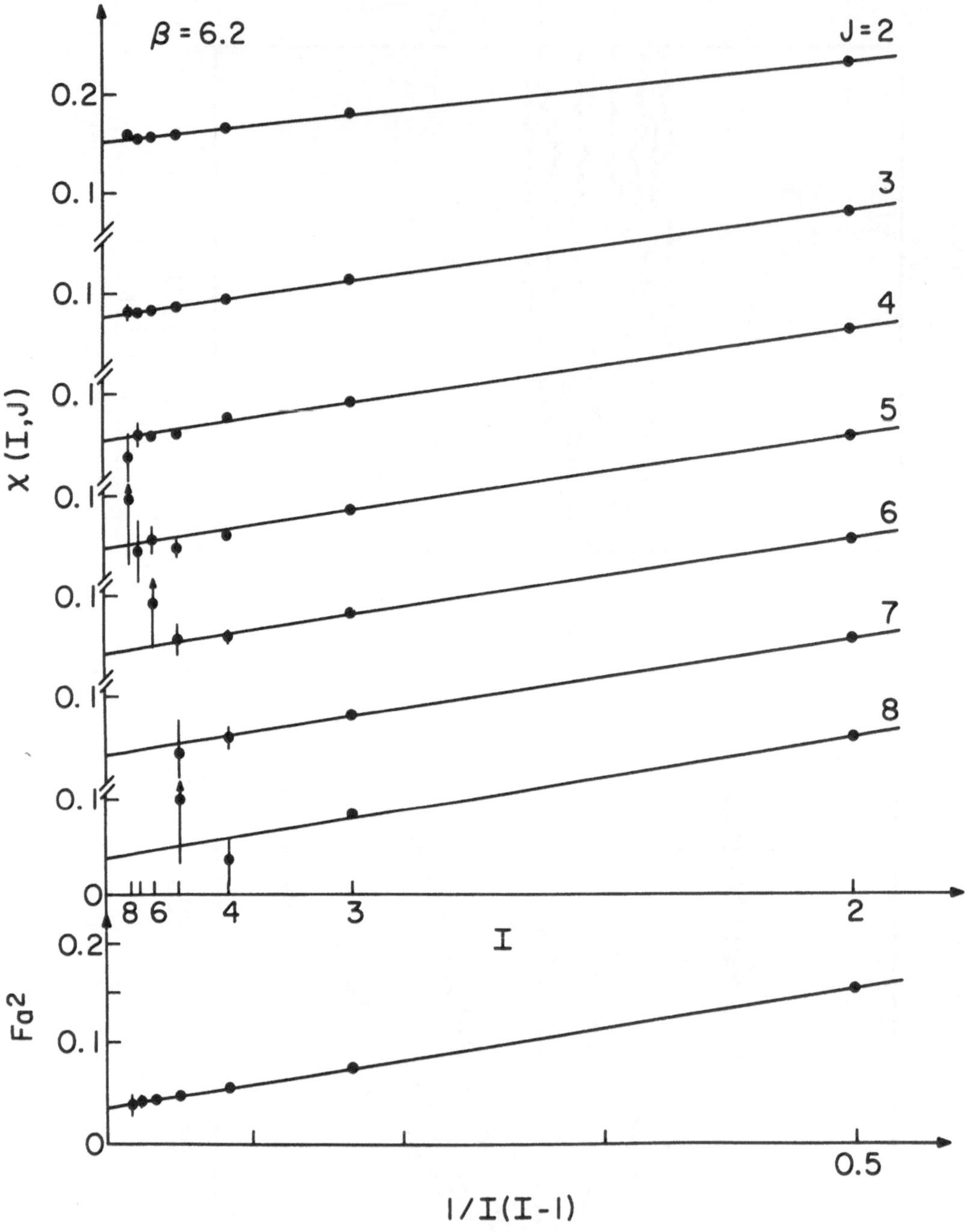

Fig.5.

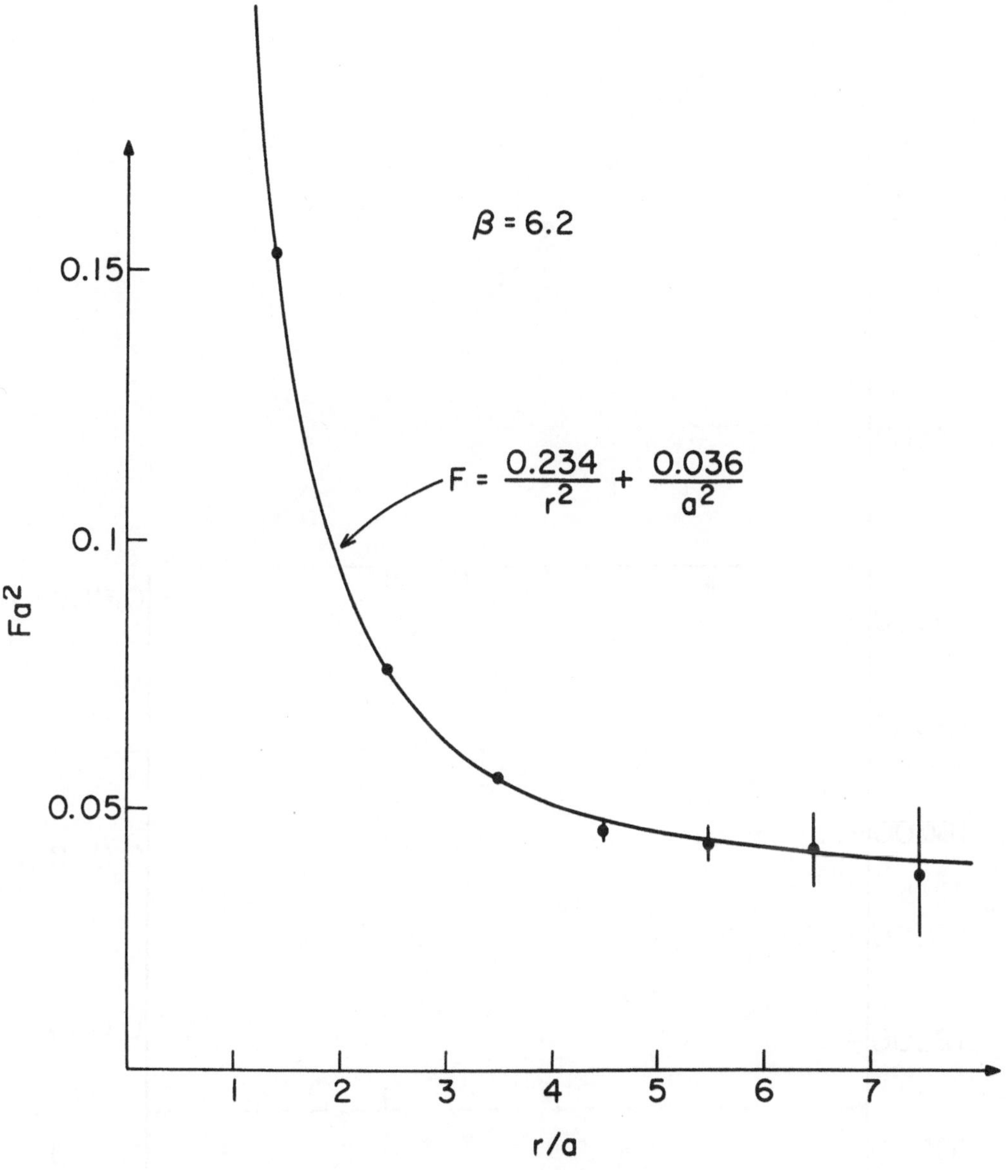

Fig.6.

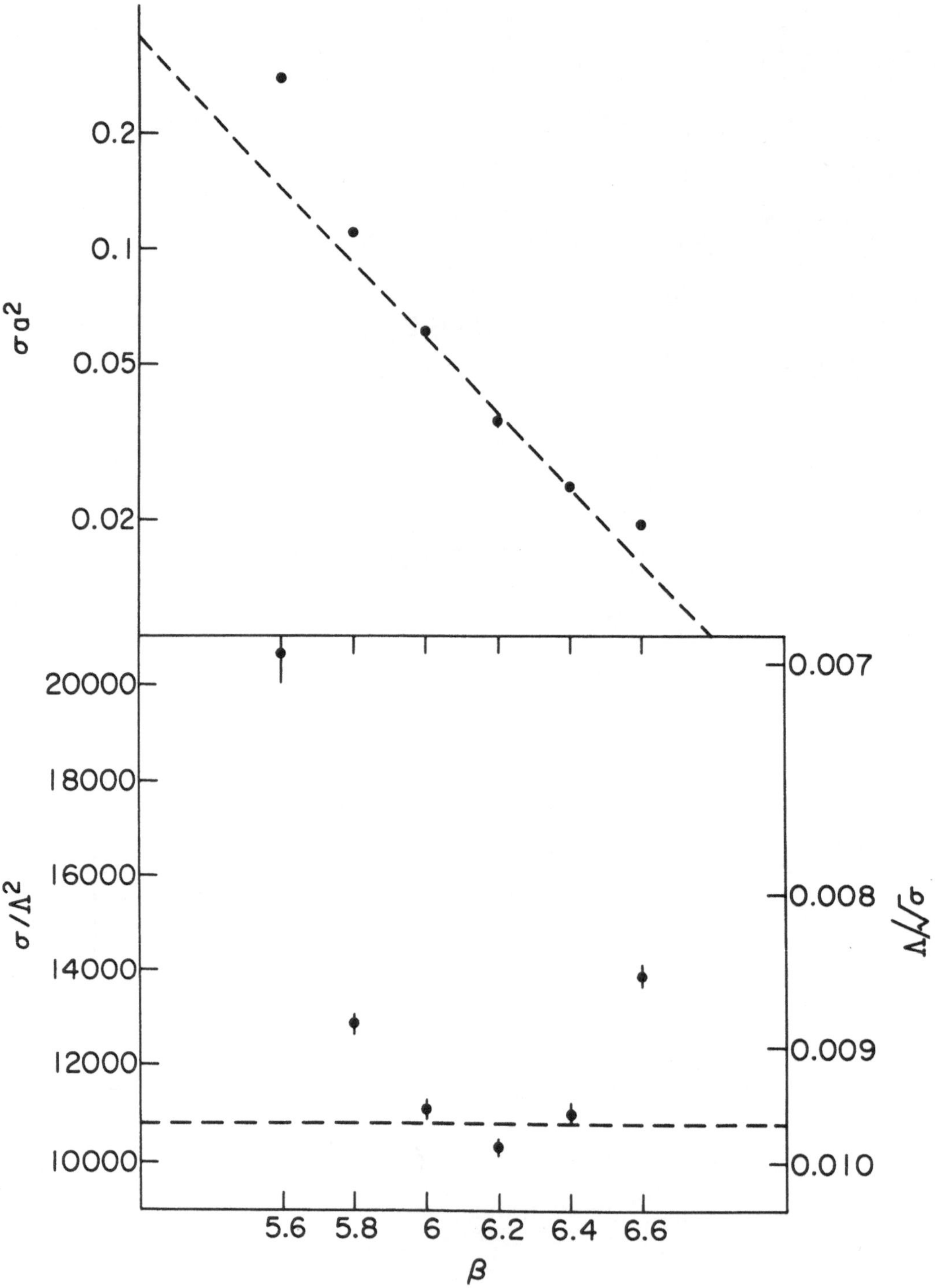
0.2
0.1
σa²
0.05
0.02
20000
18000
16000
σ/Λ²
14000
12000
10000
0.007
0.008
Λ/√σ
0.009
0.010
5.6
5.8
6
6.2
6.4
6.6
β

Fig.7.

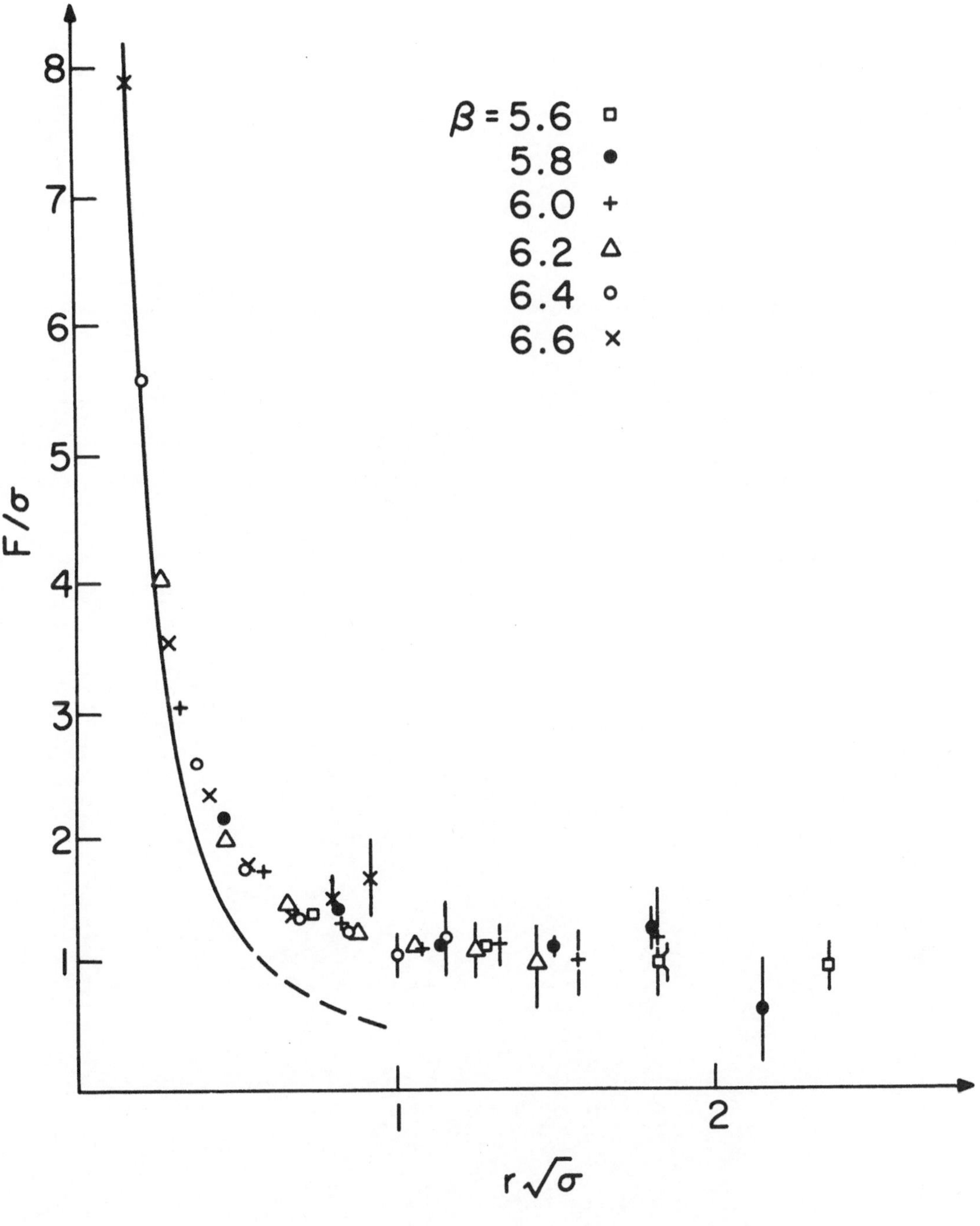

Fig.8.

Acta Physica Austriaca, Suppl. XXVI, 349–359 (1984)

THE HEAT KERNEL ON RIEMANNIAN MANIFOLDS AND LIE GROUPS+

by

T. AREDE
Fac. Engenharia - DEMEC
R. Dos Bragas
4000 Porto
Portugal

ABSTRACT

We give exact formulae for the heat Kernel on a class of Riemannian manifolds and Lie groups. These formulae express the heat Kernel in terms of lengths of geodesics of the corresponding manifolds.

1. INTRODUCTION

In this talk we discuss the heat Kernel on Riemannian manifolds and Lie groups. This correponds, via well known analytic continuation techniques, to the study of the quantum mechanical propagator for a particle moving freely on the manifold, i.e., performing, classically, a geodesic motion. The (free) heat Kernel in R^n is

+Seminar given at the XXIII. Internationale Universitätswochen für Kernphysik, Schladming, Austria, February 20-March 1, 1984.

$$K(t,x,y) = \left(\frac{1}{2\pi t}\right)^{n/2} e^{-||x-y||^2/2t} \quad ,$$

hence entirely given in terms of the geodesic distance $||x-y||$, i.e. the length of the classical path. We can then ask the question: "For which manifolds is the heat Kernel given in a exact way, for all times t, by an expression depending essentially on the geodesic distance"? In view of well known expansions for $t\downarrow 0$ or $\hbar\downarrow 0$ for the Schrödinger or heat equation [0], [4],[7], [10], [19](see also references in [1]), we shall say in the above situation that the semi-classical approximation is exact.
In this talk we exhibit classes of manifolds for which the semiclassical approximation is exact. Different aspects of the interest in these manifolds are discussed in [1]. The question of exactness of the semiclassical approximation is discussed partly heuristically, in connection with the Schrödinger equation in [8], [15]. For the heat Kernel various formulae have been obtained in special cases (see references in [1], [2]).
Here we shortly present several results for the heat Kernel, contained in [2][3] to which we refer for more details.

2. THE ELWORTHY-TRUMAN FORMULA AND APPLICATION TO HYPERBOLIC SPACES

In [9] and [10] a probabilistic formula is given for the class of Riemannian manifolds that have a pole, that is the Riemannian manifolds M of dimension $\underline{n}$ for which there exists a point, $y \in M$, such that the exponential mapping $\exp_y : T_y M \to M$, is a diffeomorphism from the tangent plane $T_y M$ to M at y, onto M. The Ruse invariant on M [21], [6] is a function on M defined by

$$\theta_y(x) = |\det(T_X \exp_y)|$$

where $T_X \exp_y$ is the tangent map of $\exp_y$ at $X = \exp_y^{-1} x$.

The Elworthy-Truman formula states that the fundamental solution of the heat equation on M

$$(\frac{\partial}{\partial t} - \frac{1}{2}\Delta) f = 0$$

- the heat Kernel on M - is given by

$$K(t,x,y) = (\frac{1}{2\pi t})^{n/2} \theta_y^{-1/2}(x) \exp \frac{-d^2(x,y)}{2t}$$

$$\cdot E\{\exp[\frac{1}{2} \int_0^t \theta_y^{1/2}(x_s) \Delta^{-1/2} \theta(x_s) ds]\} \tag{1}$$

where Δ is the Laplace-Beltrami operator on M, $d(x,y)$ is the geodesic distance between x and y, and the expectation is taken with respect to the diffusion process x_s on M, such that $x(0) = x$ and $x(t) = y$ (a bridge process) and have unit diffusion coefficient and drift:

$$Z_s(x) = \nabla_x (- \frac{d^2(x,y)}{t-s} - \frac{1}{2} \log \theta_y(x)) , \quad s < t .$$

The question of knowing when $K(t,x,y)$ can be exactly given in terms of geodesics is now reduced to the question of knowing when the expectation in (1) can be computed in an exact way and given essentially in terms of geodesics. Formula (1) applies to simply connected, complete Riemannian manifolds of nonpositive sectional curvature, and to simply connected, connected, nil-potent Lie groups. Indeed these are manifolds with a pole as it can be deduced from [16], [12].

An example of the first case is provided by the hyperbolic spaces of dimension n, H^n [18], [22]. The Ruse invariant for these spaces is given by

$$\theta_y(x) = \left(\frac{\mathrm{sh}\ d(x,y)}{d(x,y)}\right)^{n-1} ,$$

d(x,y) being the geodesic distance for the hyperbolic metric, and for n = 3 formula (1) turns to be

$$K(t,x,y) = \left(\frac{1}{2\pi t}\right)^{3/2} \frac{d(x,y)}{\mathrm{sh}\ d(x,y)}\ e^{-\frac{d^2(x,y)}{2t} + \frac{R}{12} t}$$

where $x,y \in H^3$ and R is the scalar curvature of H^3. This is then an explicit example where the semiclassical approximation is exact.

3. AN EXTENSION OF THE ELWORTHY-TRUMAN FORMULA AND THE CASE OF CLIFFORD-KLEIN SPACES

As is also pointed out in [9], the above result can be extended to the class of Riemannian manifolds for which there exists $y \in M$ such that $\exp_y : T_y M \to M$ is surjective and for all $X \in T_y M$, $\det (T_X \exp_y) \neq 0$, or in other words, y has no conjugated points along any geodesic. In these cases we need only to remark that the exponential map is a covering map between the two Riemannian manifolds ($T_y M$, $\exp_y^* g$) and (M,g), where $\exp_y^* g$ is the pull back of the Riemannian metric g of M by the exponential map. The heat Kernel on M can then be written as

$$K(t,x,y) = \sum_{X \in \exp_y^{-1} x} \tilde{K}(t,X,O) \tag{2}$$

where K(t,X,O) is the heat Kernel on $T_y M$ between its origin O and X. The latter can be given by (1) because $T_y M$ de isomorphic to R^n. We obtain

$$\tilde{K}(t,X,0) = \left(\frac{1}{2\pi t}\right)^{n/2} \tilde{\theta}_0^{-1/2}(X) \; e^{-\frac{||X||^2}{2t}}$$

$$E\{\exp[\frac{1}{2}\int_0^t \tilde{\theta}_0^{1/2}(X_s)\Delta\tilde{\theta}_0^{-1/2}(X_s)ds]\} \qquad (3)$$

where $\tilde{\theta}_0(X) = |\det T_X \exp_y|$, and the expectation is now taken with respect to a bridge process in $T_y M, X_s$, between 0 and X, with unit diffusion coefficient and drift

$$A_s(X) = \nabla_X\left(-\frac{||X||^2}{t-s} - \frac{1}{2}\log\tilde{\theta}_0(X)\right), \quad s < t .$$

This result applies to complete Riemannian manifolds with nonpositive sectional curvature and to connected nilpotent Lie groups. An example of the first class of manifolds is given by the Clifford-Klein space forms H^n/Γ [22], where H^n is the hyperbolic space of dimension n and Γ is a discrete subgroup of the Lorentz group O(n,1) acting freely and properly discontinuously on H^n. By (2) and (3) we obtain for n = 3

$$K(t,\bar{x},\bar{y}) = \sum_{\gamma\in\Gamma}\left(\frac{1}{2\pi t}\right)^{3/2} \frac{d(\gamma x,y)}{\operatorname{sh} d(\gamma x,y)} \; e^{-\frac{d^2(\gamma x,y)}{2t} + \frac{Rt}{12}} \qquad (4)$$

where $\bar{x} = x \pmod{\Gamma}$ and $\bar{y} = y \pmod{\Gamma}$, and R is the scalar curvature of H^3.

Expressions (2) and (4) can be interpreted as a sum over all geodesics between the two points; (4) gives in an exact way the heat Kernel on H^3/Γ in terms of the geodesics between $\bar{x}$ and $\bar{y}$.

4. THE EXACTNESS OF SEMICLASSICAL APPROXIMATION IN LIE GROUPS

We now apply the preceding results to Lie groups, remarking first of all, that the exponential map from a Lie algebra to the corresponding Lie group is defined using the group structure and does not coincide in general with the exponential map for manifolds that appeared in the preceding sections (see [9], [12]). Nevertheless, it is possible to consider connections on Lie groups for which the two definitions of exponential map coincide.

4.1. Compact Lie Groups

Let G be a compact Lie group and $\mathcal{G}$ its Lie algebra. Then G admits a biinvariant (invariant by left and right translations) Riemannian metric, and the associated Riemannian connection (see [17]) is such that the exponential map, exp: $\mathcal{G}\to G$ defined by Lie group structure, coincides with $\exp_e : T_e G \to G$ when G is considered as a manifold and T_eG is identified with $\mathcal{G}$ (e is the identity element of G), see [9], [12]. It is known ([5],[3]) that the heat Kernel on G is completely defined by its values on a maximal torus T of G, that is $K(t,g,g') = K(t,h,e)$, where $g,g'\in G$, $h\in T$, and gg'^{-1} is conjugated of h.
Finally we remark that exp: $\mathcal{G}\to G$ is surjective, the Ruse invariant of G is

$$\det (T_H\exp) = \left[\prod_{\alpha>0} \frac{2\sin\frac{1}{2}\alpha(H)}{\alpha(H)}\right]^2 \tag{5}$$

for $H\in \mathcal{T}$, $\mathcal{T}$ being a Cartan subalgebra of $\mathcal{G}$(the Lie algebra of T), and $\alpha>0$ are the positive roots of $\mathcal{G}$ with respect to $\mathcal{T}$ (see [12]). Moreover (5) is nonzero in a dense subset of $\mathcal{T}$[20]. Then we can apply (2) and (3) of section 3 obtaining

$$K(t,h,e) = \sum_{H\in \exp^{-1}h} \left(\frac{1}{2\pi t}\right)^{n/2} \left(\prod_{\alpha>0} \frac{\alpha(H)}{2\sin\frac{1}{2}\alpha(H)}\right) e^{-\frac{||H||^2}{2t} + \frac{Rt}{12}} \tag{6}$$

where $||.||$ is the norm in G induced by the Riemannian metric of G and R is the scalar curvature of G (constant in our case); see [3] and references therein for more details.

Equ. (6) gives then the heat Kernel on a compact Lie group in an exact way as a sum over all the geodesics between h and e, and each term of the sum depends in a essential way on the distance along the corresponding geodesic.

4.2. The Duals Of Compact Lie Groups

The dual space of a compact Lie group is a symmetric space given as a quotient of a semisimple noncompact Lie group (complexification of the compact one) by a maximal compact subgroup ([11], [12]).

Let G be a semisimple noncompact Lie group and U a maximal compact subgroup. The symmetric space G/U has a G - invariant Riemannian structure and the exponential map $\exp_O : T_O(G/U) \to G/U$ is a diffeomorphism from the tangent plane to G/U at $O = eU$ onto G/U. The Elworthy-Truman formula (1) in §2 then be applied to obtain

$$K(t,h,O) = \left(\frac{1}{2\pi t}\right)^{n/2} \prod_{\alpha>0} \left(\frac{\alpha(H)}{2\mathrm{sh}\frac{1}{2}\alpha(H)}\right) e^{-\frac{||H||^2}{2t} + \frac{Rt}{12}} \tag{7}$$

where H belongs to a maximal abelian subspace of $T_O(G/U)$, $h = \exp_O H$, $||.||$ is the norm in $T_O(G/U)$ induced by the Riemannian structure of G/U, and R is the scalar curvature of G/U. We have also that

$$\det(T_H \exp_{\mathcal{O}}) = \left[\prod_{\alpha>0} \frac{2\operatorname{sh}\frac{1}{2}\alpha(H)}{\alpha(H)} \right]^2 \tag{8}$$

where $\alpha>0$ are the positive roots of G(see also [14]). (7) gives us another situation where the semiclassical approximation is exact in the sense of §1.

4.3. The Nilpotent Lie Groups

In the case of nilpotent Lie groups we obtain similar results to the above ones for the heat equation with a left invariant second order elliptic differential operator (not necessarily a Laplace-Beltrami operator) for some metric.

For a nilpotent Lie group G with Lie algebra $\mathcal{G}$ (identified with the Lie algebra of the left invariant vector fields on G) consider a left invariant Riemannian metric (not necessarily a bi-invariant metric). The two exponentials exp: $\mathcal{G}\to G$ and $\exp_e: T_e G \to G$ will coincide if we take on G the trivial left invariant connection given by $\nabla_Y X=0$, for $X,Y\in\mathcal{G}$ (see [12]). As this connection is no more a Riemannian (or Levi-Civita) connection (its torsion tensor is different from zero) we can not use Elworthy-Truman formula as it is given in §2. Nevertheless this connection preserves the inner product. We can then still consider the stochastic development of the Brownian motion on T_eG, obtaining on G not a Brownian motion, but a diffusion process generated by a second order elliptic differential operator A given in local coordinates by

$$A = \frac{1}{2} g^{ij} \frac{\partial^2}{\partial x_i \partial x_j}$$

where g^{ij} are the components of the inverse of the Riemannian structur of G. We now extend the Elworthy-Truman formula to this new situation, see [3], and we obtain exact results for the fundamental solution K_A of the heat equation

$$(\frac{\partial}{\partial t} - A)\ f = 0.$$

Namely if G is a simply connected, connected, nilpotent Lie group we have that the exponential map is diffeomorphism, det $(T_X \exp) = 1$, $X \in G$ [12] , and

$$K_A(t,x,e) = (\frac{1}{2\pi t})^{n/2} \ e^{-\frac{d^2(x,e)}{2t}} \tag{9}$$

where d(x,e) is the distance along the geodesics given by the left invariant trivial connection.

For a non-simply connected, connected, nilpotent Lie group we have

$$K_A(t,x,e) = \sum_{X_i \in \exp^{-1} x} (\frac{1}{2\pi t})^{n/2} \quad e^{-||X_i||^2/2t} \quad , \tag{10}$$

the sum being over all geodesics between x and e.

Hence also for nilpotent Lie groups the semiclassical approximation is exact.

ACKNOWLEDGEMENTS

I am very grateful to Prof.H.Mitter and L.Pittner of the Univ. Graz and all the Organizing Committee of the Winter School of Physics, Schladming 1984, for a very kind invitation. I would also like to thank Prof.Cecile De Witt for very interesting discussions and Prof.Sergio Albeverio for a very helpful collaboration.

REFERENCES

1. S. Albeverio, Ph. Blanchard, R. Høegh-Krohn, R. Feynman, Path Integrals and the trace formula for Schrödinger operators, Commun. Math. Phys. 83 (1982) 49-76.
2. S. Albeverio, T. Arede, The relation between quantum mechanics and classical mechanics: a survey of some mathematical aspects; ZIF Preprint, Bielefeld 1984 (to be published in Proceedings of Como Conference (1983).
3. T. Arede, La geometrie du noyau de la chaleur sur les variétés, Thése de 3ème cycle, Université d'Aix-Marseille II (Luminy), 1983.
4. T. Arede, Manifolds for which the heat Kernel is given in terms of lengths of geodesics (to be published).
5. R. Azencott, H. Doss, L'equation de Schrödinger quand h tend vers zero; une approche probabiliste, to appear in Proc. of Marseille Conf. March 83, Ed. by S. Albeverio et al, Springer, Berlin (1983).
6. A. Benabdallah, Noyau de diffusion sur les espace homogénes compacts, Bull. Soc. Mat. de France 101 (1973) 263-265.
7. A.L. Besse, Manifolds all of whose geodesics are closed, 1978, Ergebnisse der Mathematik 93, Springer Verlag, Berlin.
8. C. DeWitt-Morette, A. Maeshwari, B. Nelson, Path integration in non-relativistic quantum mechanics; Physics Reports 50, no.5 (1979) 255-372, Amsterdam, North-Holland.
9. J.S. Dowker, When is the "sum over classical paths" exact? J. Phys. A. Gen. Phys, Vol.3 (1970).
10. D. Elworthy, Stochastic differential equations on manifolds, London Mathematical Society; Lect. Notes Series 70, Cambridge Univ. Press.
11. D. Elworthy, A. Truman, The diffusion equation and classical mechanics: an elementary formula, Proc. of Marseille (1981), Lect. Notes in Physics A 73, Springer-Verlag, Berlin.
12. L.D. Eskin, The heat equation and the Weierstrass transform on certain symmetric Riemannian spaces, Amer. Math. Soc. Trans. 75 (1968) 239-254.

13. S. Helgason, Differential geometry, Lie groups and symmetric spaces, Academic Press (1978).
14. N. Ikeda, S. Watanabe, Stochastic differential equations and diffusion processes; North-Holland/Kodanska (1981) Tokyo.
15. O. Loos, Symmetric Spaces, Vol.I,II; W.A. Benjamin (1969).
16. M.S. Marinov, M.V. Terentyev, Dynamics on the group manifold and path integral, Fortschr. d. Phys. 27 (1979) 511-545.
17. J. Milnor, Morse Theory, Princeton University Press (1963).
18. J. Milnor, Curvatures of left invariant metrics on Lie groups, Adv. in Math. 21 (1976) 293-329.
19. H. Ruse, A. Walker, T. Willmore, Harmonic Spaces, Ediz. Cremonese Roma (1961).
20. L.S. Schulman, Techniques and applications of path integration, Wiley, New York (1981).
21. J.P. Serre, Algebres de Lie semisimples complexes, W.A. Benjamin, New York (1966).
22. A. Walkter, Note on a distance invariant and the calculation of Ruse's invariant, Proc. Ed. Math. Soc., 2 Ser, 79 (1940) 16-26.
23. J. Wolf, Spaces of Constant Curvature, McGraw-Hill, Book Company (1976).

Acta Physica Austriaca, Suppl. XXVI, 361–364 (1984)

CONFINEMENT IN CONTINUUM QCD+

by

D. ATKINSON
University of Groningen
The Netherlands

Almost nobody has seen free quarks [1], and almost everybody believes that they are permanently confined [2], as a result of the strong exchange force generated by the gauge particles of quantum chromodynamics (QCD), the eight gluons.

In electromagnetism, the gauge particle (the photon) does not interact directly with itself, and the force lines spread out over the surface of a sphere, giving rise to the inverse-square law of Coulombic force, or equivalently to a potential that behaves like 1/r, for large separation, r. Quantum corrections affect this behaviour only to the tune of logarithms of r. In QCD, the physical idea is that the self-interaction of the gluons, a sine qua non of a non-abelian gauge theory, is so efficient that the lines of glue are essentially restricted to a one-dimensional tube between quark and antiquark (the constituents of a meson). Under these conditions, the force does not fall off at all as the separation between quark and antiquark is increased. Consequently the potential be-

---

+Seminar given at the XXIII. Internationale Universitätswochen für Kernphysik, Schladming, Austria, February 20 -March 1, 1984.

haves like r, for large r, and the increasing potential energy eventually causes the glue to break, with creation of a quark-antiquark pair. In this way one cannot manage to separate the $q\bar{q}$ constituents of a meson; rather one creates an extra meson by pumping enough energy into the system.

In momentum space, a linearly rising potential corresponds, in Born approximation, to a clothed gluon propagator that behaves like p^{-4}, as $p \to 0$, p^μ being the gluon momentum. One approach to solving the problem of such a singular infra-red behaviour is based on the Dyson-Schwinger equations, which allow non-perturbative solutions.

The work of Baker, Ball and Zachariasen (BBZ) [3], makes use of an axial gauge, in which the gluon propagator is orthogonal to a fixed vector, n^μ:

$$n^\mu D'_{\mu\nu}(p) = 0 = D'_{\mu\nu}(p) n^\nu \ . \tag{1}$$

The assumption is made that the full propagator has the form

$$D'_{\mu\nu}(p) = Z(p^2)\left[-g_{\mu\nu} + \frac{p_\mu n_\nu + p_\nu n_\mu}{(pn)} - \frac{p_\mu p_\nu n^2}{(pn)^2}\right] \times \frac{1}{p^2} \ . \tag{2}$$

Here two simplifying assumptions have been made:

a) a possible second tensor, orthogonal to n^μ, is absent, and

b) the form-factor, Z, does not depend on the scalar quantity $(pn)^2$.

Neither assumption is expected to be exactly true; but it is hoped that they constitute a reasonable approximation in the extreme infra-red.

By projecting the Dyson-Schwinger equation onto the axial vector, BBZ obtain a scalar equation that contains only the unknown $Z(p^2)$, and the three gluon vertex. For the latter they make the ansatz (where $p+q+r = 0$)

$$\Gamma'_{\lambda\mu\nu}(p,q,r) = [Z^{-1}(q^2)q_\mu - Z^{-1}(r^2)r_\mu]g_{\nu\lambda}$$
$$- \frac{Z^{-1}(q^2)-Z^{-1}(r^2)}{q^2 - r^2} [(qr)g_{\lambda\nu}-q_\lambda r_\nu](p-q)_\mu$$
$$+ \text{cyclic permutations} \quad , \qquad (3)$$

which is consistent with (although not implied by!) the Slavnov-Taylor identity, which is itself a consequence of the gauge-covariance of the theory. A nonlinear equation for Z results that has the form

$$Z^{-1}(p^2) = \frac{1+\int d^4qK(p,q)Z(q^2)}{1-\int d^4qL(p,q)Z(q^2)Z(r^2)} \quad , \qquad (4)$$

where the kernels K and L, which we do not reproduce here, are known explicitly. BBZ present a rough numerical argument to the effect that $Z(p^2)$ has a simple pole at $p^2 = 0$, so that the propagator (2) indeed behaves like p^{-4} as $p^2 \to 0$. Although much work remains to be done in order to show that (4) really does have a solution with the desired infra-red properties, some progress has been made by Atkinson and Johnson (4), with a linearized version of the equation.

In 1979, Mandelstam [5] proposed a drastically simplified gluon equation in the Landau gauge, which was analyzed [6] by functional and by numerical techniques. Although $Z(p^2)$ again exhibits the infra-red pole at $p^2 = 0$, there are impermissible complex first-sheet branch-points, whose presence is probably caused by Mandelstam's cavalier treatment of the Slavnov-Taylor identity. An improved ansatz [7], in which the clothed three-gluon vertex is assumed to be equal to the bare vertex times a scalar function, namely

$$\Gamma'_{\lambda\mu\nu}(p,q,r) = \frac{[q^2 Z^{-1}(q^2) - r^2 Z^{-1}(r^2)]}{q^2 - r^2}[g_{\mu\nu}(q-r)_\lambda + \text{cyclic permutations}] \quad (5)$$

guarantees satisfaction of the Slavnov-Taylor identity (if the Fadeev-Popov ghost is approximated by its bare value). The resulting equation has a solution free from the Mandelstam disease; but there is an unwanted pole on the spacelike real axis (in addition to the pole at $p^2 = 0$). This tachyon may be pointing to a vacuum instability of our version of the theory; and work is still in progress to clarify the anomaly.

In conclusion, there are indications that QCD has a confining infra-red singularity in the continuum form of the theory, in which no lattice discretization has been made.

REFERENCES

1. G.S. La Rue, J.D. Phillips and W.M. Fairbank, Phys. Rev. Lett. 46 (1981) 967.
2. M. Bander, Phys. Rep. 75 (1981) 205.
3. M. Baker, J.S. Ball and F. Zachariasen, Nucl. Phys. B186 (1981) 531 and 560; B226 (1983) 455.
4. D. Atkinson and P.W. Johnson, "Linearization of the Baker-Ball-Zachariasen Gluon Propagator Equation", Nucl. Phys. (to appear).
5. S. Mandelstam, Phys. Rev. D20 (1979) 3223.
6. D. Atkinson, J.K. Drohm, P.W. Johnson and K. Stam, Journ. Math. Phys. 22 (1981) 2704; D. Atkinson, P.W. Johnson and K. Stam, Journ. Math. Phys. 23 (1982) 1917.
7. D. Atkinson, H. Boelens, S.J. Hiemstra, P.W. Johnson, W.J. Schoenmaker and K. Stam, Journ. Math. Phys. (to appear)

Acta Physica Austriaca, Suppl. XXVI, 365–370 (1984)

FERMION BOUNDARY CONDITION AND θ ANGLE[+]

by

J. BALOG
Roland Eötvös University, Budapest
and
P. HRASKÔ
Central Research Institute for Physics,
Budapest, Hungary

THE PROBLEM AND MOTIVATION

There is an old problem connected with chiral symmetry breaking in gauge theories. The problem has been generated by the solution of the U(1) problem in QCD [1,2] and has been strongly emphasized by Crewther and others [3,4].

In this seminar I will first remind you briefly what this problem is and suggest a new method of calculation which may be - and we hope that it really is - free of this problem. Unfortuantely in the absence of a detailed QCD calculation I will not be able to show that this method really works in four dimensional gauge theory. However, easy calculation in a two dimensional theory, the Schwinger model, shows that our method does indeed work in

---

[+]Seminar given at the XXIII. Internationale Universitätswochen für Kernphysik, Schladming, Austria, February 20- March 1, 1984.

this case at least.

The problem is connected with the Euclidean functional integral calculation of the order parameter of the spontanous breaking of the SU(n) ⊗ SU(n) chiral symmetry (n is the number of light flavours). Successful phenomenology (based on current algebra) tells us that

$$\langle\bar{\psi}_L\psi_L\rangle \neq 0 \qquad \langle\bar{\psi}_R\psi_R\rangle \neq 0 \tag{1}$$

even in the chiral limit (for n = 2,3 at least).

If we want to reproduce (1) in QCD we can make use of the fact that the fermionic integration is Gaussian and we can formally integrate out the fermion fields:

$$\begin{aligned}\langle\bar{\psi}_L(x)\psi_L(x)\rangle &= \int \mathcal{D}A_\mu\, \mathcal{D}\psi\, \mathcal{D}\bar{\psi}\, \bar{\psi}_L(x)\psi_L(x)\, e^{-S[A]+\int d^4x\bar{\psi}\not{D}\psi} \\ &= \int \mathcal{D}A_\mu\, \mathrm{Tr}\, G_{LL}(x,x,A)\, e^{-S[A]-S_1[A]}. \end{aligned} \tag{2}$$

Here G(x,y,A) is the Green's function of the Dirac equation in the given external field A_μ and $S_1[A] = -\log\det \not{D}$.

The problem with the naive formula (2) is that because of formal chiral symmetry the matrix G(x,y,A) is off-diagonal in the Dirac indices (in the Weyl representation); therefore

$$G_{LL}(x,y,A) = G_{RR}(x,y,A) \equiv 0$$

which would give a vanishing order parameter.

One can avoid this immediate conclusion by using a more careful definition of Gaussian integrals over fermion fields. According to the works of Fujikawa and others [2,5] the correct definition is as follows.

The fermion fields $\psi(x)$, $\bar{\psi}(x)$ should be expanded in terms of a complete set of c-number spinors $\phi_n(x)$ with Grassmann-valued coefficients:

$$\psi(x) = \sum_n a_n \phi_n(x) \qquad \bar{\psi}(x) = \sum_n \phi_n^+(x) \bar{\beta}_n \ .$$

In order to maintain gauge invariance of the theory [5] the set $\{\phi_n(x)\}$ is not just an arbitrary complete set of functions but they have to be identified with the eigenfunctions of the Dirac operator:

$$\not{D}\phi_n = \lambda_n \phi_n \qquad \int d^4x \ \phi_n^+(x)\phi_m(x) = \delta_{nm} \ .$$

The integration measure is defined as

$$\int \mathcal{D}\psi \mathcal{D}\bar{\psi} = \int \prod_n (da_n d\bar{\beta}_n) \ .$$

Now, using the rules of integration over Grassmann numbers, (2) is recovered after the integration if all eigenvalues λ_n are different from zero. If, however, some eigenvalues vanish (zero modes), (2) should be modified:

$$<\bar{\psi}_L(x)\psi_L(x)> = \int \mathcal{D}A_\mu \ \phi_{OL}^+(x)\phi_{OL}(x) e^{-S[A]-S_1'[A]} \ . \qquad (3)$$

Here $\not{D}\phi_0 = 0$ and $S_1'[A] = -\log \det' \not{D}$, where the prime indicates the omission of a zero eigenvalues from the determinant.

The discreteness of the spectrum of $\not{D}$ is essential for this general prescription. This is usally achieved by the method of compactification (stereographic projection of the Euclidean space-time on the surface of a five-sphere). Using compactification the gauge configurations can be classified according to their integer topological charge ν. It can be shown that for $\nu \neq 0$ there are at least $|\nu|$ zero modes per flavour [2,6]. In particular, for $\nu = -1$ there exists a purely lefthanded zero mode $\phi_{OL}(x)$.

Now the problem is that if $n \geq 2, \psi$ can be either u,d (or s) and (3) still gives a vanishing result:

$$<\bar{u}_L u_L> = <\bar{d}_L d_L> = \ldots = \int \mathcal{D}A_\mu \; \phi^+_{OL}\phi_{OL}[\det'\not{D}]e^{-S[A]} = 0 \; , \qquad (4)$$

since det'$\not{D}$ still contains at least (n-1) zero eigenvalues even in the $\nu = -1$ case and identically vanishes for all gauge configurations.

GAUGE THEORY IN A FINITE BOX

The conclusion above expresses a very clear and disturbing contradiction between the successful results of current algebra and beautiful QCD ideas and shows that alternatives for the method of compactification are needed.

Besides of compactification the spectrum can be made discrete also by means of box-quantization. If we define our theory inside of a finite box, a boundary condition is needed to maintain the self-adjointness of $\not{D}$. For a hyper-spherical box the most general linear boundary condition conforming to (Euclidean) Lorentz symmetry is

$$e^{-\bar{\theta}\gamma_5}\gamma_r\psi = \gamma_5\psi \;\Big|_{r = R} \; , \qquad (5)$$

where $\bar{\theta}$ is an arbitrary real parameter.

Using (5) the absence of zero modes can be shown, thus (2) can be used:

$$<\bar{\psi}_L\psi_L>_R = \int \mathcal{D}A_\mu \; \mathrm{Tr}\; G^{\bar{\theta}}_{LL}(x,x,A)e^{-S[A]-S_1[A]} \; . \qquad (6)$$

Here $G^{\bar{\theta}}(x,y,A)$ is now the Green's function satisfying (5). (5) breaks chiral symmetry explicitly and acts like a "driving term" added to the Lagrangian. $G^{\bar{\theta}}_{LL} \neq 0$ for $R \neq \infty$, therefore $<\bar{\psi}\psi>_R \neq 0$. Since $G^{\bar{\theta}}_{LL} \to 0$ as $R \to \infty$, it is the question of dynamics whether some non-vanishing order parameter remains when the limit is taken after all calculations have been performed (spontanous symmetry breaking):

$$<\bar{\psi}_L\psi_L> = \lim_{R\to\infty}<\bar{\psi}_L\psi_L>_R = ? \tag{7}$$

SCHWINGER MODEL CALCULATION

One can illustrate this mechanism in QED_2 where everything can be calculated explicitly.

In the Landau gauge the gauge field configurations can be conveniently parametrized in terms of a scalar field β: $A_\mu = \varepsilon_{\mu\nu}\partial_\nu\beta$. Further [7,10]

$$S[A] = \int d^2x \, \frac{1}{4} \, F_{\mu\nu}F_{\mu\nu} = \frac{1}{2} \int d^2x(\Box\beta)^2$$

$$S_1[A] = -\log \det \not{D} = \frac{m^2}{2}\int d^2x(\underline{\nabla}\beta)^2 \qquad (m = \frac{e}{\sqrt{\pi}})$$

$$G^{\bar{\theta}}_{LL}(0,0,A) = -\frac{e^{\bar{\theta}}}{2\pi R} \, e^{2e\beta(0)} \quad . \tag{8}$$

Using (8) the (now Gaussian) functional integral (6) can be performed using the method of shifting the integration variable:

$$<\bar{\psi}_L(0)\psi_L(0)>_R = -\frac{e^{\bar{\theta}}}{2\pi R}\int \mathcal{D}\beta \; e^{-\Sigma[\beta]} = -\frac{e^{\bar{\theta}}}{2\pi R} \, e^{-\Sigma[B]} \quad . \tag{9}$$

Here $\Sigma[\beta] = S[A]+S_1[A]-2e\beta(0)$ and the configuration $B(x)$ is defined by

$$\frac{\delta\Sigma}{\delta\beta}\Big|_{\beta=B} = 0 \; . \tag{10}$$

After solving (10) for $B(x)$ we find

$$-\Sigma[B] = \ln\frac{mR}{2} + \gamma + O(e^{-mR}) \; , \tag{11}$$

which shows that the dynamics of the model enforces chiral

symmetry to be spontanously broken since the logarithmic divergence in the exponent of (9) cancels against the 1/R factor in front of the integral leading to the final result:

$$\langle\bar{\psi}_L\psi_L\rangle = \lim_{R\to\infty}\langle\bar{\psi}_L\psi_L\rangle_R = -\frac{e^{\bar{\theta}}}{2\pi}\left(\frac{1}{2}\gamma\ m\right) . \qquad (12)$$

This can be compared to the result of the Minkowski space-time operator solution of the model [8]:

$$\langle\bar{\psi}_L\psi_L\rangle = \langle\bar{\psi}_R\psi_R\rangle^{*} = \frac{e^{i\theta}}{2\pi}\left(\frac{1}{2}\gamma m\right) . \qquad (13)$$

The comparison would suggest a strong relation between our parameter $\bar{\theta}$ and the usual vacum angle θ [1,2,9] in spite of their very different origin. However, the precise nature of this connection is not clear to us and deserves further study.

REFERENCES

1. G. 't Hooft, Phys. Rev. Lett. 37 (1976) 8.
2. R. Jackiw and C. Rebbi, Phys. Rev. Lett. 37 (1976) 172.
3. R. Crewther, Phys. Lett. 70B (1977) 349.
4. R. Stacey, Phys. Rev. D26 (1982) 2134.
5. K. Fujikawa, Phys. Rev. D21 (1980) 2848.
6. B. Schroer, in Facts and Prospects of Gauge Theories (ed. P. Urban), Springer (1982) p.155.
7. N.K. Nielsen and B. Schroer, Nucl. Phys. B120 (1977) 62.
8. A.Z. Ferrari and R. Capri, Nuovo Cim A62 (1981) 273; J. Balog and P. Hraskó, KFKI preprint KFKI-1983-39.
9. A.K. Raina and G. Wanders, Ann. Phys. (N.Y.) 132 (1981) 404.
10. P. Hraskó and J. Balog, KFKI preprint KFKI-1983-95.

Acta Physica Austriaca, Suppl. XXVI, 371–380 (1984)

SEMICLASSICAL AND HIGH-TEMPERATURE EXPANSIONS FOR SYSTEMS WITH MAGNETIC FIELD+

by

D. BOLLE[++,1,2] and D. ROEKAERTS[+++,2]

[1]Zentrum für Interdisziplinäre Forschung
Universität Bielefeld, D-4800 Bielefeld 1, FRG

[2]Instituut voor Theoretische Fysica
Universiteit Leuven, B-3030 Leuven, Belgium

I. INTRODUCTION

We discuss semiclassical and high-temperature expansions for the off-diagonal thermal kernel of quantum systems with non-uniform, time-
In particular, we give an equation for the WKB-approximation for such systems. Furthermore, we provide recursion relations and an efficient diagrammatic method, based on functional integration techniques ,to calculate explicitly the coefficients in these expansions. Finally, we briefly indicate the connection between the semiclassical and high-temperature expansions.

One of the motivations to look at this problem are

+Seminar given at the XXIII. Internationale Universitätswochen für Kernphysik,Schladming,Austria,February 20-March 1,1984 .

++Onderzoeksleider N.F.W.O., Belgium

+++Aspirant N.F.W.O., Belgium

the discussions on the thermodynamical properties of nearly classical, one-component plasmas, e.g. the study of quantum effects on the Wigner crystallization of a two-dimensional system of electrons on liquid helium [1-3]. For weak magnetic fields, these quantum effects are expected to be small in the low-temperature and low-density regime of the plasma. Indeed, in this region the ratio (thermal wavelength)/(average distance of closest approach) and the ratio (thermal wavelength)/(average interparticle distance) are small such that $\hbar$ can be used as an expansion parameter. One then calculates in the literature e.g. the free energy up to order $\hbar^4$ for a uniform magnetic field [1]. In this way one finds that in three dimensions, an applied magnetic field shifts the free energy by an amount which is independent of the structure (fluid or Wigner solid) within order $\hbar^4$. So the fluid-solid transition is not displaced within that order. In two dimensions, although the free energy now depends on the structure at order $\hbar^4$, the structures of both phases are very similar near the transition. This means that in order to see the displacement of the transition by a magnetic field, one would have to go beyond $\hbar^4$. The methods presented here allow one to do this.

We remark that it is not our intention to present a rigorous study of these expansions. (For a recent survey of the relation between quantum mechanics and classical mechanics in general, we refer to [4]. We rather briefly sketch the basic arguments and results here. For more details we refer to [5-6].)

II. SEMICLASSICAL AND HIGH-TEMPERATURE EXPANSIONS: RECURSION RELATIONS

We consider the N-body d-dimensional quantum system with total Hamiltonian

$$H = \sum_{\mu=1}^{M} \frac{1}{2m} [p_\mu - A_\mu(\vec{x})]^2 + V(\vec{x}) , \tag{2.1}$$

where $M = dN$, m is the particle mass, $\vec{x} = (x^1,\ldots,x^M)$, $p_\mu = -i\hbar\partial/\partial x^\mu$, V the potential and A_μ the vector potential, both allowing a sufficient number of derivatives. Then we know that the kernel defined by

$$P(\vec{x},\vec{x}_o;\hbar,\beta) \equiv \langle\vec{x}|\exp(-\beta H(\hbar))|\vec{x}_o\rangle \quad , \qquad \beta = 1/kT , \tag{2.2}$$

satisfies the differential equation (Bloch equation)

$$\frac{\partial}{\partial\beta} P(\vec{x},\vec{x}_o;\hbar,\beta) = L(\frac{\partial}{\partial\vec{x}} ,\vec{x},\hbar)P(\vec{x},\vec{x}_o;\hbar,\beta) , \tag{2.3}$$

where

$$L \equiv \frac{\hbar^2}{2m} \partial_\mu\partial_\mu - \frac{i\hbar}{2m}[\partial_\mu A_\mu(\vec{x}) + A_\mu(\vec{x})\partial_\mu] - \tilde{V}(\vec{x}) , \tag{2.4}$$

$$\tilde{V} = \frac{1}{2m} A_\mu A_\mu + V . \tag{2.5}$$

Here $\partial_\mu \equiv \partial/\partial x^\mu$, and a sum over repeated indices is understood.

We put forward the following form for the WKB approximation:

$$P_{WKB}(\vec{x},\vec{x}_o;\hbar,\beta) = P_o(\vec{x},\vec{x}_o;\hbar,\beta)\ \exp[\frac{1}{\hbar}B(\vec{x},\vec{x}_o,\beta)+ C(\vec{x},\vec{x}_o,\beta)], \tag{2.6}$$

$$P_o(\vec{x},\vec{x}_o;\hbar,\beta) = (\frac{m}{2\pi\hbar^2\beta})^{M/2}\ \exp[-\frac{m}{2\hbar^2\beta}\Delta x^\mu\ \Delta x^\mu], \quad \Delta\vec{x} = \vec{x} - \vec{x}_o\ , \tag{2.7}$$

where B and C are in general complex. Since $P_o \sim \exp \hbar^{-2}$, we have to allow, in general, terms $\sim\exp \hbar^{-1}$ and terms independent of $\hbar$ in (2.6). We then substitute this form into Eqs. (2.3) - (2.5) and look at the different powers of $\hbar$: the terms in $\hbar^{-1}$ and $\hbar^0$ have to be equated to zero, in terms in $\hbar$ and $\hbar^2$ will provide the corrections to Eq.(2.3), since the WKB approximation is not exact in general. Higher powers of $\hbar$ are not present because H is quadratic in p (see Eq.(2.1)). In this way we arrive at

$$P_{WKB}(\vec{x},\vec{x}_o;\hbar,\beta) = P_o(\vec{x},\vec{x}_o;\hbar,\beta)\exp\{\int_o^\beta dt[\frac{i}{\hbar}\ \dot{\tilde{x}}^\mu A_\mu(\tilde{x}(t))-V(\tilde{x}(t))]\}.$$
$$.\ \exp\{-\frac{1}{2m}\int_o^\beta dt\int_o^t dt'\ \frac{t'}{t}\ \dot{\tilde{x}}^\rho\ F_{\mu\rho}(\tilde{x}(t'))\int_o^t dt''\ \frac{t''}{t}\dot{\tilde{x}}^\sigma\ F_{\mu\sigma}(\tilde{x}(t''))\} \tag{2.8}$$

and

$$\frac{\partial}{\partial\beta}\ P_{WKB}(\vec{x},\vec{x}_o;\hbar,\beta)\ -\ L(\frac{\partial}{\partial\vec{x}},\ \vec{x},\hbar)\,P_{WKB}(\vec{x},\vec{x}_o;\hbar,\beta)$$
$$=[-\hbar\ D_1(\vec{x},\vec{x}_o,\beta)\ -\ \hbar^2 D_2(\vec{x},\vec{x}_o,\beta)]\ P_{WKB}(\vec{x},\vec{x}_o;\hbar,\beta)\ . \tag{2.9}$$

Here P_o is given by (2.7) and $\tilde{x}(t)$ is the classical path between $\tilde{x}(0) = \vec{x}_o$ and $\tilde{x}(\beta) = \vec{x}$, i.e. $\tilde{x}(t) = \vec{x}_o + (t|\beta)(\vec{x}-\vec{x}_o)$. Furthermore, the quantities D_1 and D_2 are expressed in terms of the gauge-invariant field tensor $F_{\mu\nu}$ and its first derivative, and the potential V and its first and second derivative. (For the explicit forms, we refer to [5].) Expression (2.8) agrees with the one obtained in [7] by different methods. Eq.(2.9) is new.

In the case of high-temperature expansion, similar results can be obtained. E.g. the expression (2.8) for P_{WKB} is then replaced by

$$P_{HT}(\vec{x},\vec{x}_o;\hbar,\beta) = P_o(\vec{x},\vec{x}_o;\hbar,\beta)\ \exp\{\int_0^1 dt\ \frac{i}{\hbar}\ \Delta x^\mu\ A_\mu(x(t))\}\ , \tag{2.10}$$

where $x(t) = \vec{x}_o + t(\vec{x} - \vec{x}_o)$. For further details we refer to [6].

Finally, recursion relations for the higher-order coefficients in these semiclassical expansions are obtained as follows. One starts from

$$P(\vec{x},\vec{x}_o;\hbar,\beta) = P_{WKB}(\vec{x},\vec{x}_o;\hbar,\beta)\ \exp\ S(\vec{x},\vec{x}_o;\hbar,\beta)\ , \tag{2.11}$$

$$S(\vec{x},\vec{x}_o;\hbar,\beta) = \sum_{n=1}^{\infty} \hbar^n\ S_n(\vec{x},\vec{x}_o,\beta)\ . \tag{2.12}$$

Substituting these relations into (2.3) and using (2.8), (2.9), one first derives a differential equation for S. One then equates coefficients in this equation to arrive at

$$\begin{aligned}\frac{dS_n}{d\beta} &= D_1\delta_{n1} + D_2\delta_{n2} + \frac{i}{m}\ [\int_0^\beta dt\ \frac{t}{\beta}\ \dot{\tilde{x}}^\rho\ F_{\mu\rho}(\tilde{x}(t))]\partial_\mu S_{n-1}\\ &- \{\frac{1}{m}\int_0^\beta dt\ \frac{t}{\beta}\ \partial_\mu V(\tilde{x}(t)) + \frac{1}{m^2}\int_0^\beta dt\int_0^t dt'\frac{t'}{t\beta}F_{\nu\mu}(\tilde{x}(t'))\int_0^t dt''\frac{t'}{t}\dot{\tilde{x}}^\sigma F_{\nu\sigma}(\tilde{x}(t''))\\ &+ \frac{1}{m^2}\int_0^\beta dt\int_0^t dt'\frac{t'}{t}\ \frac{t'}{\beta}\dot{\tilde{x}}^\rho\partial_\mu F_{\nu\rho}(\tilde{x}(t))\int_0^t dt''\frac{t''}{t}\dot{\tilde{x}}^\sigma F_{\nu\sigma}(\tilde{x}(t''))\}\partial_\mu S_{n-2}\\ &+ \frac{1}{2m}\partial_\mu\partial_\mu S_{n-2} + \frac{1}{2m}\sum_{k=1}^{n-3}(\partial_\mu S_k)(\partial_\mu S_{n-k-2});\quad n \geq 1\ .\end{aligned} \tag{2.13}$$

This result generalizes the recursion relations of [8] (see their Eq.(4.33)) to systems with magnetic field. Employing these relations, we have calculated in [5] the off-diagonal coefficients S_1, S_2 in general and S_3 for a uniform magnetic field. For analogous results in the case of high-temperatures, we refer to [6].

III. FUNCTIONAL INTEGRATION METHODS AND DIAGONAL COEFFICIENTS

The recursion relations (2.13) can of course also be used to calculate on-diagonal coefficients. (In that case, they learn us e.g. that the expansion parameter is $\hbar^2$.) However, the disadvantage of using them in this case is that one still needs partial information about the off-diagonal form of the lower order coefficients.

To avoid this drawback we have developed a very efficient diagrammatic method based on a functional integral respresentation of $P(\vec{x},\vec{x}_o;\hbar,\beta)$, viz.

$$P(\vec{x},\vec{x}_o;\hbar,\beta) = \int_{\gamma_1(\frac{1}{2})} \mathcal{D}\vec{q}\, \exp\{-\frac{1}{\hbar^2}\int_o^\beta dt[\frac{m}{2}(\dot{\tilde{x}}^\mu + \hbar\dot{q}^\mu)^2 - i\hbar(\dot{\tilde{x}}^\mu + \hbar\dot{q}^\mu)$$

$$\cdot A_\mu(\tilde{\vec{x}}(t)+\hbar\vec{q}(t)) + \hbar^2 V(\tilde{\vec{x}}(t)+\hbar\vec{q}(t))]\}\delta(\vec{q}(0))\delta(\vec{q}(\beta)) \ . \qquad (3.1)$$

In (3.1) $\gamma_1(\frac{1}{2})$ indicates midpoint discretization [7,9]. By making an expansion around the classical path $\tilde{\vec{x}}(t)$, a representation of the quantity $\exp S(\vec{x},\vec{x}_o;\hbar,\beta) \equiv I(\vec{x},\vec{x}_o;\hbar\beta)$ is obtained that reduces on the diagonal to the following form :

$$I(\vec{x}_o,\vec{x}_o;\hbar,\beta) = (\frac{2\pi\hbar^2\beta}{m})^{M/2} \int_{\gamma_1(\frac{1}{2})} \mathcal{D}\,\vec{q}\, \exp\{\int_o^\beta dt[-\frac{m}{2}(\dot{q}^\mu)^2 +$$

$$+ \sum_{n=1}^{\infty} \hbar^n \bar{K}_n(\dot{\vec{q}},\vec{q},\vec{x}_o,\hbar)]\}\delta(\vec{q}(0))\delta(\vec{q}(\beta)) \tag{3.2}$$

with

$$\bar{K}_n = \bar{K}_n^V + \bar{K}_n^F \quad , \tag{3.3}$$

$$\bar{K}_n^V = -\frac{1}{n!}[\partial_{\rho_1} \ldots \partial_{\rho_n} V(\vec{x}_o(t))]q^{\rho_1}(t) \ldots q^{\rho_n}(t) \equiv V_n \quad , \tag{3.4}$$

$$\bar{K}_n^F = \frac{in}{(n+1)!}[\partial_{\rho_1} \ldots \partial_{\rho_{n-1}} F_{\mu\nu}(\vec{x}_o(t))]q^{\rho_1}(t) \ldots$$

$$\ldots q^{\rho_{n-1}}(t)\dot{q}^{\mu}(t)q^{\nu}(t) \equiv F_{n-1} \quad . \tag{3.5}$$

The calculation of (3.2) order by order in $\hbar$ leads to the coefficients $S_n^D(\vec{x}_o,\vec{x}_o,\beta)$. We remark that we have expanded (3.1) in $\hbar$ in order to ensure that the coefficient in front of $(\dot{q}^{\mu})^2$ in (3.2) is independent of $\hbar$.

In the diagrammatic respresentation of (3.2), S_n^D is given by the sum of all connected graphs of order $\hbar^n$. The order in $\hbar$ of a graph is the sum of the orders in $\hbar$ of the vertices $\bar{K}_n$ contained in the graph. By a simple rescaling of variables [5-6] one can also attribute an order in $\beta^{1/2}$ to the vertices as follows:

$$\bar{K}_n^V \sim \beta^{n/2+1}, \quad \bar{K}_n^F \sim \beta^{(n-1)/2+1} \quad . \tag{3.6}$$

In Table 1 we list all combinations of vertices giving rise to connected graphs of order $\hbar^n$, n = 2,4,6,in the diagrammatic representation of (3.2). Thereby we have defined

$$S_n^D = S_n^{D,V} + S_n^{D,F} + S_n^{D,NCF} \quad , \tag{3.7}$$

where $S_n^{D,V}$ contains only the potential V and its derivatives, $S_n^{D,F}$ contains the field tensor $F_{\mu\nu}$ but not its derivatives, and $S_n^{D,NCF}$ contains only derivatives of the field tensor. Note that the first contributions from the non-uniformity of the magnetic field appear in the coefficient S_4^D, and that in general S_n^D is a polynomial in β with lowest order $\frac{n}{2} + 1$ and highest order $n + 1$. Therefore the calculation of S_n^D, $n \leq n_o$, gives us also all terms in the high temperature expansion of $P(\vec{x}_o, \vec{x}_o; \not{h}, \beta)$ in powers of β up to $\beta^{n_o/2+1}$. A more systematic study of the high-temperature expansion will be presented in [6].

In [5] we have computed explicitly S_2^D, S_4^D, $S_6^{D,V}$ and $S_6^{D,F}$. (We remark that this can be done by calculating only the (basic) diagrams in Table 1 that are underlined. See [5-6] for further explanation.) The last two coefficients, as well as $S_4^{D,NCF}$, have not been obtained before.

ACKNOWLEDGEMENTS

One of the authors (D.B.) would like to thank Prof.L. Streit for his hospitality at the Zentrum für Interdisziplinäre Forschung der Universität Bielefeld. We both thank Dr.F. Peeters for informative discussions.

<u>Table 1.</u>

	β^2	β^3	β^4	β^5	β^6	β^7
$S_2^{D,V}$	$\underline{V_2}$	V_1V_1				
$S_2^{D,CF}$	F_oF_o					
$S_4^{D,V}$		$\underline{V_4}$	V_3V_1	$V_2V_1V_1$		
			$\underline{V_2V_2}$			
$S_4^{D,CF}$			$F_oF_oV_2$	$F_oF_oV_1V_1$		
			$F_oF_oF_oF_o$			
$S_4^{D,NCF}$		F_2F_o	$F_1F_oV_1$			
		F_1F_1				
$S_6^{D,V}$			$\underline{V_6}$	V_5V_1	$V_4V_1V_1$	$V_3V_1V_1V_1$
				$\underline{V_4V_2}$	$V_3V_2V_1$	$V_2V_2V_1V_1$
				$\underline{V_3V_3}$	$\underline{V_2V_2V_2}$	
$S_6^{D,CF}$				$F_oF_oV_4$	$F_oF_oV_3V_1$	$F_oF_oV_2V_1V_1$
					$F_oF_oV_2V_2$	$F_oF_oF_oF_oV_1V_1$
					$F_oF_oF_oF_oV_2$	
					$F_oF_oF_oF_oF_oF_o$	
$S_6^{D,NCF}$			F_4F_o	$F_3F_oV_1$	$F_2F_oV_1V_1$	$F_1F_oV_1V_1V_1$
			F_3F_1	$F_2F_oV_2$	$F_1F_1V_1V_1$	
			F_2F_2	$F_2F_1V_1$	$F_1F_oV_2V_1$	
				$F_1F_1V_2$	$F_1F_oF_oF_oV_1$	
				$F_1F_oV_3$		
				$F_2F_oF_oF_o$		
				$F_1F_1F_oF_o$		

REFERENCES

1. A. Alastuey and B. Jancovici, Physica 97A (1979) 349.
2. H. Fukuyama, J. Phys. Soc. Jpn. 48 (1980) 1841.
3. Mau-Chung Chang and Kazumi Maki, Phys. Rev. B27 (1983) 1646.
4. S. Albeverio and T. Arede, "The relation between quantum mechanics and classical mechanics: A survey of some mathematical apsects", University of Bielefeld, ZiF-preprint 1984.
5. D. Bollé and D. Roekaerts, "Equation for the WKB-approximation and Wigner-Kirkwood expansions for systems with magnetic fields", University of Leuven preprint KUL-TF-84/3.
6. D. Bollé and D. Roekaerts, in preparation.
7. E. Tirapegui, F. Langouche and D. Roekaerts, Phys. Rev. A27 (1983) 2649.
8. Y. Fujiwara, T.A. Osborn and S.F.J. Wilk, Phys.Rev. A25 (1982) 14.
9. F. Langouche, D. Roekaerts, E. Tirapegui,"Functional Integration and Semiclassical Expansions"(Reidel, Dordrecht, 1982).

Acta Physica Austriaca, Suppl. XXVI, 381–387 (1984)

THE GROUND STATE ENERGY OF AN INTERACTING BOSE GAS+

by

J. CONLON
Institut für Theoretische Physik
Universität Wien
A-1090 Wien, Boltzmanngasse 5,
Austria

The purpose of this article is to describe some progress in understanding the ground state energy of an interacting Bose gas. In particular we would like to understand if a formula obtained by Bogoliubov does in fact give the correct value for the ground state energy. In the early 1960's there were several papers on the Bogoliubov formula - see [4] for a review - but since then interest seems to have waned.

We consider N Bosons in a box with side of length L which are interacting via the potential $\phi(x)$. We shall assume that $\phi(x)$ is periodic on the box, is positive definite, and $\phi(0) < \infty$. Thus $\phi(x)$ has the Fourier representation

$$\phi(x) = \sum_{k \in Z^3} \nu(k)\, e^{2\pi i k\cdot x/L} \quad , \tag{1}$$

+Seminar given at the XXIII. Internationale Universitätswochen für Kernphysik, Schladming, Austria, February 20- March 1, 1984.

where $\nu(k) \geq 0$, $k \in Z^3$. Now let H_o and V be the N particle kinetic and potential energy operators respectively, so

$$H_o = \frac{1}{2}\sum_{i=1}^{N} (-\Delta_i) \; ; \qquad V = \frac{1}{2}\sum_{i\neq j=1}^{N} \phi(x_i - x_j) \; . \tag{2}$$

Letting $\rho = N/L^3$ be the particle density, we consider the Hamiltonians $\rho^\alpha H_o + V$ for various values of the parameter α. Let $\varepsilon_\alpha(\rho)$ be the ground state energy per particle of

$$\rho^\alpha H_o + V - \text{mean field energy}. \tag{3}$$

Here mean field energy (MFE) denotes the expected value of V on the constant wave function. Then we have the following result:

Theorem 1: If $\alpha > 1$ then $\lim_{\rho\to\infty} \varepsilon_\alpha(\rho) = 0$; if $\alpha < 1$ then $\lim_{\rho\to\infty} \varepsilon_\alpha(\rho) =$
$= \frac{1}{2}\nu(0) - \frac{1}{2}\phi(0)$. For $\alpha = 1$ we have

$$\lim_{\rho\to\infty} \varepsilon_1(\rho) = \frac{1}{2}\sum_{k\neq 0} \frac{1}{L^3}\{[4\pi^4(\tfrac{k}{L})^4 + 4\pi^2(\tfrac{k}{L})^2 L^3\nu(k)]^{1/2} - 2\pi^2(\tfrac{k}{L})^2 - L^3\nu(k)\} \; .$$

The formula given in theorem 1 for $\varepsilon_1(\rho)$ agrees with Bogoliubov's calculation. We shall first show how Bogoliubov makes his calculation and then indicate how the proof of theorem 1 goes. The first step is to write the operators H_o and V in second quantized form. Thus, if a_k, $k \in Z^3$, is the annihilation operator associated to the Fourier coefficient k, we have

$$H_o = \frac{2\pi^2}{L^2}\sum_{k\in Z^3} k^2 a_k^* a_k \; , \tag{4}$$

$$V = \frac{1}{2} \sum_{k,n,r} \nu(k)\; a^{*}_{n+k} a^{*}_{r-k} a_n a_r \; , \tag{5}$$

while the number operator N is given by

$$N = \sum_k a^{*}_k \, a_k \; . \tag{6}$$

Bogoliubov's basic assumption is that in the ground state most particles have zero momentum. More precisely we may state his assumption as:

$$\lim_{\rho\to\infty} <|N - a^{*}_o \, a_o|>/N = 0 \quad , \tag{B}$$

where we mean that in (B) we take the expected value of $N - a^{*}_o \, a_o$ on the ground state of the system. Based on assumption (B), Bogoliubov claimed that V should be replaced by an expression quadratic in a_k, so

$$V \simeq \frac{1}{2}\, \nu(0) N(N-1) + \frac{1}{2} N \sum_{k\neq 0} \nu(k) [a^{*}_{-k} a_{-k} + a^{*}_k a_k + a_{-k} a_k + a^{*}_{-k} a^{*}_k] . \tag{7}$$

If we take the value (7) for V then the Hamiltonian $\rho H_o + V$ is quadratic in a_k, and so one can obtain a formula for the ground state energy which is the one given in theorem 1. Of course we need to check that the original assumption (B) holds for our calculated ground state. We obtain for the proportion of particles with nonzero momentum the formula

$$<|N - a^{*}_o a_o|>/N = \frac{1}{\rho} \sum_{k\neq 0} \frac{1}{L^3} \{ [4\pi^4 (\tfrac{k}{L})^4 + 4\pi^2 (\tfrac{k}{L})^2 L^3 \nu(k)]^{-1/2} \cdot$$

$$[2\pi^2 (\tfrac{k}{L})^2 + L^3 \nu(k)] - 1 \} . \tag{8}$$

Now as $\rho \to \infty$ the expression on the right in (8) goes to zero and hence condition (B) holds. Thus we expect the formula in theorem 1 is correct in the limit $\rho \to \infty$.

We turn to the proof of theorem 1. The upper bound on $\varepsilon_1(\rho)$ is established by evaluating $\rho H_0 + V$ on the Bogoliubov ground state wave function - actually a slight variant of it. Similar calculations have been performed previously [1]. To get the lower bound we proceed as follows: For $k \in Z^3$ we put

$$A_k = \sum_{n \in Z^3} a^*_{n+k} a_n \quad . \tag{9}$$

Then it is easy to see that V is given by

$$V = \frac{1}{2} \sum_{k \in Z^3} \nu(k) \; A_k A_k^* - \frac{1}{2}[\sum \nu(k)] N \; . \tag{10}$$

Thus we may bound V below by

$$V \geq \frac{1}{2} \; \nu(0) A_0 \; A_0^* - \frac{1}{2}[\sum \nu(k)] N = \frac{1}{2}\nu(0) N(N-1) + [\frac{1}{2}\nu(0) \; - \frac{1}{2}\phi(0)] N \; . \tag{11}$$

The bound (11) has already been obtained by Lieb [3]. It gives us the lower bound for $\varepsilon_\alpha(\rho)$ in the case $\alpha < 1$.

Now let $|\psi\rangle$ be the ground state for $\rho H_0 + V$. Then we clearly have that

$$\langle\psi|\rho H_0 + V|\psi\rangle - \text{MFE} \leq 0 \; , \tag{12}$$

and from (11) we also have

$$\langle\psi|V|\psi\rangle - \text{MFE} \geq [\frac{1}{2} \; \nu(0) \; - \frac{1}{2} \; \phi(0)] N. \tag{13}$$

We conclude then from (12), (13) that

$$\langle\psi|N - a_0^* a_0|\psi\rangle / N \leq [\phi(0) \; - \nu(0)] \; L^2/4\pi^2\rho \; . \tag{14}$$

We see from (14) that the condition (B) is satisfied.

Next we wish to write V in a form similar to the Bogoliubov form (7). To do this we define S_k and T_k by

$$S_k = a_o^* a_{-k} \, ; \qquad T_k = \sum_{n+k\neq 0} a_{n+k} a_n^* \, . \tag{15}$$

Observe that A_k is related to S_k, T_k by

$$A_k = S_k + T_k^* \quad . \tag{16}$$

It is easy to see that the coefficient of $\nu(k)$ in the expression (5) is bounded below by

$$\tfrac{1}{2}[S_k^* S_k + T_k^* T_k + S_k T_k + S_k^* T_k^*] - [N - a_a^* a_o]. \tag{17}$$

We may write the kinetic energy term by means of the operators S_k by observing that

$$\rho\langle\psi|H_o|\psi\rangle = \frac{1}{L^3}\langle\psi|NH_o|\psi\rangle \geq \frac{1}{L^3}\{\sum_{k\neq 0} 2\pi^2 (\tfrac{k}{L})^2 \langle\psi|S_k^* S_k|\psi\rangle - \langle\psi|H_o|\psi\rangle\}. \tag{18}$$

Using the inequalities (17) and (18) we bound $\rho H_o + V$ below by a Hamiltonian quadratic in S_k, T_k plus some correction terms. In view of (14) we see that the correction terms disappear as $\rho \to \infty$. Thus we are left with a quadratic expression in S_k, T_k which we diagonalize just as is done for the Bogoliubov Hamiltonian. This then yields the lower bound for $\varepsilon_1(\rho)$.

Next we wish to let the dimension of the box $L \to \infty$, and $\phi(x)$ become an arbitrary positive definite function such that $\phi(0) < \infty$. Thus if

$$\phi(x) = \int_{R^3} \hat{\phi}(\xi)\, e^{2\pi i \xi \cdot x}\, d\xi \;, \tag{19}$$

then we have $L^3\nu(k) = \hat{\phi}(k/L)$. Letting $L \to \infty$ in the formula for $\varepsilon_1(\rho)$ we obtain

$$\lim_{\rho\to\infty} \varepsilon_1(\rho) = \frac{1}{2}\int_{R^3} \{[4\pi^4|\xi|^4 + 4\pi^2|\xi|^2\hat{\phi}(\xi)]^{1/2} - 2\pi^2|\xi|^2 - \hat{\phi}(\xi)\}d\xi \; , \tag{20}$$

while the formula (8) for the proportion of particles with nonzero momentum becomes

$$<|N - a_o^* a_o|>/N = \frac{1}{\rho}\int_{R^3} \{[4\pi^4|\xi|^4 + 4\pi^2|\xi|^2\hat{\phi}(\xi)]^{-1/2} \cdot \cdot [2\pi^2|\xi|^2 + \hat{\phi}(\xi)] - 1\}d\xi \; . \tag{21}$$

If we let $\rho \to \infty$ in (21) then we see that condition (B) holds and hence we expect that (20) does give the correct behaviour of $\varepsilon_1(\rho)$ as $\rho \to \infty$. Notice that the right side of (20) is finite if we substitute the Coulomb potential $\hat{\phi}(\xi) = |\xi|^{-2}$. Based on this fact Foldy [2] claimed that bosons with Coulomb interaction satisfy a $\rho^{1/4}$ law.

It is very tempting then to think that theorem 1 continues to hold for the infinite box. However this is entirely false. The reason is as follows: Notice that $\varepsilon_\alpha(\rho)$ is always larger than or equal to $-\frac{1}{2}\sum_{k\neq 0}\nu(k)$, independent of α. Now for finite L the least expensive way in terms of kinetic energy of obtaining the $\nu(k)$ term in this series is to put most particles in the a_o state and 1 particle in the a_k state. However if we allow $L\to\infty$ we can achieve it more cheaply by spreading most of our particles equally among states a_p with $|p| \leq \varepsilon L/\rho^{1/2}$ for small ε. Then we need only $\rho^{3/2}/\varepsilon^3 L^3$ particles in the state a_k to obtain the $\nu(k)$ term. Hence if we let $L\to\infty$ first and then $\rho\to\infty$ we always obtain

$$\lim_{\rho\to\infty} \varepsilon_\alpha(\rho) = -\frac{1}{2}\phi(0) \; ,$$

independent of α.

It may be objected here that the Bogoliubov approximation seems to play no role for infinite boxes, which would appear to contradict the fact that Bogoliubov's formula is important in understanding well known physical phenomena in superconductivity and superfluidity. The point is that for infinite boxes we must let L and ρ tend to infinity at rates which are related. In that case the Bogoliubov approximation does become valid provided the ratio of L^2 to ρ is correct.

REFERENCES

1. F. Dyson, Ground State Energy of a Finite System of Charged Particles, J. Math. Phys. 8 (1967) 1538-1545.
2. L. Foldy, Charged Boson Gas, Phys. Rev. 124, (1961) 649-651.
3. E. Lieb, Simplified approach to the ground state energy of an imperfect Bose gas, Phys. Rev. 130, (1963) 2518-2528.
4. E. Lieb, The Bose Fluid, Lectures in Theoretical Physics, Vol. 8C (1965) 175-224, University of Colorado Press.

Acta Physica Austriaca, Suppl. XXVI, 389–392 (1984)

FUNCTIONAL APPROACH TO A SUPERCONDUCTIVITY-TYPE QUARK-MODEL WITH BROKEN SU(4)-SYMMETRY+

by

D. EBERT
Institut für Hochenergiephysik der Akademie
der Wissenschaften der DDR
Berlin-Zeuthen, DDR

By applying path-integral techniques we recently derived a composite-meson model starting from a nonlinear quark theory with approximate chiral $SU(2)_f$ symmetry [1]. In this talk I shall discuss a suitably generalized superconductivity-type quark model with approximate chiral $SU(4)_f$ flavour symmetry [2].

Let us consider the following effective quark Lagrangian with chiral $SU(4)_f$ symmetry broken by quark masses and a global colour symmetry SU(N):

$$L(q,\bar{q}) = \bar{q}(i\gamma\partial-\hat{m}_o)q + L_{int}(q,\bar{q}),$$

$$L_{int}(q,\bar{q}) = \frac{G}{2}\sum_{i=0}^{15} [(\bar{q}\lambda_i q)^2 + (\bar{q}i\gamma_5\lambda_i q)^2] , \qquad (1)$$

where $\hat{m}_o = \mathrm{diag}\,(m_u^o, m_u^o, m_s^o, m_c^o)$ and λ_i are Gell-Mann $SU(4)_f$ matrices, $\lambda_o = \frac{1}{\sqrt{2}}\, 1$. G is a universal four-quark coupling

---
+Seminar given at the XXIII. Internationale Universitätswochen für Kernphysik, Schladming, Austria, February 20-March 1, 1984.

strength with dimension $(\text{length})^2$. It is convenient to rewrite the scalar interaction terms in L_{int} as

$$L^{s}_{int}(q,\bar{q}) = \frac{G}{2}\Big[\sum_{i=1}^{3}(\bar{q}\tilde{\lambda}_i q)^2 + \sum_{i\neq 0,8,15}(\bar{q}\lambda_i q)^2\Big] , \tag{2}$$

where now $\tilde{\lambda}_1 = \text{diag}\,(1,1,0,0)$, $\tilde{\lambda}_2 = \text{diag}\,(0,0,\sqrt{2},0)$ and $\tilde{\lambda}_3 = \text{diag}\,(0,0,0,\sqrt{2})$. The scalar interaction terms $G/2(\bar{q}\tilde{\lambda}_i q)^2$ in (2) will generate nonvanishing vacuum expectation values $\langle\bar{q}\tilde{\lambda}_i q\rangle$ so that the chiral symmetry $SU(4)_f$ will be broken both explicitly (by $\hat{m}_0$) and dynamically. Finally, in order to obtain Goldberger-Treiman relations for the singlet and 15-plet of composite 0^--mesons one has to introduce electro-weak interactions in (1). To this end, replace ∂_μ by the covariant derivative D_μ of the gauge group $SU(2)_L \times U(1)$.

To analyze the dynamical content of the nonlinear quark model (1) we consider the generating functional of Greens functions. The idea is to first introduce colour-singlet composite mesons $\sigma_i \sim \bar{q}\lambda_i q$, $\Phi_i \sim \bar{q}\gamma_5\lambda_i q$ to get an action bilinear in the quark fields. Next, performing the path integral over quark fields one gets the effective meson Lagrangian

$$L_{eff}(\{\sigma\},\{\Phi\}) = -\frac{\Lambda^2}{2}\Big(\sum_{i=1}^{3}\tilde{\sigma}_1^2 + \sum_{i\neq 0,8,15}\sigma_i^2 + \sum_{i=0}^{15}\Phi_i^2\Big) -$$

$$- i\,N\,\text{Tr}(\ln\{-S^{-1}(x,y)\})_{x=y} , \tag{3}$$

where the quark propagator S is given by

$$S^{-1}(x,y) = -\{i\gamma D - \hat{m}_0 - g\Big(\sum_{i=1}^{3}\tilde{\sigma}_i\tilde{\lambda}_i + \sum_{i\neq 0,8,15}\sigma_i\lambda_i + i\gamma_5\sum_{i=0}^{15}\Phi_i\lambda_i\Big)\}\,.$$

$$.\ \delta^4(x-y) \tag{4}$$

and we have set $G = g^2/\Lambda^2$. The distance scale Λ^{-1} turns out to be a measure characterizing the range of the short-range $q\bar{q}$-forces responsible for dynamical breaking of chiral symmetry. The scale Λ will then be used as an intrinsic cut-off for all momentum space integrals arising in the loop expansion of the quark determinant in (3).

From the stationarity condition $\delta L_{eff}/\delta\tilde{\sigma}_i = 0$ one can derive an equation for the nonvanishing vacuum expectation values $\langle\tilde{\sigma}_i\rangle = -g/\Lambda^2\langle\bar{q}\tilde{\lambda}_i q\rangle$. Introducing the matrix $\hat{m}$ of total quark masses defined as the sum of bare and dynamical masses

$$\hat{m} = \hat{m}_o + \hat{m}_{dyn} = \hat{m}_o + g \sum_{i=1}^{3} \langle\tilde{\sigma}_i\rangle\tilde{\lambda}_i \tag{5}$$

and using the notation $(m_u, m_s, m_c) = (m_1, m_2, m_3)$, we obtain

$$m_i = m_i^o + 8m_i G I_1(m_i) \tag{6}$$

where $I_1(m_i)$ is the regularized integral

$$I_1(m_i) = \frac{iN}{(2\pi)^4} \int^{\Lambda} \frac{d^4k}{k^2 - m_i^2} \quad . \tag{7}$$

The standard procedure is then to do the shift $\tilde{\sigma}_i \rightarrow \tilde{\sigma}'_i = \tilde{\sigma}_i - \langle\tilde{\sigma}_i\rangle$, expand the quark determinant in powers of $\tilde{\sigma}'_i$, σ_i, Φ_i and perform necessary renormalizations. Kinetic terms for meson fields as well as mass formulae for scalar and pseudoscalar mesons follow by evaluating quark loop diagrams with two external meson legs. One gets, for example, for the strange or charm mesons

$$M_\Phi^2 = \frac{1}{2} \cdot \left(\frac{m_i^o}{m_i} + \frac{m_j^o}{m_j}\right) \cdot g_\Phi^2 / G + (m_i - m_j)^2 \quad (\Phi = K, D, F), \tag{8}$$

where $g_\Phi = Z_\Phi^{1/2}$ is a renormalized meson-quark coupling constant.

Moreover, by considering weak ΦW_μ-transition diagrams one can derive Goldberger-Treiman relations, e.g. $F_D = (m_u+m_c)/2g_D$ etc., F_D being the pseudoscalar D-meson decay constant.

Note that our model contains five free parameters: G, Λ, m_u^o, m_s^o, m_c^o (we take N = 3). To fix them it is convenient to take the following quantities from experiment:

$$g_\pi^2/4\pi = \frac{1}{6}\, g_{\rho\pi\pi}^2/4\pi \approx \frac{1}{2}^{[1]}, \; M_\pi,\; F_\pi = 93 \text{ MeV},\; F_K \approx 1.2\, F_\pi,$$

$$m_c \approx 1.6 \text{ GeV}.$$

Using eqs.(6), (8) and the Goldberger-Treiman relations we obtain the following results:

i) $m_u^o = 5$ MeV, $m_s^o = 105\text{-}125$ MeV, $m_c^o = 1250$ MeV

$m_u = 233$ MeV, $m_s = 430$ MeV

ii) $M_K = 480$ MeV, $M_D = 2140$ MeV, $M_F = 2380$ MeV etc.

iii) $F_D \approx F_F = 140$ MeV; $g_K/g_\pi = 1.2$, $g_D/g_\pi = 2.6$ etc.

$\Lambda = 1093$ MeV. (9)

Concluding we mention that the above estimates for the bare and total quark masses roughly agree with standard estimates for current and constituent quarks or are close to them. Moreover, the predicted mass pattern for the 15-plet of 0^--mesons is in rough agreement (15 % error) with experiment.

REFERENCES

1. D. Ebert, M.K. Volkov, Z. Phys. C-Particles and Fields 16 (1983) p.205.
2. D. Ebert, JINR-Prepr. E2-83-795 (1983) Dubna.
These papers contain further references on this subject.

Acta Physica Austriaca, Suppl. XXVI, 393–399 (1984)

SOLITONS IN SOLID STATE PHYSICS+

by

H. GROSSE
Institut für Theoretische Physik
Universität Wien

ABSTRACT

We construct all reflexionless charge symmetric potentials of the Dirac equation, identify them as solitons of the MKdV equation, show the connection to the KdV solitons and indicate the relevance for polyacetylen.

I. MATHEMATICAL MOTIVATION

150 years ago J.S. Russell wrote down the first description of a solitary wave phenomenon; at the end of the last century the KdV equation for q(t,x) was written down ,

$$q_t = 6qp_x - q_{xxx} \quad , \tag{1}$$

but it was only in the late sixties possible to "solve" the initial value problem [1]. One takes q(0,x) as a potential in the one dimensional Schrödinger equation and goes over to the scattering data, which are reflection coefficient,

+ Seminar given at the XXIII. Internationale Universitätswochen für Kernphysik, Schladming, Austria, February 20-March 1, 1984.

energy eigenvalues, and bound state normalization constants $\{R(0,k), \varepsilon_\ell, c_\ell(0)\}$ at time zero. The essential step concerns the determination of the time evolution of these new variables,

$$R(t,k) = R(0,k)e^{8ik^3t}, \quad c_\ell(t) = c_\ell(0)\ e^{4\kappa_\ell^3 t}, \quad \varepsilon_\ell = -\kappa_\ell^2 \ . \quad (2)$$

It is extremely interesting that the energy levels stay time invariant as long as the potential evolves according to the KdV flow; a property which one can easily prove using the Feynman-Hellman theorem. Finally one has to go back from scattering data at time t to the new potential $q(t,x)$ by solving a Gelfand-Levitan-Marchenko type equation. This is the famous three step procedure which allows to solve the initial value problem of certain nonlinear equations.

It is remarkable that one can obtain closed expressions for the general reflexionless potentials,

$$q(t,x) = -2\frac{d^2}{dx^2} \ln \det (1 + C), \quad C_{\ell m} = \frac{c_\ell(t)\, c_m(t)}{\kappa_\ell + \kappa_m} e^{-(\kappa_\ell + \kappa_m)x}. \quad (3)$$

These are the pure soliton solutions showing the well-known stability properties. A particular n-soliton solution starts for instance from initial data $q(0,x) = -n(n+1)/ch^2x$. We also note that all solutions of the nonlinear Hamiltonian system

$$\left(-\frac{d^2}{dx^2} - \sum_{\ell}^{N} 4\kappa_\ell\ \phi_\ell^2\right)\phi_m = \varepsilon_m \phi_m \quad (4)$$

are given by reflexionless potentials; in addition it has been shown that (4) is a completely integrable dynamical system [2].

II. PHYSICAL MOTIVATION

Polyacetylen forms long $(CH)_x$ chains; since carbon has 4 electrons in the outer 2p and 2s shells and 3 of them make a sp_2 hybridization while the last one remains as a p-electron, on the one hand a full Σ band results implying the rigidity of the chain and giving rise to elastic vibrations and phonon excitations, on the other hand (due to spin) a half-filled Π band remains. The Π-electron-phonon interaction has been successfully described by the Hamiltonian [3]

$$H_N = H_{ph} + H_{el} + H_I \; , \quad H_{ph} = \sum_{n=1}^{N} \left[\frac{p_n^2}{2M} + \frac{\omega^2}{2} (u_n - u_{n+1})^2\right] \; ,$$

$$H_{el} = -t_o \sum_{n=1}^{N} (c_n^\dagger c_{n+1} + h.c.) \, , \quad H_I = \alpha \sum_{n=1}^{N} (c_n^\dagger c_{n+1} + h.c.)(u_n - u_{n+1}) \, , \tag{5}$$

where $c_n^\dagger$, c_n are creation and annihilation operators at site n, u_n gives the elongation of the (CH) complex at site n. Next one neglects the kinetic energy of the phonons; in addition one takes into account only one phonon with maximal frequency $u_n = (-)^n u$, where u acts as an order parameter. In this one dimensional system with half-filled band one expects a Peierls distortion. Expressed differently one considers a discretized Kogut-Susskind fermion field interacting with an external scalar field; due to the filled Dirac sea a dynamical symmetry mechanism implies a non-vanishing mass term (and gap) for the fermion.

In the model these ideas can be made precise. Fourier transforming $(c_n \to a_n, b_n)$ eq. (5) gives

$$H_N = 2\omega^2 u^2 N + \sum_k \{\varepsilon_k (a_k^\dagger a_k - b_k^\dagger b_k) + \Delta_k (a_k^\dagger b_k + b_k^\dagger a_k)\} \; ; \tag{6}$$

$\varepsilon_k = 2t_o \cos ka$ gives the spectrum of the free system $(\alpha \equiv 0)$ with the ground state being the phonon vacuum + filled Dirac sea, and $\Delta_k = 4\alpha u \sin ka$ is maximal at the fermi surface. The new spectrum $E_k^2 = \varepsilon_k^2 + \Delta_k^2$ has a gap with width of $2\Delta = 8\alpha u$. Filling negative energy states determines the ground state energy as a function of u; for any nonzero coupling α two degenerate ground states show up; the system chooses $u = +u_o$ or $u = -u_o$, corresponding to regular bond alternation; in addition a nontrivial n-dependence gives rise to a soliton-like structure. In the last case one observes fractional charged states [4]; in addition a zero energy bound state appears for the Dirac equation which makes it profitable for the system to choose the distorted structure. There are experimentally observed effects which can be assigned to soliton formation.

All we have said motivates a study of the Dirac equation with external soliton potentials which we describe in the next section.

III. SOLITONS OF MKdV EQUATION

All questions which have been answered for the Schrödinger equation can be posed now in connection with the Dirac equation

$$(\alpha p + \beta v(t,x))\psi = E\psi \quad , \quad \alpha = -\sigma_2 \quad , \quad \beta = \sigma_3 \; , \tag{7}$$

where σ_i are Pauli matrices. It is known [5] that the spectrum of (7) remains time invariant if v evolves according to the modified KdV equation

$$v_t = 6v^2 v_x - v_{xxx} \; . \tag{8}$$

It has also been remarked that there are no soliton solutions of (8) if $\lim_{|x|\to\infty} v(t,x) = 0$. Motivated by the physics described in part II, we modify the boundary conditions and construct all solitons with $\lim_{|x|\to\infty} |v(t,x)| = m$, corresponding to charge symmetric reflexionless potentials of (7), and map them with the help of the Miura plus Galilean transformation to the KdV solitons.

As for the direct step one follows [6] and goes over to scattering data. The time evolution follows from adjusting the AKNS-scheme [5] to the new boundary conditions. This gives [7]

$$R(t,k) = R(0,k)e^{-4it(2E^2+m^2)k}, \quad c_\ell(t) = c_\ell(0)e^{-2t\kappa_\ell(2\varepsilon_\ell^2+m^2)},$$

$$E^2 = m^2 + k^2 \quad , \quad \varepsilon_\ell^2 = m^2 - \kappa_\ell^2 \quad . \tag{9}$$

For the inverse step one has to solve a 2×2 matrix integral equation. We observe that the kernel simplifies for charge symmetric potentials and obtain all these reflexionless potentials

$$v(t,x) = \frac{d}{dx} \ln \left[\det(1 + C^+)/\det(1 + C^-)\right] + m \quad , \tag{10}$$

where the matrices $C^\pm$ are expressed in terms of scattering data ,

$$C^\pm_{\ell m} = \sqrt{c_\ell(t)\gamma^\pm_\ell c_m(t)\gamma^\pm_m}\ \frac{e^{-(\kappa_\ell+\kappa_m)x}}{\kappa_\ell + \kappa_m} \quad , \quad \gamma^\pm_\ell = 2\,\frac{m \pm \kappa_\ell}{m - \varepsilon_\ell} \quad . \tag{11}$$

The one soliton case means $v(t,x) = \text{th}(x+2t)$; special initial values leading to n-solitons are $v(0,x) = n \text{ th } x$.

All solutions (10) can be mapped to KdV solitons by applying a Miura transformation from v to $\bar{q}$ and in addition a Galilean transformation from $\bar{q}$ to q:

$$\bar{q} = v^2 - \frac{\partial}{\partial x} v \quad , \quad q(t,x) = \bar{q}(t,x - 6m^2t) - m^2 \quad . \tag{12}$$

If all $\varepsilon_\ell \neq 0$ we are able to show that a dynamical system (the analogue of (4)) is completely integrable. A related system is obtained by starting from the continuum version of the model described before (5) and taking variations of the fermionic ground state energy w.r.t. the external potentials. This leads to the Bogoliubov-de Gennes equations [8,9]

$$(\alpha p + \beta v)\psi_n = \varepsilon_n \psi_n \quad , \qquad v(x) = \lambda \sum_n{}' \bar{\psi}_n(x) \ \psi_n(x) \qquad (13)$$

where the sum runs over occupied states. It can be shown that <u>all</u> solutions of (13) are reflexionless potentials and therefore determined through (10). A further study of the relevance of these excited solitons to the physics of polyacetylen is under study.

First we have given a few well-knwon results about solitons of the KdV equation, emphasizing the deep connection to the one dimensional Schrödinger equation. In part two we discussed a model which describes the electron phonon interaction in polyacetylen. This motivated to us the study of soliton solutions of the MKdV equation with nontrivial boundary conditions, which gives the isospectral flow of the one dimensional Dirac equation. We have constructed all reflexionless charge symmetric potentials of the latter (for detail see Ref. [7]), identified them as solitons of the MKdV equation, showed the connection to the KdV solitons,and indicated a possible relevance for polyacetylen.

REFERENCES

1. C.S. Gardner et al., Comm. Pure and Appl. Math. <u>27</u> (1974) 97.
2. H. Grosse, Acta Phys. Austr. <u>52</u> (1980) 89.
3. W.P. Su, J.R. Schrieffer and A.J. Heeger, Phys. Rev. <u>B22</u> (1980) 2099.
4. R. Jackiw and J.R. Schrieffer, Nucl. Phys. <u>B190</u> (1981) 253.

5. M.J. Ablowitz et al., Phys. Rev. Lett. 31 (1973) 125.
6. I.S. Frolov, Soviet Math. Dokl. 13 (1972) 1468.
7. H. Grosse, ZiF-preprint 1984, to be published.
8. H. Takayama, Y.R. Lin-Liu and K. Maki, Phys. Rev. B21 (1980) 2388.
9. D.K. Campbell and A.R. Bishop, Nucl. Phys. B200 (1982) 297.

Acta Physica Austriaca, Suppl. XXVI, 401–408 (1984)

NORMALIZATION OF CURRENTS IN LATTICE QCD+

by

J. HOEK
Rutherford Appleton Laboratory
Chilton, Didcot, Oxon OX11 OQX
England

The definition and normalization of vector and axial vector currents in lattice QCD with Wilson's fermion method [1] are discussed. A fuller account of this work, which was performed with Rob Groot and Jan Smit, and more references can be found in [2].

On the lattice several definitions of vector and axial vector currents, all having the same naive continuum limit

$$\bar{\psi}(x)i\gamma_{\mu}\tfrac{1}{2}\lambda_{\alpha}\psi(x) \quad \text{resp.} \quad \bar{\psi}(x)i\gamma_{\mu}\gamma_{5}\tfrac{1}{2}\lambda_{\alpha}\psi(x) \quad ,$$

are possible. Most of these currents are not conserved on the lattice, even for symmetric mass terms. They are conserved in the continuum limit. In [3] it was concluded that this non-conservation leads in general to a finite renormalization of the strength of the currents which has to be

---

+ Seminar given at the XXIII. Internationale Universitätswochen für Kernphysik, Schladming, Austria, February 20-March 1, 1984.

compensated by a multiplicative constant κ. These κ's are functions of the bare coupling g^2 and are flavor independent apart from a singlet/non-singlet splitting. In the limit in which the lattice regularization is removed, g^2 goes to zero and the κ's to 1. Thus the values of the κ's for non-vanishing g^2 seem irrelevant. In Monte Carlo simulations however continuum behaviour seems to set in for values $g^2 \gtrsim 1$; in this region the κ's may give important corrections to the current matrix elements. They will be computed below up to order g^2 and for $g^2 \to \infty$.

In order to determine the κ's for all values of g, non-perturbative normalization conditions have to be given. In [3] the use of anomalies in Ward identities involving three currents was advocated. The fundamental importance of the normalization conditions lies in the fact that some of the currents enter in the electroweak interactions.

The currents we studied are the naive currents

$$\tilde{j}^{\alpha}_{\mu}(x) = \tilde{\kappa}_V \bar{\psi}(x) i\gamma_{\mu} \tfrac{1}{2}\lambda_{\alpha}\psi(x) ,$$

$$\tilde{j}^{\alpha}_{5_{\mu}}(x) = \tilde{\kappa}_A \bar{\psi}(x) i\gamma_{\mu}\gamma_5 \tfrac{1}{2}\lambda_{\alpha}\psi(x) ,$$

the point-split symmetry currents

$$j^{\alpha}_{\mu}(x) = \kappa_V \tfrac{i}{4}[\bar{\psi}(x)\gamma_{\mu}\lambda_{\alpha}U_{\mu}(x)\psi(x+a_{\mu}) + \bar{\psi}(x+a_{\mu})\gamma_{\mu}\lambda_{\alpha}U^{+}_{\mu}(x)\psi(x)] ,$$

$$j^{\alpha}_{5_{\mu}}(x) = \kappa_A \tfrac{i}{4}[\bar{\psi}(x)\gamma_{\mu}\gamma_5\lambda_{\alpha}U_{\mu}(x)\psi(x+a_{\mu}) + \bar{\psi}(x+a_{\mu})\gamma_{\mu}\gamma_5\lambda_{\alpha}U^{+}_{\mu}(x)\psi(x)] ,$$

and the conserved vector currents

$$\hat{j}^{\alpha}_{\mu}(x) = \hat{\kappa}_V i[-\bar{\psi}(x)P^{-}_{\mu}\tfrac{1}{2}\lambda_{\alpha}U_{\mu}(x)\psi(x+a_{\mu}) + \bar{\psi}(x+a_{\mu})P^{+}_{\mu}\tfrac{1}{2}\lambda_{\alpha}U^{+}_{\mu}(x)\psi(x)] .$$

In this last formula $P^{\pm}_{\mu} = \frac{1}{2}(r \pm \gamma_{\mu})$ and r is the coefficient with which the Wilson term is added to the naive fermion action ($0 \leq r \leq 1$). [3].

The symmetry currents are related to a global symmetry of the naive part ($r = 0$) of the action, and both they and the Ward identities they satisfy can be obtained by performing the gauged symmetry transformation on the generating functional. The vector currents that are conserved in the symmetry limit are similarly related to a global symmetry of the full action. The naive currents are more ad hoc, as they are not related to the quark nearest neighbour coupling in the action and we do not know of any Ward identities for them. They have current quantum numbers though and mix with the other currents.

In the weak coupling limit the use of anomalies as normalization conditions is awkward as $O(g^2)$ effects come in only at two loop level. In this region we have used the standard normalization of the quark axial vector charge where $O(g^2)$ effects come in at 1-loop level. This normalization is expected to be equivalent to the anomaly normalization (at least up to $O(g^2)$) provided that the chiral limit is taken before removing the infrared regulating gluon mass (an infrared ambiguity exists in the axial vector quark form factor). The quark form factor at zero momentum transfer is given by

$$\langle\vec{p}|j^{\alpha}_{5_\mu}(0)|\vec{p}\rangle = \bar{u}(\vec{p})i\gamma_\mu\gamma_5\,\tfrac{1}{2}\lambda_\alpha u(\vec{p})\kappa_A\,\frac{Z_2}{Z_1^A}$$

and similarly for the other currents. Here Z_2 is the fermion wave function renormalization constant and the Z_a's are vertex renormalization constants. Requiring $\kappa Z_2/Z_1 = 1$ in the symmetry limit determines the κ's. These computations have been independently performed by three groups [2,8,9] and are all in agreement. With the parametrization

$$\kappa = 1 - g^2 C_2 d(r) + O(g^4), \qquad C_2 = \frac{N^2-1}{2N} ,$$

where N is the number of colors, our results for r = 1 are:

$$d^V(1) = .061972(1) \ , \qquad d^A(1) = .054857(1) \ ,$$

$$\tilde{d}^V(1) = .130561(2) \ , \qquad \tilde{d}^A(1) = .100030(2) \ .$$

The bracket indicates the one standard deviation error in the last decimal place. Note that $\hat{\kappa}_V = 1$.

In the strong coupling region the VVA - anomaly can be used as a normalization condition. Lattice QCD can be reformulated in this region in terms of an effective action for meson and baryon fields [4 - 7]. Large N considerations support the use of this effective action in the tree graph approximation in the mesonic sector. However, in this approximation the AAA - anomaly vanishes and one other condition is called for in order to determine both κ_A's and κ_V's. Non-perturbative conditions can be obtained by imposing formfactors for vector currents between true one-particle states. In the strong coupling region the pion or nucleon can be used, but also the ρ which is stable in the quenched approximation.

The currents are conveniently introduced by adding external field terms for each of them to the original action,

$$\sum_{x,\mu} (j^\alpha_\mu(x) v^\alpha_\mu(x) + j^\alpha_{5_\mu} a^\alpha_\mu + \tilde{j}^\alpha_\mu \tilde{v}^\alpha_\mu + \tilde{j}^\alpha_{5_\mu} \tilde{a}^\alpha_\mu + \hat{j}^\alpha_\mu \hat{v}^\alpha_\mu) \ .$$

In the effective action the effect of this is a shift of the sourceterm for the meson field M for the naive currents,

$$J(x) \to J(x) - \frac{1}{2}\lambda_\alpha i\gamma_\mu [\tilde{\kappa}_V \tilde{v}^\alpha_\mu(x) + \tilde{\kappa}_A \gamma_5 \tilde{a}^\alpha_\mu(x)] \ ,$$

and a shift of the $P^\pm_\mu$ in the kinetic term for the other currents,

$$P_\mu^\pm \rightarrow P_\mu^\pm + \frac{1}{4}\lambda_\alpha i\gamma_\mu[\kappa_V v_\mu^\alpha(x) + \kappa_A\gamma_5 a_\mu^\alpha(x)] \pm iP_\mu^\pm \frac{1}{2}\lambda_\alpha\hat{\kappa}_V\hat{v}_\mu^\alpha(x) .$$

The expressions for the currents in terms of the internal fields of the effective action formalism are found by isolating the terms linear in the external fields v_μ^α, a_μ^α, ... after the shifts have been performed. We take $r = 1$ from now on. Then the vacuum expectation value of the meson field is $v = \frac{1}{2}$ if a flavor independent mass term is assumed such that the pion mass vanishes. The fluctuations of M around v are parametrized as

$$M(x) = v\ 1 + \frac{i}{\sqrt{N}}\ \phi^\alpha(x)\ \frac{\lambda^\alpha}{\sqrt{2}} \quad ,$$

$$2\phi^\alpha = S^\alpha + P^\alpha\gamma_5 + A_\mu^\alpha\ i\gamma_\mu\gamma_5 + V_\mu^\alpha\gamma_\mu + \frac{1}{2}T_{\mu\nu}^\alpha\sigma_{\mu\nu} .$$

In terms of these fields the expression for the naive currents is

$$\tilde{j}_\mu^\alpha(x) = \tilde{\kappa}_V(2N)^{1/2}\ V_\mu^\alpha(x) \ , \qquad \tilde{j}_{5_\mu}^\alpha(x) = -\ i\tilde{\kappa}_A(2N)^{1/2}A_\mu^\alpha(x) \ ,$$

and the low energy form of the other currents is

$$j_\mu^\alpha(x) = \kappa_V[v(2N)^{1/2}V_\mu^\alpha(x) - \frac{1}{4}\ f_{\alpha\beta\gamma}P^\beta(x)\partial_\mu P^\gamma(x) + \ldots] \quad ,$$

$$j_{5_\mu}^\alpha(x) = \kappa_A[v(2N)^{1/2}(\frac{1}{2}\partial_\mu P^\alpha(x) - iA_\mu^\alpha(x)) + \ldots] \ ,$$

$$\hat{j}_\mu^\alpha(x) = \hat{\kappa}_V[-\ \frac{1}{2}\ f_{\alpha\beta\gamma}P^\beta(x)\partial_\mu P^\gamma(x) + if_{\alpha\beta\gamma}P^\beta(x)A_\mu^\gamma(x) + \ldots] \quad .$$

The naive currents have a pure vector dominance form whereas the symmetry currents have bilinear and higher order terms in the fields in addition to a suppressed vector dominance term. The conserved vector currents lack a vector dominance term in the infinite coupling limit (they acquire one in $O(1/g^2)$). This implies an infinite ρ-decay constant and is a major reason to consider vector currents other than the conserved one. They may interpolate more smoothly between the strong and weak coupling domains.

First the κ_V's are determined by requiring the usual matrix elements of the vector charges between one pion states, using as interpolating pion fields $P^\alpha(x)$. To lowest order in the external momenta this leads to

$$<\vec{p}\beta|\tilde{j}^\alpha_\mu(0)|\vec{q}\gamma> \simeq - if_{\alpha\beta\gamma}(p+q)_\mu \tilde{\kappa}_V \frac{16}{5} ,$$

so the standard normalization implies $\tilde{\kappa}_V = \frac{5}{16}$. Similarly one finds $\kappa_V = \frac{10}{21}$ and $\hat{\kappa}_V = 1$ as it should be. We could use rho-meson states instead and normalize to

$$<\vec{0}\lambda\beta|j^\alpha_4(0)|\vec{0}\sigma\gamma> = - if_{\alpha\beta\gamma}\vec{e}^{(\lambda)*}\cdot\vec{e}^{(\sigma)}2im_\rho F_1 .$$

Imposing $F_1 = 1$ leads to the same κ's, so the operator currents appear to be conserved in the low momentum limit.

The κ_A's are subsequently found by requiring the standard Adler-Bardeen [10-12] form for the VVA anomaly. For the current correlation function of the naive currents we find in leading order in the external momenta

$$\sum_{xy} e^{-ipx-iqy}<\tilde{j}^\beta_\mu(x)\tilde{j}^\gamma_\nu(y)\tilde{j}^\alpha_{5_\lambda}(0)> \simeq \frac{1}{4} Nd_{\alpha\beta\gamma}\tilde{\Gamma}_{\mu\nu\lambda}(p,q) + ct ,$$

$$\tilde{\Gamma}_{\mu\nu\lambda}(p,q) + ct = 16\tilde{\kappa}_V^2 \tilde{\kappa}_A[\tfrac{2}{3}\varepsilon_{\lambda\mu\nu\rho}(p_\rho - q_\rho) - \frac{14}{15}\frac{k_\lambda}{k^2}\varepsilon_{\mu\nu\rho\sigma}p_\rho q_\sigma] ,$$

where $k = -p-q$ and the contact terms "ct" (polynomials in p and q) are to be chosen such that the vector Ward identities are satisfied,

$$p_\mu\tilde{\Gamma}_{\mu\nu\lambda}(p,q) = q_\nu\tilde{\Gamma}_{\mu\nu\lambda}(p,q) = 0 .$$

Thus

$$\tilde{\Gamma}_{\mu\nu\lambda}(p,q) = - 16\tilde{\kappa}_V^2 \tilde{\kappa}_A \frac{14}{15}\frac{k_\lambda}{k^2}\varepsilon_{\mu\nu\rho\sigma} p_\rho q_\sigma .$$

Requiring the standard Adler-Bardeen form for the axial Ward identity ,

$$(p + q)_\lambda \tilde{\Gamma}_{\mu\nu\lambda}(p,q) = \frac{2}{2\pi^2}\,\varepsilon_{\mu\nu\rho\sigma} p_\rho q_\sigma \ ,$$

and using the value found above for $\tilde{\kappa}_V$ we find $\tilde{\kappa}_A = \frac{12}{35\pi^2} = .035$. Similarly $\kappa_A = .048$.

In conclusion, we find that the κ corrections as extrapolated from the weak coupling to the region $g \approx 1$ are small but non-negligible (8-18%). In the strong coupling limit the deviations of the κ's from 1 are substantial for the vector currents ($\kappa_V = .48$, $\tilde{\kappa}_V = .31$) and large for the axial vector currents ($\kappa_A = .048$, $\tilde{\kappa}_A = .035$). This large deviation from 1 at strong coupling might be accompanied by a rapid crossover to weak coupling behaviour. It is therefore hazardous to make predictions involving currents in the intermediate coupling region without knowledge of the κ's. At strong coupling the κ's reduce the discrepancies with the experimental values of the decay constants. Defining the pion decay constant f_π by

$$\langle j^\alpha_{5\mu}\ j^\beta_{5\nu}\rangle \approx \delta_{\alpha\beta}\ \frac{p_\mu p_\nu}{p^2}\ f^2_\pi$$

we find $f_\pi = \frac{\sqrt{5}}{2}\kappa_A$ and $\tilde{f}_\pi = \frac{2}{\sqrt{5}}\tilde{\kappa}_A$. Converting to MeV with $m_\rho = .867/a = 776$ MeV (a is the lattice distance) gives $f_\pi = 50$ MeV (1000 MeV without κ) and $\tilde{f}_\pi = 29$ MeV (800 MeV without κ), to be compared with the experimental value of 93 MeV. Similarly, we find for f_ρ, defined by [13],

$$\langle \vec{\partial}\lambda\beta | j^\alpha_k | 0\rangle = i\delta_{\alpha\beta} e^k_\lambda\ \frac{m^2_\rho}{f_\rho} \quad ,$$

the values $f_\rho = 1.71$ (.81) and $\tilde{f}_\rho = 1.30$ (.41), still quite below the experimental value of 5.3.

REFERENCES

1. K.G. Wilson, in New phenomena in subnuclear physics, ed A. Zichichi (Plenum,1977).
2. R. Groot, J. Hoek and J. Smit, Normalization of currents in lattice QCD, University of Amsterdam preprint ITFA-83-6.
3. L.H. Karsten and J. Smit, Nucl. Phys. B183 (1981) 103.
4. N. Kawamoto and J. Smit, Nucl. Phys. B192 (1981) 100.
5. J. Hoek, N. Kawamoto and J. Smit, Nucl. Phys. B199 (1982) 495.
6. H. Kluberg-Stern, A. Morel, O. Napoly and B. Petersson, Nucl. Phys. B190 [FS3] (1981) 504.
7. H. Kluberg-Stern, A. Morel and B. Petersson, Nucl. Phys. B215 [FS7] (1983) 527.
8. B. Meyer and C. Smith, Phys. Lett. B123 (1983) 62.
9. G. Martinelli and Zhang Yi-Cheng, Phys. Lett. B123 (1983) 433; Phys. Lett. B125 (1983) 77.
10. S.L. Adler, Phys. Rev. 177 (1969) 2426.
11. S.L. Adler and W.A. Bardeen, Phys. Rev. 182 (1969) 1517.
12. W.A. Bardeen, Phys. Rev. 184 (1969) 1848; Nucl. Phys. B75 (1974) 246.
13. O. Dumbrajs, R. Koch, H. Pilkuhn, G.C. Oades, H. Behrens, J.J. de Swart and P. Kroll, Nucl. Phys. B216 (1983) 277.

Acta Physica Austriaca, Suppl. XXVI, 409–413 (1984)

STOCHASTIC QUANTIZATION AND SUPERSYMMETRY+

by

R. KIRSCHNER
Sektion Physik
Karl-Marx-Universität Leibzig, GDR

Since the paper by Parisi and WU [1] there is increasing interest in stochastic quantization.

The convergence of the stochastic relaxation process in the large-time limit to the corresponding quantum theory can be shown using the Fokker-Planck equation [2-5]. In the framework of perturbation theory the validity of stochastic quantization has been shown analyzing the stochastic graphs generated by the Langevin equation [6-8].

Consider a system with the action

$$S = \int d^n x \; L(\phi(x)) \tag{1}$$

and couple it to a white-noise random force $\eta(x,t)$. The behaviour of the system is described by the Langevin equation with initial conditions at $t = 0$. We extend the range in time to the full axis defining, that for $t < 0$ a process takes place obtained from the process at $t > 0$ by time reflection.

The generating functional $Z(j,t)$ of the stochastic equal-time correlation functions can be written as the

+Seminar given at the XXIII. Internationale Universitätswochen für Kernphysik, Schladming, Austria, February 20- March 1, 1984.

generating functional of a quantum superfield theory [8] on the superspace consisting of n space dimensions, the additional time dimension t and the Grassmann dimensions θ and $\bar{\theta}$ with a restricted form of the current

$$J_t(x,t',\theta,\bar{\theta}) = j(x)\,\delta(t-t')\,\delta(\bar{\theta})\,\delta(\theta) \ . \tag{2}$$

The superfield Lagrangian is given by

$$L(\Phi) = L_K(\Phi) + i\, L(\Phi) \ ,$$

$$L_K(\Phi) = \Phi(\partial_\theta \partial_{\bar{\theta}} - \frac{i}{2}\,\mathrm{sgn}\, t(\partial_t \theta \partial_\theta - \partial_\theta \theta \partial_t))\Phi \ . \tag{3}$$

$\phi(x,t)$ is the lowest component of the superfield Φ. This superfield theory has a supersymmetry of the type of supersymmetric quantum mechanics (transforming t, θ and $\bar{\theta}$).

The generating functional $Z(j,t)$ is invariant with respect to a further supertransformation. This transformation leaves the following combination of the superfield coordinates invariant :

$$\tau = t - \frac{i}{2}\,\mathrm{sgn}\, t\ \bar{\theta}\theta \ . \tag{4}$$

We call this symmetry reduction supersymmetry. It is not a symmetry of the superfield Lagrangian. The invariance of $Z(j,t)$ under this transformation is due to the special form of the current J_t, eq. (2). The variation of the Lagrangian can be compensated by a change of variables in the functional integral.

Applying a time translation by an amount proportional to $\bar{\theta}\theta$ it can be shown that the reduction supersymmetry coincides with the supertransformation invariance found in[7] by analyzing the stochastic graphs.

Besides of the stochastic quantization by relaxation processes discussed so far there is a similar quantization procedure working with two additional space dimensions. Starting with the same theory, eq. (1), we consider the action with the field $\phi(x)$ replaced by a field $\tilde{\phi}(\tilde{x})$ living in n+2 dimensions ($\tilde{x} = (x, x^{(n+1)}, x^{(n+2)})$). Correspondingly, one has to replac the Laplacian involved in $L(\phi)$ by the n+2 dimensional Laplacian. We couple the resulting system to a white-noise stochastic force $\tilde{\eta}(\tilde{x})$.

Also in this case the generating functional of the stochastic correlation functions $\tilde{Z}(j)$ can be expressed as the generating functional of a quantum superfield theory. The superfield Lagrangian has the form

$$\tilde{L}(\tilde{\Phi}) = L(\tilde{\Phi}) - \frac{1}{2} \tilde{\Phi} \partial_\theta \partial_{\bar{\theta}} \Phi \; . \tag{5}$$

For stochastic quantization we need the correlation functions with their arguments restricted to the submanifold $x^{(n+1)} = x^{(n+2)} = 0$. Thus the current in the generating functional has the special form

$$J^{(n)}(\tilde{x}, \theta, \bar{\theta}) = j(x)\delta(x^{(n+1)})\delta(x^{(n+2)})\delta(\bar{\theta})\delta(\theta) \; . \tag{6}$$

As established by Parisi and Sourlas [9] this superfield theory is symmetric with respect to superspace rotations. The only invariant in the superspace of these rotations is given by

$$\nu^2 = \tilde{x}^2 + 4\,\bar{\theta}\theta \; . \tag{7}$$

Unlike the reduction supersymmetry of relaxation systems the superrotations leave the Lagrangian eq. (5) invariant.

The mechanism of stochastic quantization in the scheme with two additional space dimensions is based on the superrotation symmetry [9]. Cardy has proposed a non-perturbative proof for this case [10].

The scheme of this proof can be extended to the case of

quantization by stochastic relaxation processes. Here the quantization mechanism is based on the reduction supersymmetry.

We divide the superfield Lagrangian into two parts and introduce a parameter λ. In the case of relaxation processes we have

$$L_{\lambda}(\Phi) = [\lambda-(1-\lambda)\frac{i}{2}\delta(\bar{\theta})\,\delta(\theta)\,(\delta(t'-t)+\delta(t'+t))]i\ L(\Phi)+L_{K}(\Phi). \tag{8}$$

We consider the more general functional $Z_{\lambda}(j,t)$ constructed now with L_{λ} instead of L.

The proof consists of two steps. First one shows that $Z_{\lambda}(j,t)$ coincides for $\lambda = 1$ with the generating functional of the stochastic correlation functions and for $\lambda \to 0$ with the generating functional of the quantized theory eq.(1) in n dimensions. In the second step one shows that the connected Green functions derived from $Z_{\lambda}(j,t)$ are independent of λ in the large time limit. In this step the reduction supersymmetry is essential. A condition on the long-distance behaviour has to be imposed which is equivalent to the requirement that in the large-time limit the equal-time correlation functions become independent of the initial conditions.

The results of the two steps prove that under the mentioned condition the stochastic correlation functions converge in the large-time limit to the Green functions of the quantized theory eq.(1).

REFERENCES

1. G. Parisi, Y. Wu, Scientia Sinica 24 (1981) 483.
2. L. Baulieu, D. Zwanziger, Nucl. Phys. B193 (1981) 163.
3. J.D. Breit, S. Gupta, A. Zaks, preprint, Inst. Adv.

Study, Princeton (1983).

4. C. Bender, F. Cooper, B. Freedman, Nucl. Phys. B219 (1983) 61.
5. E. Floratos, J. Iliopoulos, Nucl. Phys. B214 (1983) 392.
6. W. Grimus, H. Hüffel, Zeitschr. f. Phys. C18 (1983) 129.
7. H. Nakazato, M. Namiki, I. Ohba, K. Okano, Progr. Theor. Phys. 70 (1983) 298.
8. E.S. Egorian, S. Kalitsin, Phys. Lett. 129B (1983) 320.
9. G. Parisi, N. Sourlas, Phys. Rev. Lett. 43 (1979) 744; Nucl. Phys. B206 (1982) 321.
10. J.L. Cardy, Phys. Lett. 125B (1983) 470.

Acta Physica Austriaca, Suppl. XXVI, 415–421 (1984)

QUANTUM FIELD THEORY IN GRAVITATIONAL BACKGROUND[+]

by

R. HAAG and U. STEIN
II. Institut für Theoretische Physik
Universität Hamburg

H. NARNHOFER
Institut für Theoretische Physik
Universität Wien

I. INTRODUCTION

This is a short survey of results contained in [1]. We study the influence of gravitation on quantum field theory, insofar, that gravity is considered as background field, changing Minkowski space into a general Riemannian manifold. Considerations in this respect were started in [2]. The propagator for the Weyl algebra was found in [3]. Interest in the model increased when in [4] thermal radiation was predicted as consequence of a reasonable state in a Schwarzschild metric (the choice of this state was based on arguments on the time evolution of the black hole). We will restrict the allowed physical states by local conditions and deduce their consequences in the tangent space. They determine already the temperature of the black hole.

[+]Seminar given at the XXIII. Internationale Universitätswochen für Kernphysik, Schladming, Austria, February 20-March 1, 1984.

II. THE ALGEBRA OF THE FREE SCALAR FIELD

In [3] it is shown that on a globally hyperbolic manifold the Cauchy propagator that satisfies

$$[-\partial_\mu g^{\mu\nu}\sqrt{-g}\partial_\nu + m^2\sqrt{-g}]G(x,x') \equiv (-\Delta_x + m^2)G(x,x') =$$

$$= (-\Delta_{x'} + m^2)G(x,x') = 0 ,$$

$$G(x,x') = -G(x',x) , \qquad G(\, , \, ') = 0 \text{ for } x \text{ spacelike to } x' , \tag{1}$$

is unique. Thus G(x,x') defines a symplectic form (though degenerate)

$$\sigma(f,g) = -\sigma(g,f) = \int f(x)G(x,x')g(x')d\Omega_x d\Omega_{x'} , \quad f \in C_0^\infty(M) , \tag{2}$$

that can be used for the construction of the Weyl algebra [5] where

$$W(f)W(g) = e^{-i\sigma(f,g)/2}W(f+g) , \qquad W(f)^\dagger = W(-f) ,$$

$$||W(f)|| = 1 , \qquad ||W(f) - 1|| = 2 , \qquad \neq 0 . \tag{3}$$

Thus there is no ambiguity in quantizing a free scalar field.

III. STATES ON THE WEYL ALGEBRA

The states on the Weyl algebra are by far too numerous as we know from statistical mechanics, where states with infinitely many particles in a finite region are mathematically possible but are excluded for physical reasons.

Some special states are the quasifree (with trivial

truncated functions) states. Those that correspond to an irreducible representation are given by an operator J, $J^2 = 1$, $\sigma(f,Jg) = -\sigma(Jf,g)$,

$$\omega_J(W(f)) = e^{-\sigma(f|JF)} \quad . \tag{4}$$

This operator allows to define creation and annihilation operators by

$$\Phi(f) = -i \frac{d}{d\lambda} W(\lambda f)$$

$$a_J(f) = \frac{\Phi(f) + i\Phi(Jf)}{\sqrt{2}} \quad , \qquad a_J^{\dagger}(f) = \frac{\Phi(f) - i\Phi(Jf)}{\sqrt{2}} \quad . \tag{5}$$

The state ω_J is the vacuum state for the a_J and $a_J^{\dagger}$. Other quasifree states are

$$\omega(W(f)) = e^{-\sigma(f|A^t JAf)} \quad .$$

They give rise to an irreducible representation iff $A^t = A^{-1}$. As in statistical mechanics we state, that only those states are physically realizable that, when restricted to a finite region, give rise to a certain standard representation. We must look for conditions that fix this standard representation. It is evident that these conditions should mimick as much as possible Minkowski space.

IV. THE WIGHTMAN DISTRIBUTION IN TANGENT SPACE

In differential geometry the difference between Minkowski space and general manifolds disappears in the tangent space. Therefore we assign to a quantum field on the Riemannian manifold a quantum field on the tangent space by a scaling procedure and demand that on the tangent space they are equal to the one obtained for the Minkowski space.

Quantum field theory is given by the Wightman functional

$$W(f,g) = \langle\Phi(f)\Phi(g)\rangle \ . \tag{6}$$

The quantum field theory of the tangent space in the point x is determined by the Wightman functional $W_x(f,g)$ given in the following way:

Let ξ be a map from $T_x \to M$ satisfying

$$\xi(0) = x, \qquad \left.\frac{d\xi(sz)}{ds}\right|_{s=0} = z, \qquad z \in T_x \ .$$

Then

$$W_x(f,g) = \lim_{s\to 0} N(s)\int W(\xi(sz_1),\xi(sz_2))\bar{f}(z_1)g(z_2)d\mu(z_1)d\mu(z_2) \ , \tag{7}$$

where $N(s)$ has to be chosen such that the limit is not trivial, namely $N(s) = s^{-6}$. It can be shown that the limit $W_x(f,g)$ does not depend on the special choice of ξ. If $W(\xi(sz_1),\xi(sz_2))$ are the Wightman functions of a free field with mass m on Minkowski space ,

$$W(x,y) = D_m^+(x-y) = \frac{1}{(2\pi)^3} e^{i\vec{p}(\vec{x}-\vec{y})-ip^o(x-y)^o} \cdot$$

$$\cdot \ \delta(\vec{p}^{\,2}+m^2-p^{o2})\Theta(p^o)d^4(p) \ ,$$

then

$$W_x(f,g) = \int\bar{f}(y_1)g(y_2)D_o^+(y_1 - y_2)d^4y_1d^4y_2 \ ,$$

thus the Wightman functionals of a free massless field independent of the point x. In general one can see that

$W_x(f,g)$ has to be Poincaré invariant. Therefore we can define $\tilde{W}_x(p)$ by

$$W_x(z_1,z_2) = \int \tilde{W}_x(p)\ e^{ip(z_1-z_2)} d\mu(p) \ ,$$

and we demand that $\tilde{W}_x(p)$ has support in $p^o \geq 0$, $(p,p) \leq 0$. By a scaling argument

$$\tilde{W}_x(\lambda p) = \lambda^{-2}\tilde{W}_x(p) \ .$$

It follows that

$$W_x(z_1,z_2) = D_o^+(z_1 - z_2) \ . \tag{8}$$

The parameter that is left free by the scaling argument is fixed by the commutation relations of the field.

V. CONSEQUENCES FOR THE SCHWARZSCHILD METRIC

We start with the Schwarzschild manifold with

$$ds^2 = -(1 - \frac{r_o}{r})dt^2 + (1 - \frac{r_o}{r})^{-1} dr^2 + r^2 d\Omega \ , \tag{9}$$

but assume that the manifold can be extended to a larger one, though it is not necessary to fix how. All we assume, that points on the horizon are interior points such that (8) has to be satisfied for $x = (t,r_o,\Omega)$. Schwarzschild time is a killing vector field. Thus the construction of creation and annihilation operators can be done in complete correspondence to Minkowski space, finding the solution to

$$(-\frac{\partial^2}{\partial \xi^2} + \frac{\ell(\ell+1)}{r^2}\frac{r-r_o}{r} + m^2\frac{(r-r_o)}{r} + \frac{r_o(r-r_o)}{r^4})\psi_{\ell,\omega} = \omega^2\psi_{\ell,\omega} \ , \tag{10}$$

where ξ is the tortoise coordinate $\xi = r + r_o \ln(r-r_o)$. Close to the horizon ($\xi \to -\infty$) it reduces to

$$\left(-\frac{\partial^2}{\partial\xi^2} + e^{\xi/r_o}(k^2 + m^2) - \omega^2\right)\psi_{\omega,k} = 0 \quad . \tag{11}$$

From the region $\xi \to +\infty$ it is separated by a potential barrier that increases like $k^2 (\simeq \ell(\ell+1))$ for $k \to \infty$. Introducing $\rho = e^{\xi}$ we obtain the equation for the Rindler space

$$\left(\frac{d^2}{d\rho^2} + \frac{1}{\rho}\frac{d}{d\rho} - (k^2 + m^2) + \frac{\omega^2 r_o^2}{\rho^2}\right)\psi = 0 \quad . \tag{12}$$

Here we know the explicit solutions, the modified Hankel functions. If we first assume that the state has a well defined temperature with respect to Schwarzschild (or Rindler) time, we can calculate

$$\langle\Phi(t_1,x_1)\Phi(t_2,x_2)\rangle_\beta \simeq \frac{1}{(2\pi)^3}\int d\omega \, \frac{e^{\beta\omega}}{e^{\beta\omega}-1} e^{-i\omega(t_1-t_2)} \sinh \pi\omega \; , \tag{13}$$

$$K_{i\omega}(\mu\rho_1) K_{i\omega}(\mu\rho_2) e^{ik^{\perp}(x_1^{\perp} - x_2^{\perp})} \, d^2k^{\perp} \, , \qquad \mu = (k^{\perp 2} + m^2)^{1/2} \quad .$$

If we stay away from the horizon, only $\omega \to \pm\infty$ becomes relevant and the dependence on β disappears. If we approach the horizon ,

$$K_{i\omega}(\mu\rho_1) \to \pi\delta(\omega) \quad , \qquad \frac{e^{\beta\omega}}{e^{\beta\omega}-1}\,\delta(\omega) = \frac{1}{\beta}\,\delta(\omega) \quad , \tag{14}$$

and due to our condition (8) only $\beta = \pi r_o$ is admissible, which is just the Hawking temperature. However it should be noted that in our context there is no reason why we should restrict ourselves to temperature states. In fact, another state that occurs in literature, i.e. the state in which only the outgoing waves have a thermal distribution, whereas the incoming waves have the vacuum distribution, is also admissible, because the incoming waves are reflected on the angular momentum barrier and do not contribute to the state on the tangent space of points on the horizon.

REFERENCES

1. R. Haag, H. Narnhofer, U. Stein, On Quantum Field Theory in Gravitational Background, DESY preprint (1984).
2. R.U. Sexl, H.K. Urbantke, Phys. Rev. 179 (1969) 1247.
3. A. Lichnerowith, in "Relativity, Groups, Topology", B.S. DeWitt and C.M. DeWitt, eds. (Gordon & Breach, New York, 1964).
4. S.W. Hawking, Comm. Math. Phys. 43 (1975) 199.
5. O. Bratteli, D.W. Robinson, "Operator Algebras and Quantum Statistical Mechanics II" (Springer, New York, 1981).
6. C. Moreno, Rep. Math. Phys. 17 (1980) 333.
7. W.G. Unruh, Phys. Rev. D14 (1976) 870.

Acta Physica Austriaca, Suppl. XXVI, 423–426 (1984)

THE JOSEPHSON POTENTIAL AS A STATISTICAL PHENOMENON+

by

A. RIECKERS
Institut für Theoretische Physik
Universität Tübingen
D-7400 Tübingen, BRD

The theoretical description of collective phenomena requires the passage to the thermodynamical limit. In the operator algebraic formulation of infinite quantum systems [1] a spontaneous symmetry break down (in virtue of a phase transition) is reflected by a non-trivial barycentric decomposition of the limiting Gibbs state (as a state on the quasi-local C^*-algebra) into factor states of lower symmetry. This so-called central decomposition constitutes a classical part of the quantum statistical ensemble theory and expresses a filtering into sub-ensembles by fixing all values of the classical observables. These sub-ensembles have been successfully interpreted as pure phase states.

Only in the calculation of the limiting Gibbs state ϕ^β of the BCS-model by direct methods [2] the mechanism for breaking gauge invariance and the meaning of the central decomposition

+Seminar given at the XXIII. Internationale Universitätswochen für Kernphysik, Schladming, Austria, February 20- March 1, 1984.

$$\phi^{\beta} = \int \phi^{\beta\theta} d\theta/2\pi \quad , \tag{1}$$

which brings into play the macroscopic phase angle $\theta \in [0,2\pi)$, has been clarified: the equipartition of the pure phase states is the conservation law for the "Cooper pairs at infinity". Reversely stated signifies every deviation from the equidistribution over θ an exchange of these particles with the surroundings. In the GNS-reconstructed quantum mechanics over ϕ^{β} these "condensed Cooper pairs" have a field operator of the form

$$s(\beta) = \text{st-op} \lim_{\Lambda} m_{\Lambda} \tag{2}$$

with

$$m_{\Lambda} := \sum_{k \in \Lambda} b_k / |\Lambda| \quad . \tag{3}$$

Here b_k is the pair-annihilation operator of momentum k (in the representation) and $|\Lambda|$ denotes the cardinality of the finite set of momenta Λ.

In order to form a microscopic model for the Josephson junction we interprete the weak coupling between two BCS-superconductors a and b in a twofold manner: also for the infinite system only a finite number of electrons tunnels with a normal frequency, but there is also a global tunneling which involves asymptotically infinitely many pairs with vanishing frequency. By the requirements of Hermiticity, total particle number conservation and symmetry in a and b the latter exchange has the unique lowest order potential

$$h_{\Lambda a,\Lambda b} = g(m^{*}_{\Lambda a}\, m_{\Lambda b} + m_{\Lambda a}\, m^{*}_{\Lambda b}) \quad . \tag{4}$$

By means of an all order perturbation theory in the interaction (4) it is demonstrated in [3] and [4] that the limiting Gibbs state ω^{β} of the coupled model exists and has the central decomposition

$$\omega^{\beta} = \iint \phi^{\beta\theta a} \otimes \phi^{\beta\theta b} \; d\mu^{\beta}(\theta a, \theta b) \tag{5}$$

with

$$d\mu^{\beta}(\theta a, \theta b) = c \exp[-d \cos(\theta a - \theta b)] d\theta_a d\theta_b \; . \tag{6}$$

The positive constants c and d in (6) depend on the natural temperature β, and d vanishes if not both critical temperatures β_{ca} and β_{cb} are smaller than β. From (5) we observe that the pure phases of the coupled and uncoupled composed systems are the same.
Using again exact perturbation theory one can treat the limiting Heisenberg dynamics in the representation over ω^{β} and obtains the same as in the non-interacting case.

We observe that the deductive treatment of the infinite quantum system provides us with a precise but rather subtle answer to Josephson's question "whether there can be any behaviour of two superconductors that is intermediate between those characteristic of complete separation and complete uion" [5]. The picture that comes out in the thermodynamical limit depicts complete separation in what concerns the pure phases and the Heisenberg dynamics. The interaction by means of Josephson's cosine-potential manifests itself solely by a correlation term in the statistical distribution for the pure phase states. Since the interaction (4) leaves the total number of particles in the composed system invariant, the non-equidistribution over the macroscopic phase angles θ_a and θ_b indicates an exchange of condensed pairs. By means of these statistical arguments one is led to the existence of a supercurrent in spite of the Heisenberg dynamics not transferring condensed particles from one subsystem to the other.

If a weak normal interaction is added, the pure phase states in the central decomposition are coupled with each other, but the reasoning for the condensed particles remains unaltered.

In [4] it is worked out how the Josephson current may be calculated by an essential extension of the dynamical formalism; the present discussion supplements some interpretational background.

REFERENCES

1. O. Bratteli and D.W. Robinson, Operator Algebras and Quantum Statisctical Mechanics I (1979) and II (1981), Springer, Berlin.
2. W. Fleig, Acta Physica Austriaca 55 (1983) 135.
3. M. Ullrich, Calculation of the Limiting Gibbs States for Weakly Coupled Macroscopic Quantum Systems with Application to the Josephson Oscillator, preprint, Tübingen (1984).
4. A. Rieckers and M. Ullrich, Microscopic Derivation of the Finite Temperature Josephson Relation in Operator Form, preprint, Tübingen (1984).
5. B. Josephson, thesis, Cambridge University (1964).

Acta Physica Austriaca, Suppl. XXVI, 427–433 (1984)

GENERATING FUNCTIONAL AND QUANTUM STABILITY IN FIELD THEORY MODELS WITH SOLITONS+

by

L. ROSZKOWSKI
Institute for Theoretical Physics
Warsaw University, Warsaw, Poland

For many years it has been known that the existence of particle-like solutions of classical field equations has nontrivial implications in quantum theory. In this seminar I will propose a new representation of generating functional and a new concept of quantum stability in field theory models with solitons.

We shall consider a class of field theory models of a multiplet $\vec{\Phi} = (\phi_1, \ldots, \phi_N)$, $N = 1, 2, \ldots$, of real scalar fields in the Minkowski space-time M [3,1] , defined by the total action

$$S[\vec{\Phi}] = \int d^4x L(\vec{\Phi})(x) = \int d^4x [\tfrac{1}{2}\partial_\mu \vec{\Phi} \partial^\mu \vec{\Phi} - \tfrac{1}{2}\mu_0^2 \vec{\Phi}^2 + L_I(\vec{\Phi})](x) \qquad (1)$$

where $x = (x^\mu) = (t, \vec{x})$, $\partial_\mu = (\partial_t, \nabla)$ and $c = 1$.

For the classical non-linear field equations

+Seminar given at the XXIII. Internationale Universitätswochen für Kernphysik, Schladming, Austria, February 20- March 1, 1984.

$$(\Box + \mu_o^2)\phi_k - \frac{\partial L_I(\vec{\Phi})}{\partial \phi_k} = 0 \quad , \quad k = 1,\ldots,N, \ \Box = \partial_t^2 - \Delta \ , \quad (2)$$

we assume in the considered models the existence of soliton (particle-like) solution $\vec{\Phi}_o$, i.e. such an object that (in its rest frame):

i) it is "localized" around some point $\vec{x}_o$ (usually $\vec{x}_o = 0$),
ii) the energy functional $E[\vec{\Phi}_o]$ is finite.
Such objects have been found analytically or numerically in many models in 2,3 or 4 dimensions.

A NEW FORMULA FOR GENERATING FUNCTIONAL

In order to study "quantum implications" of the existence on the classical level of particle-like objects we consider a generating functional $J[\vec{f}]$ and utilize its representation in terms of Feynman path integrals over all classical fields:

$$J[\vec{f}] = Z^{-1}\int_\Delta e^{i\vec{\Phi}(\vec{f})+\frac{i}{\hbar}S[\vec{\Phi}]}D\vec{\Phi} \quad , \quad Z = \int_\Delta e^{\frac{i}{\hbar}S[\vec{\Phi}]}D\vec{\Phi} \quad . \quad (3)$$

We expand (3) around $\vec{\Phi}_o$ and, after shifting $\vec{\Phi}-\vec{\Phi}_o\to\vec{\Phi}$, we obtain

$$J[\vec{f}] = e^{i\vec{\Phi}_o(\vec{f})}Z^{-1}\int_\Delta e^{i\vec{\Phi}(\vec{f})}\exp\{\frac{i}{\hbar}\int d^4x[-\frac{1}{2}\vec{\Phi}\hat{K}\vec{\Phi}+ L_I^{eff}(\vec{\Phi})](x)\}D\vec{\Phi} \quad (4)$$

where

$$\hat{K} \equiv (\Box + \mu_o^2)I - \left|\left|\frac{\partial^2 L_I(\vec{\Phi})}{\partial\phi_i\partial\phi_j}\right|\right|_{\substack{i,j=1,\ldots N \\ \vec{\Phi}=\vec{\Phi}_o}} \quad , \quad (5)$$

and $L_I^{eff}(\vec{\Phi})$ is the Taylor expansion of $L(\vec{\Phi})$ after the subtraction of terms of order lower than three. However, it turns out that the operator $\hat{K}$ has a number of normalizable eigenvectors with the zero eigenvalue - this forbids us to construct the propagator directly,

$$\hat{G}(x,y) = (\hat{K}-i\varepsilon)^{-1}(x,y) \ .$$

These eigenvectors, usually called zero translation modes (ZTM), are of group theoretical origin. To see this notice that although the symmetry group in the considered models is $G = \Pi \times SO(N)$ (where Π is the Poincaré group and SO(N) is the internal symmetry group), the x-dependent solution $\vec{\Phi}_o(x)$ has only a proper subgroup $G_o = \Pi_o \times SO(N)_o$, $G_o \subset G$, as its invariance group. Denote by $G^{\perp} = \Pi^{\perp} \times SO(N)^{\perp}$ the set of all one-parameter subgroups in G which are not in G_o, then $G = G^{\perp} G_o$. Let X_a, $a = 1, \dots, \dim G^{\perp}$, denote generators of $G^{\perp}$ and $U(\alpha_a)$-a group element generated by X_a and a parameter α_a. Now, since $\vec{\Phi}_o$ satisfies the field equations (2), then also does this $U(\alpha_a)\vec{\Phi}_o$. Differentiating the field equations for $U(\alpha_a)\vec{\Phi}_o$ with respect to α_a and setting $\alpha_a = 0$ one obtains

$$[(\Box+\mu_o^2)I - \left|\left|\frac{\partial^2 L_I(\vec{\Phi})}{\partial\phi_i \partial\phi_j}\right|\right|_{\vec{\Phi}=\vec{\Phi}_o}](X_a\vec{\Phi}_o) = \hat{K}(X_a\vec{\Phi}_o) = 0 \quad .$$

Thus (the non-normalized) vectors $X_a\vec{\Phi}_o$ are the ZTM in the theory.

In order to define $\hat{G}(x,y)$ correctly one has to remove all ZTM from the domain of $J[\vec{f}]$. I will propose here the following procedure which allows to perform this [1]:

i) Choose in the subspace of ZTM a set of orthonormal eigenfunctions $\vec{u}_a^{\,o}$ of $\hat{K}$, $a=1,\dots, d=\dim \tilde{G}^{\perp}$ ($\tilde{G}^{\perp}$ is the collection of those one-parameter subgroups in $G^{\perp}$ which form a complete set of orthonormal ZTM) - in every so far considered model with volume cutoff this has been always possible.

ii) Construct the following identity:

$$\int_{\tilde{G}^{\perp}} d\alpha \prod_{a=1}^{d} \delta[\int d^4z \vec{u}^o{}_a(z)(U(\alpha)\vec{\Phi})(z)] I(U(\alpha)\vec{\Phi}) = 1$$

where $\alpha = (\alpha_1,\ldots,\alpha_d)$, $U(\alpha) = \prod_{a=1}^{d} U(\alpha_a)$, and $I(U(\alpha)\vec{\Phi})$ is a Jacobian ,

$$I(U(\alpha)\vec{\Phi}) = \det||\int d^4z \vec{u}^o{}_a(z) \frac{\partial}{\partial\alpha_b}(U(\alpha)\vec{\Phi})(z)||_{a,b=1,..,d.}$$

iii) nsert this identity into (3) and expand $J[\vec{f}]$ around $\vec{\Phi}_o$. This makes the domain Δ of $J[\vec{f}]$ free of ZTM and finally we obtain the following new representation for the generating functional in models with solitons:

$$J[\vec{f}] = \hat{P}\tilde{J}[\vec{f},\vec{\Phi}_o] \quad , \qquad \hat{P} = \int_{\tilde{G}^{\perp}} d\alpha U(\alpha)$$

$$\tilde{J}[\vec{f},\vec{\Phi}_o] = Z^{-1} \int_{\Delta'} e^{i(\hbar^{1/2}\vec{\Phi}+\vec{\Phi}_o)(\vec{f})} \exp\{i\int d^4x[-\tfrac{1}{2}\vec{\Phi}\hat{K}\vec{\Phi}+\hbar^{-1}L_I^{eff}(\hbar^{1/2}\vec{\Phi})](x)\}\tilde{I}(\hbar^{1/2}\vec{\Phi}+\vec{\Phi}_o)D'\vec{\Phi} \quad . \qquad (6)$$

(Here the prime means that the ZTM have been removed.) The final formula (6) shows that the contributions to the effective generating functional give all those solutions $U(\alpha)\,\vec{\Phi}_o$ which can be obtained from the arbitrary chosen solution $\vec{\Phi}_o$ by use of all transformations in $\tilde{G}^{\perp}$. In that manner the elimination of ZTM implies the averaging over all dinstinct solutions which can be obtained from $\vec{\Phi}_o$.

Now one can develop a new perturbation theory or an $\hbar$-expansion in terms of generalized Feynman diagrams. The $\hbar$-expansion is also useful in studying of quantum stability problem in models with solitons.

QUANTUM STABILITY IN FIELD THEORY MODELS WITH SOLITONS

The conventional approach, where the instability implies that the second-quantized energy operator spectrum is unbounded from below, is very complicated. Thus, we will propose here a simpler concept of stability: we say that a quantum theory defined by Lagrangian density L is stable if mass^2 -singularities of one-particle propagator $\tau^{(2)}(p_1,p_2)$ are non-negative [1].

In our case $\tau^{(2)}_{rs}(x,y)$ -functions, $r,s = 1,\ldots,N$, are given by

$$\tau^{(2)}_{rs}(x,y) = i^{-2} \frac{\delta^2 J[\vec{f}]}{\delta f_r(x)\delta f_s(y)}\bigg|_{\vec{f}=\vec{0}} \quad , \; r,s = 1,\ldots,N \; , \tag{7}$$

and we use here the new representation (6) for $J[\vec{f}]$.

In what follows, for the sake of clarity of presentation, we shall restrict our analysis to the case N=2 and assume stationary form of solutions $\vec{\Phi}_o$,

$$\vec{\Phi}_o(t,\vec{x}) = \binom{\cos \omega t}{-\sin \omega t} u(\vec{x}) \quad , \qquad u^*(\vec{x}) = u(\vec{x}) \tag{8}$$

(ω-parameter). One can show [1], that in the rest frame $p_1 = (M,\vec{0})$, $M \neq 0$, we have

$$\tau^{(2)}_{rs}(p_1,p_2) = (\text{const.})\,\delta(M+E_2)\,\delta^{(3)}(\vec{p}_2)\,\delta_{rs}\hbar \sum_n{}' \sum_{k=1,2} \cdot$$

$$\cdot \frac{\tilde{\vec{\Psi}}_{nk}(\vec{0})\tilde{\vec{\Psi}}^*_{nk}(\vec{0})}{\omega_n^2(M)-M^2-i\varepsilon} + O(\hbar^2) \quad , \quad r,s = 1,2 \; . \tag{9}$$

Here, $\tilde{\vec{\Psi}}_n(\vec{0})$ is the Fourier transform of a vector $\vec{\Psi}_n(\vec{x})$ at

$\vec{p}_1 = \vec{p}_2 = 0$.
The vectors $\vec{\Psi}_n(\vec{x})$ are the eigenvectors of the Hamiltonian H with the eigenvalues ω_n^2,

$$\hat{H}(M)\vec{\Psi}_n(\vec{x}) = \begin{bmatrix} -\Delta+\mu_o^2-\omega^2-2L_I'(K)\,|_{K=u^2(\vec{x})} - 4KL_I''(K)\,|_{K=u^2(\vec{x})}, & -2i\omega M \\ 2i\omega M, & -\Delta+\mu_o^2-\omega^2-2L_I'(K)\,|_{K=u^2(\vec{x})} \end{bmatrix} \vec{\Psi}_n(\vec{x}) = \omega_n^2(M)\vec{\Psi}_n \tag{10}$$

We see that the mass^2-singularities of $\tau^{(2)}(p_1,p_2)$ are, up to $O(\hbar^2)$, determined by eigenvalues ω_n^2 which in turn are in general functions of the parameter ω.

The given QFT model is stable if there exists such a domain of values of ω for which all ω_n^2 are positive.

For an illustration we consider a Rosen model [2] given by

$$L(\vec{\Phi}) = \frac{1}{2}\partial_\mu\vec{\Phi}\partial^\mu\vec{\Phi} - \frac{1}{2}\mu_o^2\vec{\Phi}^2 + \frac{\alpha}{2}\vec{\Phi}^2\ln(\vec{\Phi}^2), \qquad \alpha > 0, \ N = 2 \ . \tag{11}$$

This model possess the soliton solution in the form (8) where

$$u(r) = \exp[-\tfrac{1}{2}\alpha r^2 + \tfrac{1}{2\alpha}(\mu_o^2-\omega^2)+1], \qquad r = |\vec{x}|, \tag{12}$$

and energy and charge are finite. The eigenvalue problem for $\hat{H}$(10) is now analitically solvable and one obtains

$$\overset{\pm}{\omega}{}_n^2 = \alpha(2n + \kappa - 1 \sqrt{2n\kappa+(\kappa-1)^2}), \ \kappa \equiv \frac{2\omega^2}{\alpha}, \ n = 0,1,2,\ldots \tag{13}$$

For $|\omega| > \sqrt{\frac{\alpha}{2}}$ all $\overset{\pm}{\omega}{}_n^2$ are positive and the model is stable. (Let us note here that on the classical level one can consider the Liapunov stability [1,3] and in the Rosen model the domains of stability are exactly the same in both cases.)

One can also study quantum stability in a general case with $N \geq 2$ [1]. However, since the operator $\hat{H}$ is non-diagonal, in general methods of estimation of its eigenvalues or computer programme must be used. Even the solutions $\vec{\Phi}_o$ may be found in most cases only numerically. We have studied in this manner a model of π-mesons with interaction Lagrangian density $L_I(\vec{\Phi}) = \alpha(\vec{\Phi}^2)^{3/2}+\lambda(\vec{\Phi}^2)^2$, N=3, and obtained the trajectories of particles with a good quantitative agreement with the experimental data (for the details see [1]).

REFERENCES

1. R. Raczka, L. Roszkowski, "Analysis of Green's functions and stability problem in models of quantum field theory with solitons", ICPT preprint IC/83/190, Trieste, 1983.
2. R. Rosen, Phys. Rev. 183 (1969) 1186.
3. A. Kumar, V.P. Nisichenko and Yu.P. Rybakov, Int. J. Theor. Phys. 18 No.6 (1979) 425.

Acta Physica Austriaca, Suppl. XXVI, 435–439 (1984)

STOCHASTIC QUANTIZATION OF THE LINEARIZED GRAVITATIONAL FIELD+

by

H. RUMPF
Institut für Theoretische Physik
Universität Wien
A-1090 Vienna, Austria

The stochastic quantization scheme of Parisi and Wu [1] exhibits a Euclidean quantum field $\psi(x)$ as the stationary limit with respect to a fictitious time t of the stochastic relaxation process defined for $t \geq 0$ by the Langevin equation

$$\frac{\partial}{\partial t} \psi(t,x) = - \frac{\delta S[\psi(t,x)]}{\delta\psi(t,x)} + \eta(t,x) \quad . \tag{1}$$

Here $x = (x^a)$, $a = 0,..,3$, S is the classical action, and η is a Gaussian white noise with correlation function

$$<\eta(t,x)\eta(t',x')>_\eta = 2\cdot\hbar\delta(t-t')\delta^{(4)}(x-x') \quad . \tag{2}$$

The Euclidean Green'sfunctions are obtained as the "equilibrium limit" of the correlation functions of the process:

+Seminar talk given at the XXIII. Internationale Universitätswochen für Kernphysik,Schladming,Austria,February 20-March 1, 1984.

$$<\psi(x_1)\ldots\psi(x_n)> = \lim_{t_1=\ldots=t_n\to\infty} <\psi(t_1,x_1)\ \ldots\psi\ (t_n,x_n)>_\eta\ . \tag{3}$$

A most interesting aspect of the Parisi-Wu method for continuum field theories is that in gauge theories the perturbative calculation of the right hand side of eq. (3) may be based on the classical action S alone, i.e. no gauge-fixing term and associated Faddeev-Popov ghosts are necessary.

One would like to exploit this advantage also for the gravitational field. But it is well-known that the Einstein-Hilbert action

$$S_{EH} = (2\kappa)^{-1}\int d^4x\,(g)^{1/2}R[g_{ab}]$$

is not positive definite in the Euclidean regime. Therefore the process (1) does not describe an "approach to equilibrium" in this case (rather the correlation functions diverge exponentially in t). This difficulty turns up already at the linearized level, where (with $g_{ab} = \delta_{ab} + 2(\kappa)^{1/2}\psi_{ab}$)

$$S_{EH}^{(o)}[\psi] = \frac{1}{2}\int d^4x\ \psi\ V\ \psi \tag{4}$$

with the operator V given in momentum space by

$$V = k^2(P^2 - 2P^{o\prime})\quad . \tag{5}$$

Here P^2 and $P^{o\prime}$ are members of a complete set of spin projection operators $P^o, P^{o\prime}, P^1, P^2$ acting on the space of symmetric tensor fields [2]. As is well known, the action (4) describes a massless helicity-2 field in Minkowski space and is invariant under the abelian gauge transformations

$$\psi_{ab} \to \psi_{ab} + k_a\Lambda_b + k_b\Lambda_a\quad . \tag{6}$$

The corresponding gauge-invariant "field strength" is the linearized Riemann tensor

$$R_{abcd} = 4k_{[a}\psi_{b][c}k_{d]} \ . \tag{7}$$

In the following we modify the Parisi-Wu prescription so as to make it applicable to the action (4) and to reproduce the results of its standard quantization. Recall that the standard procedure consists in imposing a gauge condition, e.g.

$$C_a^{(\lambda)} \equiv k_c\psi_{ac} - \lambda k_a\psi = 0 \ , \qquad \lambda \neq 1 \ , \tag{8}$$

formally obtained by adding a gauge-fixing term $\alpha^{-1}C_a^{(\lambda)}C_a^{(\lambda)}$ to the Lagrangian in (4). This yields an invertible operator $V^{(\alpha,\lambda)}$, the corresponding propagator being given by

$$K_{abcd}^{(\lambda,\alpha)}(k) = k^{-2}(\delta_{(a|c}\delta_{b)d} - \frac{1}{2}\delta_{ab}\delta_{cd}) + L_{abcd}^{(\lambda,\alpha)} \ . \tag{9}$$

The gauge-dependent part $L^{(\lambda,\alpha)}$ does not contribute to $\langle R\ldots(x)\ R\ldots(x')\rangle$ and vanishes for $\lambda = 1/2$, $\alpha = 1$. Now the propagator $K^{(0,0)}(k)$ can be obtained by considering the limit $m^2 \to 0$ of the "naive" massive extension of (4) defined by adding a mass term $m^2\Pi$ to V. (This differs from the so-called Fierz-Pauli mass term, hence the extension is unphysical, containing a spin-0 tachyon besides a massive spin-2 particle, and thus does not contradict the van Dam-Veltman [3] mass discontinuity.) The propagator of the massive field is

$$K(m^2;k) = \frac{P^2}{k^2+m^2} + \frac{P^{0'}}{m^2-2k^2} + \frac{P^0+P^1}{m^2} \ . \tag{10}$$

The first two terms on the r.h.s. of (10) give exactly $K^{(0,0)}(k)$

if $m^2 = 0$. Introducing a mass term is, via Fourier transform, equivalent to introducing a fictitious time parameter s, which is sometimes called "proper time" and was extensively used by Schwinger and others. Consider the Fourier transform of (10),

$$\tilde{K}(s;k) = (2\pi)^{-1} \int_{-\infty}^{\infty} d(m^2)\, e^{im^2 s} K(m^2;k) \tag{11}$$

with the poles circumvented in such a way that the integration contour can be rotated counterclockwise into the imaginary m^2-axis. Since the residues are projectors, one has

$$K^{(0,0)}(k) + \frac{p^0+p^1}{m^2} \cdot \infty^2 = \int_{-\infty}^{\infty} ds\, \tilde{K}(s,k) =$$
$$= 2\left(\lim_{s\to\infty} \int_0^s d\sigma - \lim_{s\to-\infty} \int_s^0 d\sigma\right) \tilde{K}(s-\sigma,k)\tilde{K}(s-\sigma,k). \tag{12}$$

But $\tilde{K}$ is related to the (resolvent) Green's function G of the deterministic part of the Langevin equation (1) via $G(t,k) = i\tilde{K}(it,k)$. Therefore the integrals appearing at the r.h.s. of eq. (12) may be interpreted as correlation functions of processes defined by (1), but with η replaced by $\eta_{\pm t}$, where $\eta_{\pm t}$ are Gaussian white noise in the intervals $(0,t)$ and $(-t,0)$, respectively, but zero otherwise. Thus we have arrived at the following modification of equation (3):

$$\langle\psi(x_1)\ldots\psi(x_n)\rangle = \left(\lim_{t\to\infty} - \lim_{t\to-\infty}\right) \langle\psi(t,x_1) \ldots \psi(t,x_n)\rangle_{\eta_t} . \tag{13}$$

Together with the asymptotic fall-off condition (in the fictitious time) implied by the resolvent Green's function the prescription (13) is fully equivalent to the standard quantization of the free linearized gravitational field, as far as gauge-invariant quantities are concerned. The appearance of the "Landau gauge" $\lambda = \alpha = 0$ in the finite part of the propagator is completely analogous to what happens in the stochastic quantization of gauge theories. Different values of $\alpha > 0$ (but not λ) may be obtained by prescribing certain initial or final probability distributions for ψ.

Recently "stochastic gauge fixing" has been introduced into the Parisi-Wu scheme [4]. In the gravitational context we define "generalized gauge transformations" (GGT) by letting Λ_a in (6) depend on t. The Langevin equation (1) is not invariant under GGT, but the change affects only the operator V. For calculations in perturbation theory it may be advantageous to introduce a "nonholonomic" (i.e. involving $\partial\psi/\partial t$) "gauge constraint" (NHGC) with the property that the GGT from the space of solutions of (1) to the stochastic constraint surface change the operator V into an invertible operator W, thus yielding a finite propagator. But although there is a natural correspondence between the linear, local, covariant gauge conditions (8) and NHGC's, the propagators obtained are different, except for $\lambda = 0$. The reason is that W is in general not self-adjoint (and hence not derivable from an action). However the gauge-invariant part (cf. eq. (9)) of all the propagators is the same, and therefore, at least in the free case, all the generalized gauges are physically equivalent.

Finally we note that the modification (13) of the Parisi-Wu prescription is a very conservative one. A more radical modification may be necessary for a consistent perturbative treatment beyond the linearized level. More details will be found in a forthcoming paper being prepared in collaboration with Helmuth Hüffel.

REFERENCES

1. G. Parisi and Wu Yong-Shi, Sci. Sinica 24 (1981) 483.
2. P. van Nieuwenhuizen, Nucl. Phys. B60 (1973) 478.
3. H. van Dam, M. Veltman, Nucl. Phys. B22 (1970) 397.
4. L. Baulieu, D. Zwanziger, Nucl. Phys. B193 (1981) 163; see also E. Seiler, lectures at this school.

Acta Physica Austriaca, Suppl. XXVI, 441–446 (1984)

ROTATION-INVARIANT REGULARIZATION OF QUANTUM CHROMODYNAMICS IN STRONG COUPLING+

by

H. SCHLERETH
The Niels Bohr Institute
University of Copenhagen
Blegdamsvej 17, DK-2100 Copenhagen
Denmark

We present a regularization of strong coupling QCD which is rotation-invariant and suggest its Hamiltonian formulation to be used in practical calculations. Dynamical fermions are introduced. The fermion doubling problem is seemingly avoided.

The intention of this project is to improve on the prospects of the strong coupling expansion in quantum chromodynamics (QCD). Usually the use of this expansion is hampered by the occurrence of complex singularities [1] and the roughening singularity [2]. We present here an approach which is rotation- and translation-invariant from the start and thus avoids the latter type of singularity,

+Seminar given at the XXIII. Internationale Universitätswochen für Kernphysik, Schladming, Austria, February 20- March 1, 1984.

while the role of complex singularities must be investigated separately in this approach. In the lattice regularization [3] there are attempts to use lattices with more symmetry axes than the hypercubic lattice has, which has led to some improvement in the scaling behaviour of the string tension[4].

Here we write what might be called an effective block spin theory. Ideas of [5] are used, but the final formulation is different. Furthermore the introduction of dynamical fermions is discussed.

We assume that the theory in question has some non-locality which is used for ultraviolet (U.V.) regularization. We also require it to have the classical "continuum" limit.

The basic idea of U.V. regularization [5] in brief is the following: take some "effective" link variable U(x,y), to be more closely defined below, and subdivide Euclidean space of dimension d into cells of volume a^d, and take the points x resp. y out of disjoint cells. The theory is U.V. regularized by use of a delocalization parameter b, which in general should obey b>>a. For the action then one writes the general expression

$$S = \frac{1}{g_o^2} \int dx\, dy\, dz\, dv\, K(x,y)K(y,z)K((z,v)K(v,x) \cdot U_{\square}(x,y,z,v), \tag{1}$$

where

$$U_{\square}(x,y,z,v) = :U(x,y)U(y,z)\bar{U}(z,v)\bar{U}(v,x) ;$$

$$K(x,y) =: \left(\frac{1}{b^2}\right)^2 \cdot e^{-\frac{(x-y)^2}{b^2}} \xrightarrow[b \to 0]{} \delta^{(d)}(x-y) \quad . \tag{2}$$

Furthermore, since we take the limit a → 0, we have replaced

$$a^4 \cdot \sum_{\text{cells}} \longrightarrow \int d^4x \equiv \int dx \quad . \tag{3}$$

(We take d = 4 in the sequel,and tr is over some representation of the underlying group.)
Under a local gauge transformation $\Omega(x)$ we have

$$U'(x,y) = \Omega(x)U(x,y)\Omega^{-1}(y) \quad , \tag{4}$$

so that (1) is gauge invariant. Beyond this (1) is obviously also rotation- and translation-invariant in d = 4 Euclidean space. (For a numerical treatment one must take $a > 0$, but $a \ll b$ still approaches these symmetries.)

The smearing kernels K(x,y) in (2) are taken as in [5], their role in the introduction of fermions will be elucidated below.

The choice of U(x,y) is to a high degree arbitrary, apart from the requirement that the classical limit $b \to 0$ of (1) should give $S = 1/g^2 \int d^4x \; tr(F_{\mu\nu})^2$.

In [5] a choice is made which for SU(2) amounts to writing

$$U(x,y) = \Phi(x,y) \; V(x,y) \quad , \tag{5}$$

where $\Phi(x,y)$ is a bilocal scalar, and V(x,y) is a group element of SU(2). This leads to the fact that the leading term in the strong coupling expansion is entirely based on the $\Phi(x,y)$ and is not sensitive to the group element V(x,y)$^+$ (for details see [5]). Also the definition of the

---

$^+$Furthermore the continuation to weaker coupling will depend on how fast $\Phi(x,y)$ takes on its long wave length behaviour, where it is supposed to go to a constant (so that the correct "continuum limit" at $b \to 0$ is obtained).

functional measure is unclear.

We therefore prefer to take U(x,y) itself to be a group element. Gauge field averages of this type are considered in [6].

The partition function is then defined using the Haar measure by

$$Z = \int DU(x,y)e^{-S}. \tag{6}$$

The introduction of fermions is accomplished by adding to (1) the part

$$S_F = \int dx\ dy\ K(x,y)\bar{\Psi}(x)\gamma_\nu \frac{(x-y)^\nu}{b^2} U(x,y)\psi(y) \quad . \tag{7}$$

(The notation is obvious.) Using (2) and

$$U(x,y) = e^{iA_\nu(\frac{x+y}{2})\cdot(x-y)^\nu + O[x-y]^2} \quad , \tag{8}$$

then by Taylor expansion at fixed x+y/2 around (x-y) the usual continuum Dirac equation containing interaction with a four potential $A_\nu(x)$ is gotten taking $b \to 0$.

Before commenting further on the fermions let us point out that in its present form the formalism contains a difficulty. Namely K(x,y) is also smearing out the time variable, which in fact corresponds to having derivatives in time of infinite order. So one worries about the existence of a transfer matrix in this scheme [7]. To avoid this difficulty it is preferable to use for the rotation-invariant regularization from the beginning a Hamiltonian formulation. A Hamiltonian with the correct classical limit $H = \int d^3x(E^2 + B^2)$ as $b \to 0$ is

$$H \doteq \frac{g^2}{2} \int d^3x\ d^3y\ K(\vec{x},\vec{y})E^a(\vec{x}\vec{y})^2 - \frac{1}{g^2 b}\int d^3x\ d^3y\ d^3z\ d^3v\ K(\vec{x},\vec{y})\cdot$$

$$K(\vec{y},\vec{z})K(\vec{z},\vec{v})K(\vec{v},\vec{x})\,\mathrm{tr}\,U_{\Box}(\vec{x},\vec{y},\vec{z},\vec{v}) \tag{9}$$

with

$$K(\vec{x},\vec{y}) \simeq \frac{1}{b^3}\, e^{-\frac{(\vec{x}-\vec{y})^2}{b^2}} \xrightarrow[b \to 0]{} \delta^{(3)}(\vec{x}-\vec{y}) \quad ; \tag{10}$$

and $U_{\Box}(\vec{x},\vec{y},\vec{z},\vec{v})$ is the spacelike analogue of the first line in (2). Furthermore $E^a(\vec{x},\vec{y})$ (the color electric field operator, a≡ color index) and $U(\vec{x},\vec{y})$ have to obey special commutation relations and the physical states must satisfy the Gaussian law [8]. For the present model this will be discussed elsewhere in more detail [9].

Again (9) can be extended to include dynamical fermions and the system can be treated in strong coupling by e.g. Raleigh-Ritz methods.

The system stands somewhat in between the Random Lattice [10] and the original Wilson lattice [3], with the small exponential non-locality not allowing the system to loose confinement. Also (9) is rotation-invariant in 3-space, so that no roughening singularity [2] should come up.

To discuss the fermion spectrum let us look at the free Dirac equation in 1+1 dimensions (for simplicity),

$$i\psi(x,t) = i\alpha\int dy\; K(x,y)\, \frac{x-y}{b^2}\psi(x,t) \tag{11}$$

$(\alpha=\gamma_0\gamma_1)$. The dispersion law one finds is the following:

$$E(k) = \alpha\; k\; \exp(-k^2\, b^2/4)\;. \tag{12}$$

There are two low energy solutions [11], $k^2 = 0$ and $k^2 = \infty$. However the last one at infinite momentum cannot be exited in our case.

The formalism presented is designed for analytic strong coupling calculations. However numerical and mean field methods using (6) with nonzero a-cells and only a resticted time smearing over a few neighbouring cells can also be considered [9].

ACKNOWLEDGEMENT

I thank H. Bohr, H. Dass, E. Katznelson and H.B. Nielsen for many helpful discussions and comments, and the ICTP/ Trieste, where part of this work was done, for its kind hospitality.

REFERENCES

1. M. Palcioni, E. Marinari, M.L. Paciello, G. Parisi and B. Tagliente, Phys. Lett. 102 B (1981) 270.
2. M. Lüscher, Nucl. Phys. B180 (1981) 317.
3. K.G. Wilson, Phys. Rev. D14 (1974) 2455.
4. W. Celmaster, Phys. Rev. Lett. 52 (1984) 403.
5. J.M. Drouffe, Nucl. Phys. B 218 (1983) 89.
6. T. Balaban, J. Imbrie and A. Jaffe, HUTMP - Preprint 83/ B 149 (1983).
7. See also: M. Lüscher and P. Weisz, DESY - Preprint 84-018.
8. See e.g. J. Kogut, Rev. Mod. Phys. 55 (1983) 775.
9. H. Schlereth, in preparation.
10. N. Christ, R. Friedberg and T.D. Lee, Nucl. Phys. B 202 (1982) 89; and ibid. 210 (1982) 310,337.
11. H.B.Nielsen and M. Ninimiya, Nucl. Phys. B 185 (1981) 20.

Acta Physica Austriaca, Suppl. XXVI, 447–452 (1984)

MONTE CARLO SIMULATION OF THE PROCESS $e^+e^- \to \tau^+\tau^-(\gamma)$ +

by

Z. WAS
Jagellonian University, Cracow, Poland

The purpose of this talk is to discuss some results obtained from the Monte Carlo (M.C.) simulation of the combined production and decay process $e^+e^- \to \tau^+\tau^-(\gamma)$, $\tau \to \nu\ e(\mu,\pi,\rho\ldots)$ based on the $O(\alpha^3)$ QED calculations, including Z_0 exchange (in the low energy approximation). The effects due to finite mass of τ, and to spin polarization of $e^\pm$ and $\tau^\pm$ are also included. In this way all possible effects which would produce systematic variation of the cross section, down to 1 % level, are under control. Thus, accuracy level is assured from the τ production threshold up to the highest PETRA/PEP energies ($\sqrt{s} \sim 45$ GeV).

The M.C. program based on the above QED calculations is being prepared for publication [2]. It generates randomly the four momenta of $\tau^+\tau^-$, the momentum of the radiative photon (if present) and also the momenta of the decay products, all over the entire phase space (including hard photons).

The production and decay parts of the algorithm program are clearly separated allowing thus to change easily

---

+Seminar given at the XXIII. Internationale Universitätswochen für Kernphysik, Schladming, Austria, February 20-March 1, 1984.

the decay part, if necessary. Spin effects are included in the following way: In the first part of the algorithm the momenta of particles are generated assuming that electrons are not polarized. The decay of each τ is simulated in its rest frame, assuming that τ is not polarized. Next, effects due to $e^{\pm}$ polarization and to spin effects in τ-decays are introduced by event rejection. As usually in the M.C. type of calculations the user of the program may introduce his own selection criteria (cut-offs) and the apparatus efficiency by means of event rejection. Since τ decay is included in the simulation, the rejection may be done using the momenta of the decay products as in the real experiment.

The following discussion of the numerical results will be rather general. More details can be found in [1]. The effects due to radiative corrections and in particular from hard bremsstrahlung are essentially the same as in muon production [3], especially at higher energies. Radiative corrections contribute to the total cross section and to angular charge asymmetry. Rather trivial but new, with respect to muon production, are effects due to finite mass of τ. They decrease strongly with growing energy, but even for $\sqrt{s} \sim 35$ GeV the terms of order $4\, m_{\tau}^{2}/s$ are larger then the aimed 1 % accuracy level. In the hard bremsstrahlung cross section (and spin density matrices) there are also much larger terms of order $2\, m_{\tau}/\sqrt{s}$, but the measurable quantities seem to depend weakly on them.

Spin phenomena are characteristic to τ pair production process. Since τ is not observed directly and τ decay is sensitive to its polarization, spin effects cannot be eliminated from the process and even for unpolarized $e^{\pm}$ they are present in form of spin correlations in the final state. Energy spectrum of τ decay products depend on τ-spin, therefore, these correlations show up as correlations among the momenta of the τ decay products. For example in $e^{+}e^{-} \to \tau^{+}\tau^{-}(\gamma)$, $\tau^{\pm} \to \pi^{\pm}\nu$ process one observes the surplus of the fast-fast and slow-slow $\pi^+\pi^-$ pairs over fast-slow pairs as compared to the uncorrelated decays by 25 % ($\sqrt{s} >> 2m_\tau$). Radiative corrections make pions softer but the pattern of the correlations remains unchanged. See fig. 1 for results from M.C. simulation. In the case of both τ's decaying into $\pi\nu$, the cross section and the measured branching ratio may be affected by up to 5 % due to spin correlations if the typical cut-off on the visible energy $E(\pi^+) + E(\pi^-)$ is used to select events.

In addition to the longitudinal correlations discussed above there are also the corrections among the transverse momenta of decay products with respect to the direction of decaying τ. For example it was shown [1] that the plane defined by momenta of the decay products ($\pi^+\pi^-$) tends to be perpendicular to the reaction plane ($e^+\pi^+$). Even though this effect is sometimes of order of 5 % [1] it does not influence strongly the quantities measured in typical experiment.

The polarization of single τ is zero in $O(\alpha^2)$ QED ($e^\pm$ unpolarized). The radiative corrections (box and magneting vertex) produce small nonzero polarization. Another, more interesting, source of the polarization is the Z_0 exchange. The latter polarization causes the energy spectrum of the decay product (π,μ...) to be dependent on the scattering angle. The Z_0 -τ vector coupling constant may be estimated [4] or with better statistics extracted from the above dependence. Such measurement may be the unique experimental source for the vector coupling constant before the SLC/LEP era. For more quantitative discussion of this and other spin effects see [1].

Summarizing, I conclude that the presented calculation algorithm embodied in the M.C. program constitutes a powerful tool, very useful in the analysis of the high statistics data on the τ-pair production process down to 1 % level.

FIGURE CAPTIONS

Fig.1 The energy distributions $d\sigma/dE(\pi^+)dE(\pi^-)$ at $\sqrt{s}$ = 44 GeV from M.C. simulation in $O(\alpha^3)$ QED, a) with, b) without spin correlations.

REFERENCES

1. S. Jadach and Z. Was, QED $O(\alpha^3)$ radiative corrections to the reaction $e^+e^- \to \tau^+\tau^-$ including spin and mass effects, TPJU-14/83, Jagellonian Univ. preprint, submitted to Acta Phys. Pol..
2. S. Jadach and Z. Was, Monte Carlo simulation of the process $e^+e^- \to \tau^+\tau^-$ including radiative $O(\alpha^3)$ QED corrections, mass and spin effects, in preparation.
3. F.A. Berends and R. Kleiss, Nucl. Phys. B177 (1981) 237.
4. Tau branching ratios and polarization limits..., CELLO preprint, DESY 93-019.

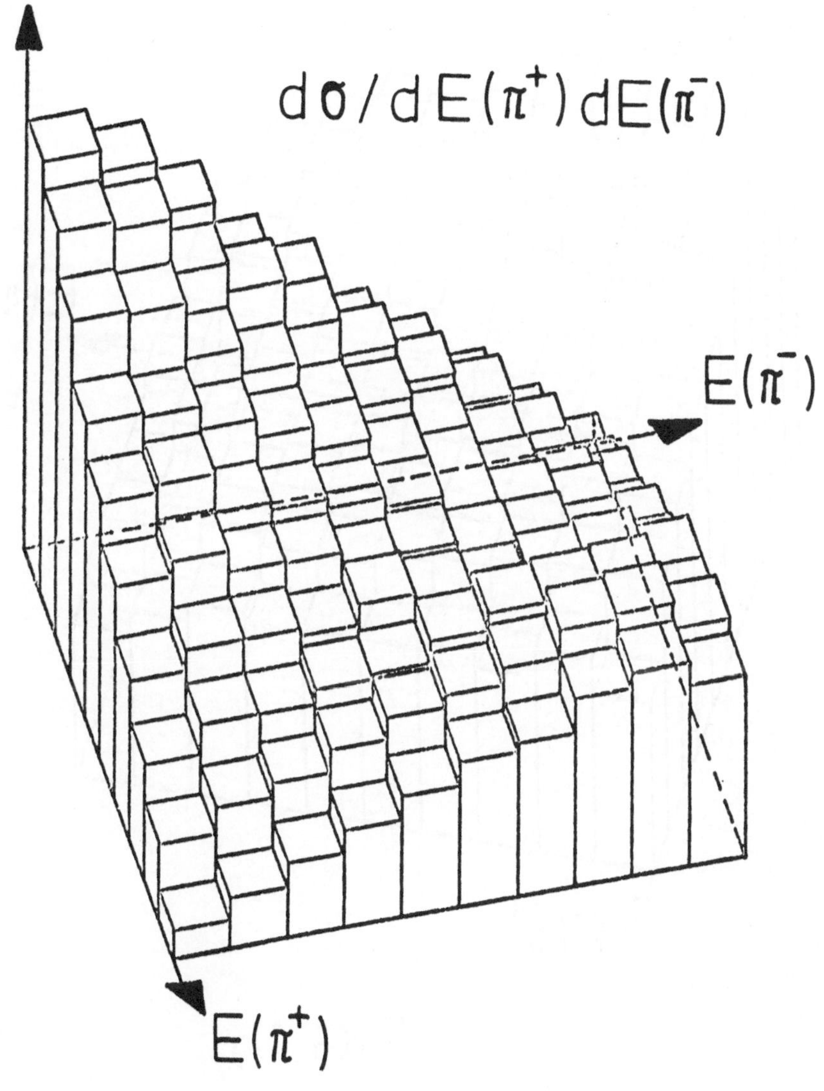

FIG. 1a

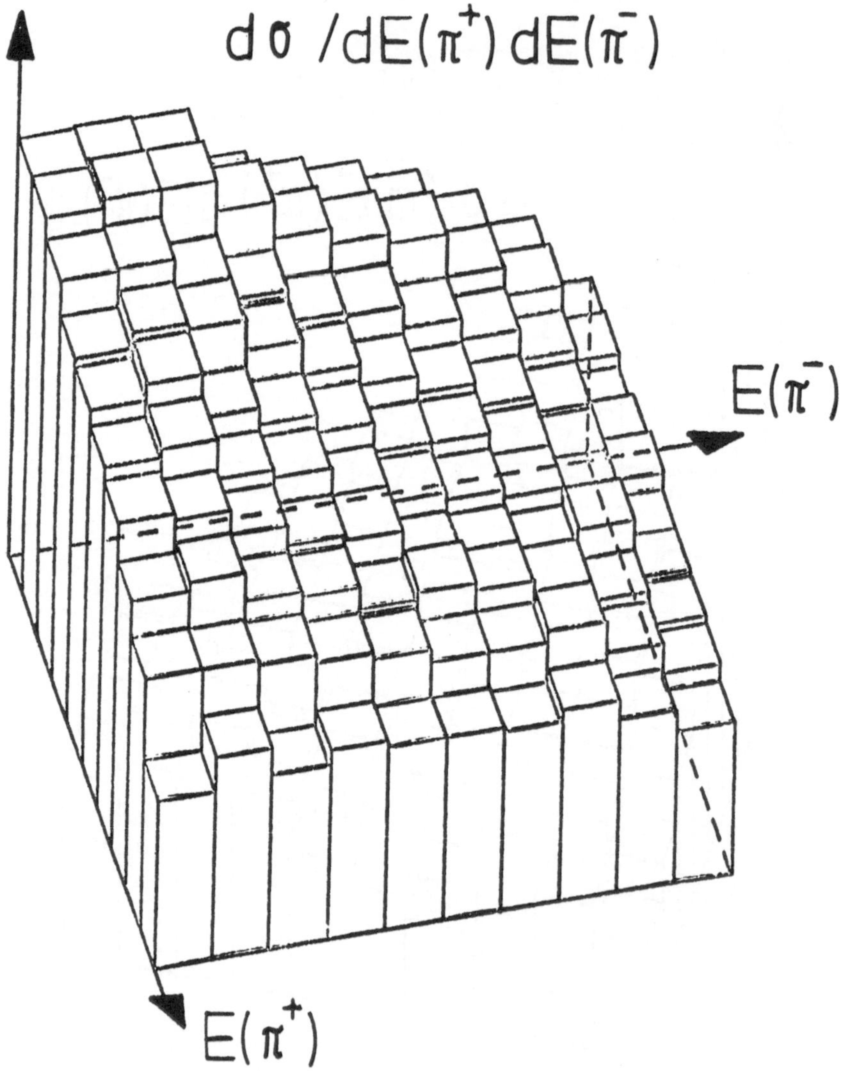

FIG.1b

Druck: Novographic, Ing. Wolfgang Schmid, A-1230 Wien.